Environmental Footprints and Eco-design of Products and Processes

Series Editor

Subramanian Senthilkannan Muthu, Chief Sustainability Officer, Green Story Inc., Tung Chung, Hong Kong

Indexed by Scopus

This series aims to broadly cover all the aspects related to environmental assessment of products, development of environmental and ecological indicators and eco-design of various products and processes. Below are the areas fall under the aims and scope of this series, but not limited to: Environmental Life Cycle Assessment; Social Life Cycle Assessment; Organizational and Product Carbon Footprints; Ecological, Energy and Water Footprints; Life cycle costing; Environmental and sustainable indicators; Environmental impact assessment methods and tools; Eco-design (sustainable design) aspects and tools; Biodegradation studies; Recycling; Solid waste management; Environmental and social audits; Green Purchasing and tools; Product environmental footprints; Environmental management standards and regulations; Eco-labels; Green Claims and green washing; Assessment of sustainability aspects.

Kishore Kumar · Susela Devi Suppiah ·
Preeti Bhaskar · Seda Yıldırım
Editors

AI-Driven Circular Economy for Carbon Neutrality in Emerging Economies

Springer

Editors
Kishore Kumar
School of Management
Bennett University
Greater Noida, Uttar Pradesh, India

Susela Devi Suppiah
Faculty of Business and Accountancy
Universiti Selangor
Petaling Jaya, Selangor, Malaysia

Preeti Bhaskar
Department of Marketing and Management
College of Banking and Financial Studies
University of Technology and
Applied Sciences
Muscat, Oman

Seda Yıldırım
Department of Business Administration
Tekirdag Namık Kemal University
Tekirdağ, Türkiye

ISSN 2345-7651 ISSN 2345-766X (electronic)
Environmental Footprints and Eco-design of Products and Processes
ISBN 978-3-032-30297-7 ISBN 978-3-032-30298-4 (eBook)
https://doi.org/10.1007/978-3-032-30298-4

This Springer imprint is published by the registered company Springer Nature Switzerland AG
The registered company address is: Gewerbestrasse 11, 6330 Cham, Switzerland

Contents

1 **Governing Artificial Intelligence for Circular Economy Transitions: Policy Architecture and Emerging Economy Realities** 1
Kishore Kumar, Susela Devi Suppiah, Preeti Bhaskar, and Seda Yıldırım

Part I Ethics, Governance, and Responsible AI

2 **Ethical Leadership in the Age of AI: Fostering Sustainability, Building Trust, and Accountability in Data-Driven Organizations** 9
Krupa B. Nair, P. C. Radhika, and Bhagyalakshmi S. Pai

3 **AI-Powered Decision-Making and Responsible Approaches to Ethical Dilemmas** 25
Amit Kumar, Nitendra Kumar, and Jasreen Kaur Kakar

4 **Sustainability and Ethics in the Digital Economy: Exploring Corporate Responsibility Among Nifty 50 Companies** 37
Akancha Srivastava, Shefali Mishra, and Dipanjan Bagchi

5 **Corporate Responsibility: Sustainability and Ethics in the Digital Economy** 57
Anil Kumar, Neha Nupoor, and Mukesh Babu Gupta

6 **Artificial Intelligence, Sustainability, and the Future of Human Resources in the Organisation** 75
Devika Agrawal, Teena Bagga, and Mahanish Panda

Part II Digital Transformation and Green Growth

7 Dynamic Capabilities for Socially Sustainable Digital Circular Transformation: A Four-Pillar Framework for Emerging Economies 93
Hosam Alden Riyadh, Shahid Latif, Maher Ali Al-Shmam, Ali Malallah Al-Sendy, and Sinan Zuhair Mohammed Jameel

8 Reimagining Growth: ICT-Driven Green Growth Transition in Emerging BRICS Economies 111
Shikha Daga, Remy Jonkam Oben, and Kyla L. Tennin

9 From Algorithms to Action: AI's Contribution to Achieving the SDGs 127
Yasser Mahfooz, Obaidur Rahman, and Manik Batra

10 The Role of Technology and Innovation in Carbon Accounting and Climate Policy: A Bibliometric Mapping of Knowledge Evolution and Structural Fragmentation 143
M. I. M. Riyath and Sushma Maurya

11 Conceptualizing Climate Change Vulnerability and Adaptation: A CIMO-Based Strategic and Policy Perspective 165
Priyanka Verma, Arif Hasan, Mini Agrawal, and Shakeel Ahmad Sofi

Part III AI and Circular Economy Innovations

12 Roadmap for AI-Powered Circular Value Chains for Carbon Neutrality and Corporate Sustainability in Emerging Economies 183
Sanjana Santra, Kishore Kumar, and Ankita Sharma

13 Beyond Today: Strategic Foresight in AI-Powered Circular Economies for Environmental Sustainability 199
Shubhra Mishra

14 If 'AI Waste-Sorting Robots Are the Future of Circular Economy,' How Are Emerging Economies Responding? 217
Sanjana Santra, Kishore Kumar, and Rubee Singh

15 Integrating Generative AI for Sustainable and Resilient Supply Chains 231
Vivek Gupta and Ravi Shankar

16 The Greta Effect and AI-Driven Circular Practices in Sustainable Hospitality: A Pathway to Guest Engagement and Innovation 247
Praveen Srivastava, Shelly Srivastava, and Gautam Shandilya

Part IV AI, Finance, and the Digital Economy

17 Automated Decisions and Market Fairness in Decentralised Finance (DEFI): Exploring the Ethical Dimensions of AI-Powered Trading 269
Animesh Singh

18 The Ethical Landscape of DeFi: Challenges and Opportunities in a Tokenised Economy for Emerging Economies 285
Justine Mbudaya, David Chikwere, Kundayi Chinyongo, and Munyaradzi Chibaro

19 Time-Varying Interactions Between Artificial Intelligence and Sustainable Finance in the Post-pandemic Era 303
Mohit Verma, Mukhtaruddin Mukhtaruddin, and Bikram Paul Singh Lehri

20 Generative AI a Boon or a Threat: Rising Impact of Deepfake on Circular Economy in Southeast Asia 317
Sonal Tikku and Ritesh Dwivedi

Part V Circular Economy Transitions, Carbon Markets, and Global Sustainability Policy

21 Strategizing Urban Sustainability for Emission Reduction: A Best–Worst Method (BWM) Approach 337
Kiran Yadav, Noman Arshed, Bhavna Yadav, and Shubham Shukla

22 Carbon Markets in the Global South: Neoliberal Foundations Amid Neocolonial Critiques 355
Sana Fatima and Nabeel Ahmad

Index 375

About the Editors

Dr. Kishore Kumar is Faculty of Sustainability Management and Area Chair (General Management) at the School of Management, Bennett University (Times of India Group). He also serves as Chair of the Bennett University Centre for Sustainable Development and UN PRME Faculty Lead. He brings strong expertise in sustainability reporting, ESG, and Business Responsibility and Sustainability Reporting (BRSR), along with extensive experience in sustainability training and executive capacity building. His academic and professional work spans corporate sustainability, ESG integration, CSR, and climate change. Dr. Kumar has published over 50 research articles in high-impact journals indexed in Scopus, ABDC, and Web of Science, garnering more than 1750 citations, with an H-index of 18 and an i10-index of 24 (Google Scholar). He is the editor of several Scopus-indexed books on sustainability and climate change and holds multiple editorial roles with reputed international journals. A GRI-certified Sustainability Reporting professional and UGC-NET/JRF qualifier, he is deeply committed to advancing sustainability-driven business education, executive training, and ESG-informed decision-making.

Prof. Dr. Susela Devi Suppiah is an Honorary Professor at Universiti Selangor (UNISEL), Malaysia, an adjunct Professor at Sunway Business School and a leading scholar in sustainability accounting, ESG governance, and regulatory reform in emerging economies. With over four decades of academic and professional engagement, her work examines the intersection of digital transformation, AI governance, accounting professionalization, and sustainable development. She served as Director of the Sunway Centre for ESG and has led multiple initiatives on integrating ESG, artificial intelligence, and digital competencies into professional and higher education frameworks. Her research critically explores regulatory capture, institutional reform, and the role of professional bodies in advancing accountability and carbon transition strategies in developing economies. Professor Susela serves as Associate Editor for leading international journals and has previously served as Chief Editor of the Asian Journal of Business and Accounting. She is actively involved in regional and international benchmarking initiatives on professional competency frameworks aligned with global standards. Her editorial contribution to AI-Driven Circular Economy for

Carbon Neutrality in Emerging Economies reflects her commitment to advancing ethically grounded, technology-enabled sustainability transitions across emerging markets.

Dr. Preeti Bhaskar is an Assistant Professor at the College of Banking and Financial Studies, Muscat, with over 14 years of academic experience in Human Resource Management. Her research spans Human Resource Management, the adoption of emerging technologies in higher education, and sustainability-oriented practices. She places particular emphasis on digital transformation, responsible innovation, and sustainable institutional development. Her work has been published in reputed Scopus-indexed and ABDC-ranked journals. She has authored books and case studies, and her research excellence has been recognized through Best Research Paper awards. In addition to her research contributions, she actively engages in faculty and student development initiatives at both national and international levels.

Prof. Dr. Seda Yıldırım currently works at Tekirdag Namık Kemal University in Turkey. She has a doctoral degree (PhD) in marketing. She studies on sustainable development, 2030 sustainable development goals, greenwashing, sustainable consumption patterns, consumption patterns, environmental quality and food security. She is married and has two children. She has over 1900 citations by Google scholar. She has many publications in journals indexed by Web of Science (SSCI, SCI-Exp, ESCI and others) and SCOPUS. She published several book chapters in IGI Global publishing. She won Emerald literati award-Outstanding Paper award 2022.

Chapter 1
Governing Artificial Intelligence for Circular Economy Transitions: Policy Architecture and Emerging Economy Realities

Kishore Kumar, Susela Devi Suppiah, Preeti Bhaskar, and Seda Yıldırım

Abstract A carbon–neutral economy has become the buzzword in recent years. Countries across the world are under immense pressure to reduce carbon emissions to achieve a circular economy. Advanced technologies such as artificial intelligence (AI) and Blockchain have emerged as a potential solution to achieve carbon neutrality and implement a circular business model. There is a need for a global framework, mechanisms and policies for carbon–neutral strategies. This chapter outlines the summary of contribution in this book. It includes ethical leadership for developing an AI-based smart infrastructure to shift to a digital circular economy. It also discusses the significance of green growth transition and policy actions based on AI-powered solutions to achieve various Sustainable Development Goals (SDGs). Further, it also highlights the important concepts discussed in the book, like structural climate governance, circular economy-related reforms based on smart technologies, AI-powered financial system promoting ethics, fairness and market integrity.

Keyword Circular economy · Artificial intelligence · Climate change · Ethical leadership · Sustainable development goals

K. Kumar (✉)
BU Centre for Sustainable Development (BUCSD), School of Management, Bennett University, Greater Noida, India
e-mail: akishore001@gmail.com

S. D. Suppiah
Universiti Selangor (UNISEL), Shah Alam, Selangor, Malaysia
e-mail: suseladevi@unisel.edu.my

P. Bhaskar
College of Banking and Financial Studies, Muscat, Oman
e-mail: preeti.bhaskar52@gmail.com

S. Yıldırım
Department of Business Administration, Tekirdag Namık Kemal University, Tekirdağ, Turkey
e-mail: sedayil1@gmail.com

K. Kumar et al. (eds.), *AI-Driven Circular Economy for Carbon Neutrality in Emerging Economies*, Environmental Footprints and Eco-design of Products and Processes,
https://doi.org/10.1007/978-3-032-30298-4_1

1 Reframing Carbon Neutrality as an Institutional Transformation

The global discussion on the multi-stakeholder commitment to achieving carbon neutrality has become a major imperative of this century. As the world moves forward with the mission of building a sustainable future for the people and planet, a need for reconfiguration of the structural transformations for impactful progress has been observed. In this scenario, reframing carbon neutrality as an essential practice with institutional transformations, such as supply chain and logistics, financial systems, industrial productions and societal behaviour is necessary to understand the impact of decarbonization. In the emerging economies' contexts, these challenges are more complex with multidimensional issues, such as eradicating environmental externalities, examining social barriers and sustaining economic development.

In this context, the vision of the circular economy offers the pathway to merge all these aspects under one conceptual understanding. Not only by offering the systemic 4R framework, stating reduction, reuse, recycling and recovery as the process of production, circulation and consumption, but in its holistic understanding, it defines sustainable institutional practices for better environmental, social and economic development. As circularity is operationally a complex process, more technological advancements with real-time data insights, predictions and forecasting, systems optimization and impact measurements are required. With the advent of Artificial Intelligence, this regard has become a disruptive tool that may help in achieving these functions to enable the governance infrastructure to perform optimally.

AI has revolutionized the circular product and process designing but also enabled automatic and remote monitoring of the efficiency during the manufacturing process, by providing transparent reporting and accounting practices. In enhancing climate sustainability, this helps in better carbon accounting, climate financial budgeting, waste management optimization and sustainable urban remodelling. However, in developing economies, there are a multitude of challenges, keeping the effective adoption and implementation of AI, creating a fragmented, uneven impact across these domains.

The volume addresses these caveats while building on a central proposition: AI-driven circular economy transitions are fundamentally institutional and governance challenges rather than merely technological implementations and innovations.

2 The Policy Orientation

The volume presents twenty-one chapters that collectively engage with the policy implications of AI-driven circular economy growth and transitions. As the proposition positions circularity as a governance challenge, it is imperative to focus on the emerging policy debates in this intersection. The policy discourse in this book presents four principal dimensions. First, the chapters articulate normative foundations, engaging with the discussions on ethical standards, stakeholders' responsibilities and leadership and accountability mechanisms in AI implementation. Second, the studies identify the institutional gaps in this implementation: the regulatory fragmentation, digital divides and structural constraints come forward. Third, the volume proposes strategic policy frameworks to bridge these gaps by proposing structured pathways and roadmaps for adopting AI within the existing practices of circular economy and carbon neutrality strategies. Finally, the studies offer multiple empirical and modelling-based insights that present emerging analytical perspectives to support evidence-sensitive policymaking strategies. While being methodologically robust and conceptually diverse, the volume keeps a unifying thread of policy relevance. Each contribution, either theoretical, review-based, empirical, or analytical, engages explicitly or implicitly with governance implications.

3 Methodological Composition

The academic and industrial research at the intersection of AI, carbon neutrality and circular economy is still at a nascent stage, especially in the emerging economies. Mapping long-term AI-driven sustainability outcomes through empirical datasets and designing regulatory frameworks are still in the adaptation phase. With the pace of rapid technological innovation, more research on this front is needed with diverse methods. In this purpose, the chapters in this volume engage with structured literature synthesis, theoretical framework construction, Bibliometric mapping approaches, strategic foresight analysis, case studies and normative policy designing. A few studies also engage deeply with quantitative deduction, interactional analysis and sectoral cross-case examination. The conceptual and theoretical integration with the emerging concepts is at the central focus with structured analytical frameworks, risk identification and governance guidance. The volume, therefore, presents highly policy-oriented discussions with conceptual coherence while also identifying the areas for future empirical expansion.

4 Thematic Contributions to Policymaking

4.1 Ethical Leadership and Digital Infrastructure

The convergence of corporate responsibility and ethical leadership has come forward in several chapters, articulating the need for responsible digital transformation and algorithmic accountability for building a sustainable institutional ecosystem. These contributions incorporate AI governance with underscoring transparency and explainability standards, safeguards against algorithmic bias, human-centric design principles and corporate accountability mechanisms. These insights are very important in the stakeholders' participation, highlighting the need to embed ethical oversight within the Circular economy procurement policies, regulatory codes and corporate governance standards.

4.2 Development Transitions and Digital-Circular Capabilities

A substantial discussion has been undertaken, examining the dynamic capabilities in AI-enabled growth and its contributions to social sustainability through defining pathways to carbon–neutral and circular economies. In emerging BRICS countries, the green growth transitions through ICT-driven solutions are discussed, exploring what the actionable solutions are to achieve the Sustainable Development Goals (SDGs). These chapters exclusively align circular economy frameworks with industrial policies, while addressing digital inequality through targeted capacity building by integrating sustainability objectives into national digital strategies. It emphasizes that AI-adoption must not be undertaken in an isolated manner but be inclusive of the contextual socio-economic disparities.

4.3 Carbon Accounting, Adaptation and Measurement Systems

Chapters building on the role of ICT in carbon accounting, strategic adaptation modelling and AI-driven value chain present the decarbonization pathways to offer governance-related insights. For the emerging economies, the carbon accounting practices are often criticized for non-transparent data protocols, inadequate carbon reporting standards and inaccurate AI-auditable measurement systems. To address these issues, several chapters have provided actionable solutions and important insights into mapping this structural fragmentation and help in knowledge evolution for building credibility in global climate governance.

4.4 Financial Systems, DeFi and Digital Risk

The concepts of decentralized finance, AI-driven trading systems, sustainable finance budgeting and risk management with generative AI take a central space in the discussion of the circular economy in this volume. Starting from the post-pandemic era with time-varying interactions to understanding automated decisions and market fairness in decentralized finance extends the discussion of sustainable finance in circular economy imperatives. The financial policies in this regard focused on the supervision of AI-powered training systems, mitigating systemic risks in decentralized markets and preventing algorithmic manipulation to balance innovation incentives with managing market integrity. The discussion not only reinforces the necessity of updating financial regulatory capacity in the emerging economies' contexts but also provides insights into building financial capability-building mechanisms in digitally intensive sustainability transitions.

4.5 Structural Climate Governance and Market Reform

The final thematic cluster of the volume addresses the issues regarding carbon emissions reductions and carbon market legitimacy in the developing economies. These particular chapters exclusively focus on the concepts of equity, sovereignty and institutional trust concerns. Here, the authors of the respective chapters focus on situating the AI-driven technical process within the concepts of sociocultural and economic ideologies, such as how the practices in the carbon markets in global south will redefine the neoliberal foundations amid neocolonial critiques. Here, the focus is also shifted to sustainable urban development with an exploration of the emissions reductions. Here, the policy implications examine the carbon market inequity, designing transparent prioritization frameworks while strengthening multi-layered governance systems.

5 Editorial Integrity and Scholarly Development

Throughout the editorial process, attention was towards enhancing the diverse methodological aspects while ensuring a transparent and consistent methodical approach across the chapters. As per the central proposition of this volume, the focus is explicitly on the articulation of the policy implications in every thematic cluster. This is to ensure that the governance and institutional practices in the circular economy are grounded in policy imperatives. A terminological consistency has been maintained throughout the chapters to keep the discussion grounded on the central theme of the volume. A special focus has been taken to contextualize the realities of the emerging economies by analysing the social and industrial cases that are working towards this cause. While isolated variation remains inherent in the edited volumes, a rigorous revision of the chapters has strengthened the analytical clarity and governance relevance. However, the chapter does not claim empirical finality but rather contributes towards a structured policy discourse at a formative stage of AI-driven sustainability and circular economy research.

6 Towards Integrated Governance Structure

Across the contributions, the primary insights that come forward are the need for alignment across four policy domains, such as digital governance and data regulation, environmental and climate policy, financial supervision and innovation strategies. These areas require major focus while implementing AI-driven transitions for circular economy designs. The chapters highlight that the global south still lacks the robust and long-term strategies to enhance research capacity, sustainable investments into research facilities and skills, as well as evidence-based policymaking. Hence, the urgent call for action that comes through the volume is to prioritize AI-driven capacity building, policy advice and informed financial decisions for more sustainable outcomes for the developing economies in future.

Part I
Ethics, Governance, and Responsible AI

Chapter 2
Ethical Leadership in the Age of AI: Fostering Sustainability, Building Trust, and Accountability in Data-Driven Organizations

Krupa B. Nair, P. C. Radhika, and Bhagyalakshmi S. Pai

Abstract With most of the companies using artificial intelligence in recent times, ethical issues and responsible leadership call for awareness of the developments occurring around. Leaders have obligations to uphold the ethical guidelines of the company in order to protect the values and integrity. Since the companies depend on artificial intelligence to guide decisions, promote sustainability, be responsible, open, fair and human-centered leadership becomes absolutely vital. Ethical leadership is very important in guiding companies across this technological change by including moral responsibility and ethical governance into the design and execution of intelligent systems. This chapter provides significant new perspectives on how ethical leadership enables data-driven companies to build dependable and open artificial intelligence ecosystems. It looks at leadership strategies including stakeholder participation, values-driven decision-making, data ethics education, and corporate integrity promotion to help ethical artificial intelligence to be adopted. The chapter puts forward the necessity to coordinate the leadership activities with ethical principles to reduce the adverse impact of algorithmic bias, preserve the privacy of data, and create a sense of trust in artificial intelligence systems among people. It also notes that leaders must be proponents of conscientious innovation and a culture of responsibility should be fostered. This chapter provides insightful advice on integrating ethical governance into the value system of the organization through a qualitative research on the views of CFOs, CHROs, and senior managers in various fields. The implications of the findings are to clarify the reason why ethical leadership is capable of balancing technological development and long-term interests of society and the

K. B. Nair (✉) · P. C. Radhika · B. S. Pai
Department of Management and Professional Studies, Rajagiri College of Social Sciences (Autonomous), Kalamassery, Kochi, India
e-mail: krupa@rajagiri.edu

P. C. Radhika
e-mail: radhika@rajagiri.edu

B. S. Pai
e-mail: bhagyalakshmi@rajagiri.edu

K. Kumar et al. (eds.), *AI-Driven Circular Economy for Carbon Neutrality in Emerging Economies*, Environmental Footprints and Eco-design of Products and Processes,
https://doi.org/10.1007/978-3-032-30298-4_2

environment to guarantee resilience and legitimacy in the era of intelligent machines. This chapter enables the creation of ethical leadership models that suit the digital age because it is centered on sustainability, trust, and accountability.

Keywords Ethical leadership · AI governance · Sustainability-oriented leadership · Trustworthy AI · Responsible innovation · Algorithmic ethics

1 Introduction

Algorithms and data analytics such as artificial intelligence (AI) have become the pervasive elements of modern business practices, which have also generated more sophisticated ethical concerns. Implementing AI in the decision-making process is an enormous change to the conventional corporate practice. The AI can revolutionize the strategic decision, automate the routine activities, and personalize the executive communication with the stakeholders. However, such innovations raise the issue of algorithmic bias, the necessity of transparency of AI technologies, and the potential loss of such key leadership qualities as empathy and ethical custodianship (Frimpong, 2025). The consequences of AI-driven leaders are unlikely to take independent leadership roles as autonomous robots instead, programmers and the following decisions produced by AI systems might be controlled by the AI leader. This means that regulatory frameworks are required that would regulate the relationship between human and machine in collaboration. The difference between actual leadership and control by managers in this kind of setting is not clear, which requires further research into the proper supervisory systems (Smith & Green, 2018).

The AI technology suggests possibilities to increase the efficiency of individuals, make informed decisions, and automatize the working process. However, concerns related to algorithmic bias, privacy loss, and non-transparency of decision-making remain, which explains the fact that it is necessary to follow the reviewed ethical standards (Floridi, 2023; Osasona et al., 2024). These issues require the creation of leaders who can show a balance between ethical requirements and technological progress. Uddin (2023) explains that the digital era requires leaders who apply compassion, trust, and long-term strategic vision in AI governance. There is a lot of literature on the adaptations of leadership in the digital environment. Indicatively, Abasaheb and Subashini (2023) note that AI is transforming business processes and changing the leadership behavior.

Heyder et al. (2023) theorize the idea of ethical leadership in organizations, which are enabled by AI, as the administration of sociotechnical relations, not solely the management of technology. The focus of their review is that the leaders are to align AI deployment with human values by interfering with responsibility to rules and ethical judgment subject to virtues. In this perspective, ethical leadership will be the key aspect of human-AI interaction and, therefore, will provide accountability, transparency, and value alignment in organizational decision-making.

Modern leaders should transform themselves into digital mentors who are skilled at managing as well as deliberating on technology as an ethical concern. On the same note, Madanchian et al. (2024) state that AI-oriented environments demand leaders able to change the organizational culture, encourage ethical decision-making, and foster flexibility of the workforce. The concepts also reveal that ethical leadership requires the existence of trust. The extent of trust people put in AI depends mostly on its visibly exposed structure and the responsibility of its managers (Díaz-Rodríguez et al., 2023). Ethical leaders develop this trust through having strong regulatory frameworks, involving all players, and creating open discourse. The case under analysis is by Petrucci and Rivera (2018), who discuss leadership transformation in digitally enabled organizational settings where the authors believe that an effective leader needs to combine the conventional leadership skills with technological skills and data-intensive practices. Their discussion highlights that digital leaders are turning to information systems and feedback mechanisms which are real time to drive decision-making and organizational development. They provide an example of the current organizations that have implemented digital feedback as a replacement of the traditional annual performance appraisal, thus allowing the leaders to develop a culture of meritocracy and improve the performance outcome using continuous data streams.

According to Srinivasan and González (2021), the concept of empathy and person-centered focus are two inseparable components of AI accountability. As a result, ethical behavior becomes a highly important quality of leaders. Furthermore, sustainability issues emerge since organizations are moving towards AI-focused operations; leaders should make sure that their AI plans are consistent with organizational goals and the larger societal needs to transform the situation in the world. Mahmood and Abbas argue that responsible innovation is important because the ethical and technical abilities of the leadership can have a strong impact on the ability of a firm to be resilient. The conceptualization of leadership proposed by Samani et al. (2012) is based on the idea that leadership is a process of behavior and relationships that progressively integrate intelligent technological agents to promote coordination and support decision-making. Their study emphasizes the development of distributed human–machine forms of leadership and reminds of the unchanged need of human ethical responsibility regarding the decision-making contexts that are still technologically mediated. Such an attitude strengthens the need to have leadership models that combine technological consciousness and ethical accountability.

The intersection of technology innovation and sustainability is not an easy challenge that organizational leaders are dealing with, especially in the data-intensive environment. The overall aim is to educate the decision-making processes that enhance workforce adaptation, which will help the human capital to efficiently react to the changes in the technological environment. In these scenarios, ethical leadership will play a crucial role in terms of gaining confidence of stakeholders, creating a sense of accountability and matching the organizational strategies with the long-term social and environmental well-being. To this end, the current chapter outlines the mechanisms that can be used to promote the sustainability, credibility and accountability of artificial intelligence-based businesses by ethical leadership. The topicality of the present study is broadened to the creation of the ethical governance frameworks, which would suit the digital age contextually.

It is essential that the study can be applied to the creation of ethical governance frameworks that are appropriate in the digital age through the analysis of qualitative information gathered among top executives in various sectors. It helps organizations to bridge the gap between theory and practice, thus making it possible to create ethically responsible AI ecosystems. The modern practice lays stress on the fact that leaders should demonstrate ethical behavior to enable the stakeholders to trust them. The leaders also need to enforce accountability and ensure that the organizational initiatives are not harmful to the society or the environment in the long run.

In this line, this chapter explores the possibility of sustainability, trustworthiness, responsibility of AI companies through ethical leadership. The research contributes to the development of ethical governance models that can be used in the digital era based on the empirical findings of senior leaders. The research enables capacity building of organizations in ethical, responsible, and resilient AI systems by reducing the theory–practice gap.

2 Context and Theoretical Background

Artificial intelligence is disrupting organizational processes and thus it requires a corresponding adaptation in leadership. Decision-making is no longer an activity of human beings; now computers, algorithms, and various technological artefacts are involved. Leaders have the responsibility of ensuring that these technologies are not applied in a manner that is unjust, exclusionary, and protective of the future (Díaz-Rodríguez et al., 2023; Floridi, 2023). In the old paradigm, quality leadership was equated to profitability, efficiency of operations and job satisfaction. The modern environment requires leaders to consider data ethics, algorithmic fairness, and inclusion of stakeholders and long-term implications in their reasoning (Brown & Treviño, 2006). They need to understand how technology works and set ethical frontiers. Since AI is becoming an intervention factor in decision-making, leaders are in charge of the moral consequences of the automated decisions. This change extends to a technical change and goes into a philosophical change (Osasona et al., 2024; Srinivasan & González, 2021). AI and other technologies force one to re-evaluate power, equality, and discrimination. Computational systems do not have any moral cognition as human agents do; moral values have to be programmed intentionally. As a result, data-driven organizations cannot just oversee the algorithms; their leaders need to actively consider ethics when designing, using, and assessing such systems (Madanchian et al., 2024; Smith & Green, 2018).

Current cross-cultural studies show that the expectations of how artificial intelligence should be ethically governed vary significantly in the institutional, cultural, and regulatory contexts. Unlike Western paradigms, where transparency and individual responsibility are of paramount importance, frameworks in emerging economies are more likely to prophesy collective good and social influence (Mahmood et al., 2024;

Uddin, 2023). These differences demonstrate that ethical-oriented leadership frameworks, which are responsive to context as opposed to prescriptive across the board, are required, which explains why adaptive ethical leadership is relevant in AI-driven organizations. (Alasiri & AlKubaisy, 2022). Besides, within the economically integrated global environment, the choices of AI in a single set of jurisdictions may resonate throughout the societies, economies, and ecological systems of the whole world. Leaders should thus take into consideration local organizational needs and global ethics acknowledging the different perceptions of right and wrong across different cultures (Díaz-Rodríguez et al., 2023). The theoretical basis of the present research is a synthesis of the views concerning exemplary leadership, stakeholder inclusivity, and responsible innovation. Good leaders put ethical principles in place; they mobilize other people to follow suit. Stakeholders are not only the proprietors but also the employees, customers, local communities and regulatory bodies (Frisch & Huppenbauer, 2013). Responsible innovation implies active anticipation, retrospective evaluation of the old culture and responsiveness to the new technologies. Samani et al. (2012) uses a futuristic in technology perspective to study how leadership qualities of society in which smart machines and robotics are involved in decision-making process develop. They argue that robots do not replace human leadership, leadership has to change in situations where technology plays roles that are traditionally taken by human leaders. The combination of the mentioned factors helps to understand the all-encompassing nature of leadership in AI-assisted businesses, combining technical factors, human agency, and the need to foster transparency and responsibility. In this way, the chapter does not consider ethics as a subsidiary factor in regard to AI governance. Rather, it pre-empts ethical consideration as a fundamental responsibility of leadership, which calls on leaders to thoughtfully consider the moral implications of AI strategy and day-to-day operations. This focus is essential in the modern environment, where organizational image, legal and social behavior are determined by the moral use of technology (Mahmood et al., 2024).

In the existing literature, there is an ongoing conflict between the use of AI in performance and ethical responsibility. Although data-driven optimization is claimed to be efficient and scalable, researchers fear a decline in ethical values due to the delegation of decision-making power to opaque algorithms in large volumes (Anupama Prasanth et al., 2023; Osasona et al., 2024). This conflict highlights a central position of leadership in balancing the process of innovation and ethical responsibility. The chapter argues that ethical leadership serves as the means of integration that relates artificial intelligence governance to sustainability. Ethical leadership provides normative basis, AI governance provides structural control, and sustainability ensures the long-term value creation to the society and organization. Together, these constructs are more of a mutually reinforcing system and not separate theoretical domains. The current discourse describes the applicability of ethical leadership within data-driven environment, which provides a paradigm within which to explain further empirical research. The chapter contributes to empowering both scholars and practitioners with a subtle perspective of ethical AI leadership in contemporary organizations by basing the argument on existing literature.

3 Methodology

Gioia analytical framework was especially useful in deriving subtle information about verbal and non-verbal communicative attributes of informants, and hence, offer a solid platform through which theoretical constructs can be built. The current study employed a purposive sampling approach, which selected the participants who have a lot of knowledge in areas that are related to the integration of artificial intelligence, ethical governance, and strategic leadership. Two different groups were included to guarantee a high level of validation: (1) the scholars that study technology ethics, artificial intelligence, and leadership studies; and (2) the industry professionals, such as chief financial officer, chief human resources officer, and senior managers working in AI-utilizing companies. The interviews were semi-structured and a rather small sample was used, the task of which was to not only test the already existing hypotheses but also come up with new ones in connection with ethical leadership in AI-oriented organizations. This approach decision was based on the ability to obtain profound thoughts about the leadership experiences of the participants, the current ethical culture in their organizations, and the adoption of AI strategies.

Twenty-three people took part in semi-structured interviews; half of them were academics and half were professionals. The interview guide covered issues including making decisions based on values, getting stakeholders involved in AI governance, making algorithms more transparent, setting up accountability systems, and making sure that tech-driven plans are long-lasting. Each interview lasted between 45 and 60 min and was recorded so that it could be analyzed later. Even though twenty-three professionals agreed to take part in the study at the beginning, only nine of them filled out the entire semi-structured interview guidelines. Scheduling and professional obligations were also credited with the attrition. Following the standards of qualitative research and the Gioia approach, nine educated participants were enough to reach the theoretical saturation.

We used a three-tier coding structure based on the Gioia approach. First-order codes recorded terms and phrases that were focused on the informant. Then, these were put into second-order themes that showed the main ideas and new patterns that were starting to show up. Finally, these concepts were put into larger groups that helped create a conceptual framework for ethical leadership in AI settings. The inductive technique made it possible for a grounded theoretical model to come out of real-world activities and experiences. Member verification and peer debriefing were used to make the information more reliable and trustworthy. This method not only helps to confirm conceptual notions, but it also gives leaders who want to make sure that AI use is in line with moral obligations useful information. This part of the methodology explains how the research was set up to answer the main question: How can ethical leadership in data-driven enterprises foster sustainability, trust, and accountability in AI governance? The Gioia method made it easy to see and follow a strict process for turning real-life experiences into useful theoretical contributions.

3.1 Conducting Interviews

Despite the fact that the ethical AI principles have plenty of support, their practical implementation can be a challenging issue. The most common issues that are reported by the leaders include the shortage of resources that can be explained by the existence of the legacy information systems, unclear accountability processes, and the deficiency of AI literacy. The need of quantitative indicators, i.e., transparency indices, AI incident reporting systems, and stakeholder trust indices, was emphasized by the respondents as the means of moving ethical commitments to tangible governance practices. The interview guide was thoroughly designed to rely on a literature review that is exhaustive in terms of ethical AI leadership, stakeholder governance, digital transformation, and sustainable management. The next step involved the creation of a preliminary collection of thirty salient questions that were created to expand the discussions around the value-based leadership, the establishment of trust on an AI system, information management, the role of the AI system, and the application of the sustainability principles in digital leadership. The questions were shaped by the existing studies on AI ethics and responsible leadership and models of governance (Díaz-Rodriguez et al., 2023; Floridi, 2023). The last phase in designing the interview guide was an evaluation by three academic specialists in AI governance, corporate ethics, and strategic leadership, who rated the quality and applicability of the interview guide.

The final semi-structured interview guide, incorporating the feedback of the panel, consisted of twenty-five central elements, which were organized in a methodical way around all the most important elements of the conceptual model, and thus formed a solid analytical framework,

- Ethical Governance and Accountability.
- Trust and Transparency in AI Deployment.
- Sustainability in Digital Leadership.

Each person was asked to think about their organization's procedures, leadership decisions, and perceptions in each of the theme categories. The last questionnaire had both open-ended questions and evaluation prompts that used a Likert-type scale (1 to 4) to measure how clear, aligned, and important certain ethical leadership practices were thought to be in AI environments. The first interaction with participants was an email inviting them to take part in the study. The last group had 9 people in it. There were 2 from academics and 7 senior professionals, such as CHROs, CFOs, and top-level managers, who agreed to take part in video-conference interviews. Before the meeting, each participant got the interview guide and a short idea brief to help them get used to the material and think about it ahead of time. According to qualitative research standards, a sample size of 9 is appropriate because it is between 5 and 25, which is enough to get topic saturation. We got permission to record all of the interviews and then transcribed them for later analysis using the Gioia approach.

The demographic characteristics of the respondents including gender, age, educational qualification, years of experience and occupation are presented in Table 1.

Table 1 Demographic profile of respondents

Demographic summary	Categories	Respondents
Gender	Male Female	5 4
Age	25–34 35–44 45–54 55 and above	1 4 3 1
Educational qualification	Bachelors Degree Post Graduate PhD	1 6 2
Years of experience	6–10 years 11–15 years 16–20 years 21 and above	2 3 3 1
Occupation	CFO CHRO CIO/CTO Academician/Researcher AI/Data Professional	3 2 2 1 1

The study included nine people from a mix of academic, executive, and professional backgrounds to make sure that the understanding of ethical leadership in AI-driven enterprises was balanced and rich in context. Table 1 presents the summary of their demographic traits

This distribution reflects a mature and experienced cohort, offering informed insights on ethical leadership practices, AI governance, and sustainability principles. The representation of both academic and practitioner viewpoints ensures credibility and diversity in perspectives for qualitative analysis.

3.2 *Data Analysis*

3.2.1 Ethical Governance and Accountability

The initial analysis of Ethical Governance and Accountability defines current attitudes and behaviors related to ethical issues in AI leadership. The respondents reiterated the need to explicitly regulate AI application to enhance fairness, compliance, and accountability. They mentioned the constructs like compliance culture, ethical training programs, leader with values and open decision-making process. These constructs were organized into Leadership Responsibility, Ethical Oversight and Governance Structures. It has been stressed that ethics should be part of the AI strategy at the very beginning and not an incidental factor. Altogether, these reactions show that ethical leadership is critical in successful AI policy formulation and risk reduction. Figure 1 shows the Data Structure (Gioia Methodology) of the Ethical Governance and Accountability Dimension.

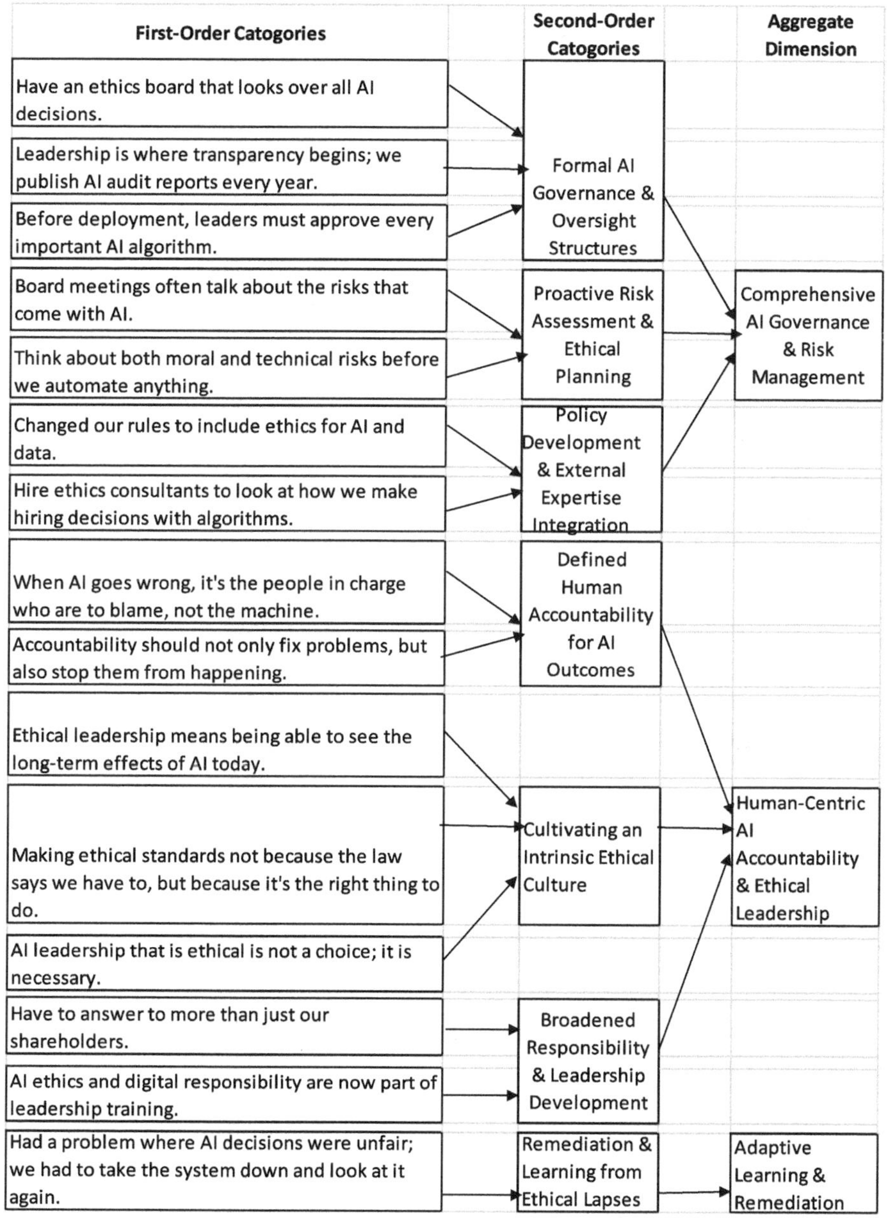

Fig. 1 Data structure (Gioia methodology) of the ethical governance and accountability dimension. *Source(s)* Authors' own creation

3.2.2 Trust and Transparency in AI Deployment

The next section, Trust and Transparency in AI Deployment, explores processes by which heads develop trust in AI projects at their companies. It is shown in interviews that some salient elements that help in building confidence are transparency on the topic of algorithmic decision-making, proactive communication, and productive stakeholder engagement. Some of the most common themes are explaining the results of AI, auditing bias, participatory, and the interpretability of data. Our participants repeatedly emphasized the point that openness is not only about the technology itself but about other organizational practices, which requires an articulate conversation and action. Therefore, the reliability of AI depends on functional operation, ethical implementation, and the level of inclusion in the policy process and risk control that the leaders allow stakeholders to partake in. Figure 2 shows the Data Structure (Gioia Methodology) of the Trust and Transparency in AI Deployment Dimension.

3.2.3 Sustainability in Digital Leadership

Figure 3 shows the Data Structure (Gioia Methodology) of the Sustainability in Digital Leadership Dimension. Sustainability in Digital Leadership is the third area of focus and it deals with the sustenance of digital transformations by the leaders without causing negative societal, environmental, and fiscal consequences. Interviewees describe its concurrence with responsible behavior and see AI as a way to make the world and the planet better. These aspects were grouped into categories like Earth-Friendly Innovation, Inclusive Technological Leadership, and Ethical AI Skill Development. According to the leaders, modern AI leadership requires a two-way benefit, which promotes digital initiatives in a progressive way. This point of view highlights the view of a holistic leadership approach that integrates ethical responsibility, human wellbeing, and creative problem-solving, beyond the scope of financial rewards. On the basis of the discourse above, there is a series of findings based on the expert opinions, which are organized in the Gioia data structure as well as the results of the evaluations in terms of clarity, coherence, and relevance. The expression of such findings has been deeply revised and then slightly modified. Results of the Gioia Methodology will be included here and the results will be given below.

4 AI-LEAD Framework (*AI Leadership for Ethics, Accountability, and Development*) Model

The AI-LEAD model is based on inductive development of the framework based on the stories stated by the participants. The first-order concepts such as explainable algorithms, ethical audits and human oversight were clustered into second-order themes of transparency, accountability and learning. The collective themes form the

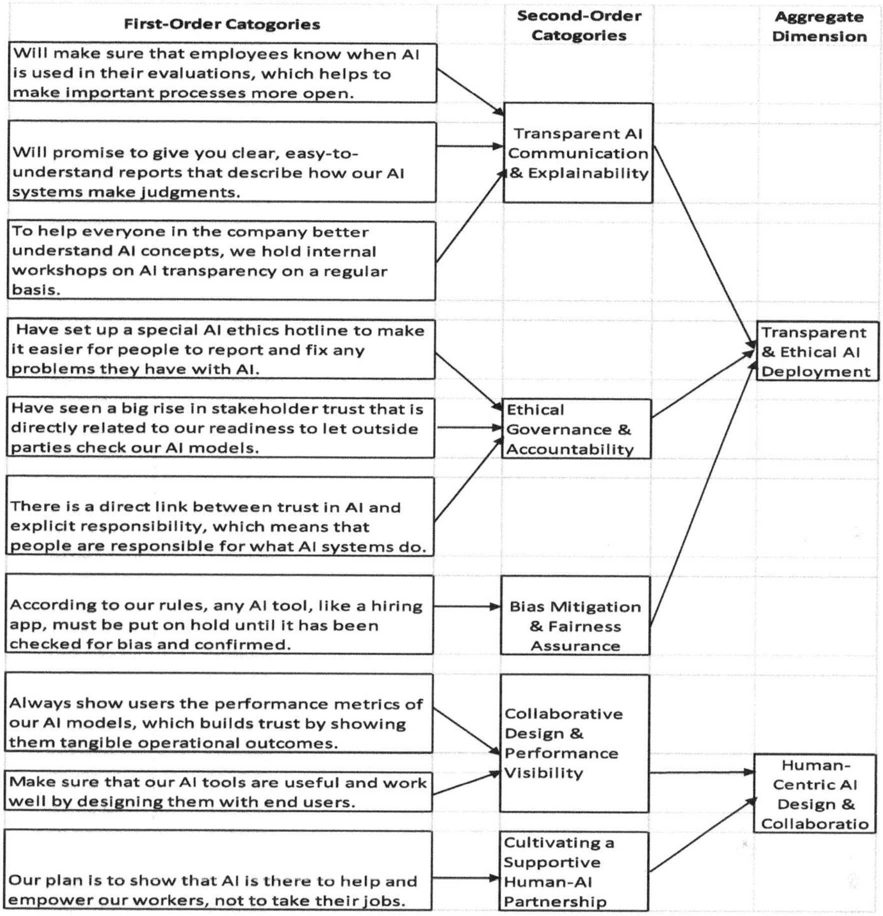

Fig. 2 Data structure (Gioia methodology) of the trust and transparency in AI deployment dimension. *Source(s)* Authors' own creation

basis of the composite dimensions which form the AI-LEADs framework such that the model is grounded on empirical knowledge. Table 2 presents the findings for the AI-LEAD Framework by aggregate dimension.

The paper is a contribution to the ethical theory of leadership because it broadens the scope of the theory to include the context of AI-mediated decisions. It adds to the existing body of knowledge on AI governance by showing how values of leadership are transformed into structural governance practices. In addition, it adds to the sustainability theory in a conceptualization of ethical AI as a long-term value-creation approach, as opposed to a compliance measure.

The AI-LEAD framework serves as a diagnostic instrument and a guideline to implementation to practitioners. It can be used by executives to build ethics boards, to incorporate accountability metrics, and to equate AI initiatives with ESG goals. The

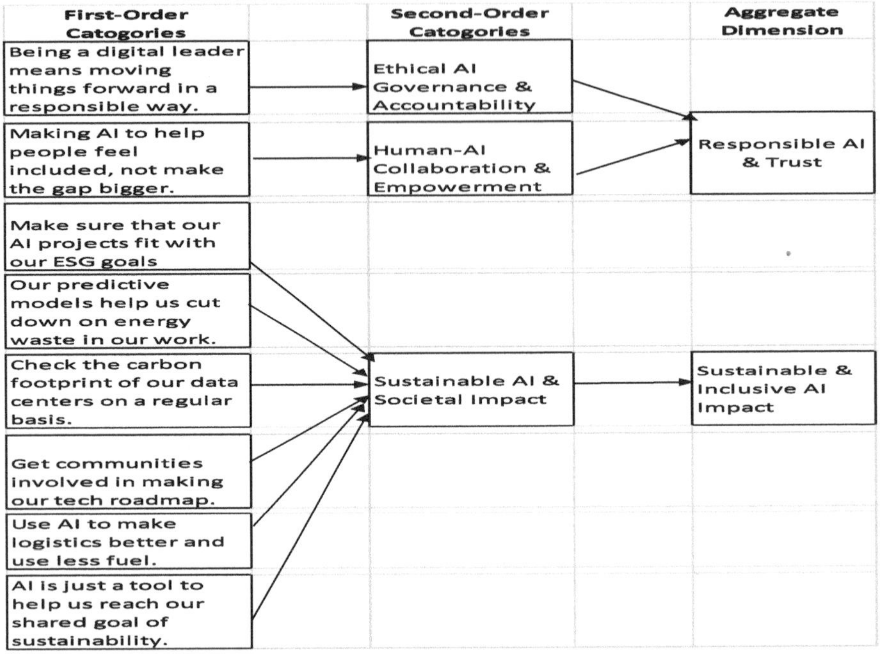

Fig. 3 Data structure (Gioia methodology) of the sustainability in digital leadership dimension. *Source(s)* Authors' own creation

framework also helps in the policy formulation process since it transforms abstract ethical principles into tangible leadership practices.

5 Discussion and Concluding Remarks

The application of AI to the work of the company is turning into a trend, and it implies that the leaders will have to consider some new ethical aspects. This paper examines how one can be an ethical leader to balance the enhancement of technology with key concepts like trust, responsibility, and doing what is sustainable. Based on the interviews with the big bosses and experts, the study developed a leadership plan, which considers the ethics in the way AI is operated, the clarity of things, the level of trust people have, and the willingness of the company to become sustainable.

One of the important things to note is that ethical management and responsible leaders are very important in planning. All the interviewed indicated that it is necessary to watch AI when making major decisions. Avoiding ethical leadership can only imply doing more than merely obeying the regulations. It is also associated with planning, scrutinizing things continuously, and being comfortable with the idea

Table 2 Findings for the **AI-LEAD framework** by aggregate dimension

Model dimension	Model subcategory	Aggregate dimension Gioia	Changes to be incorporated
Strategic leadership and governance	Strategic oversight	Comprehensive AI governance and risk management	• Set up AI ethical boards that include people from different departments • Keep track of AI risks in strategic evaluations Set up AI compliance KPIs Add AI provisions to data policies
	Reflective leadership	Human-centric AI accountability and ethical leadership	• Use case-based learning to teach leaders about AI problems • Include ethics in leadership scorecards Promote talking about how to use AI Support the values of the organization
	Inclusive development	Sustainable and inclusive AI impact	• Keep records of AI incidents Include comments from stakeholders in ethical updates • Use AI dashboards that let you give feedback
	Stakeholders trust	Responsible AI and trust	• Make AI tools explainable by law Make sure there are reports that explain Have other parties check for fairness Put the results in ESG/ CSR reports
Operational excellence and innovation management	Continuous feedback	Adaptive learning and remediation from ethical incidents	• Include a wide range of users in the design of AI Use ethical scenario planning Don't let too much automation happen in human jobs • Make sure design is in line with human rights

(continued)

Table 2 (continued)

Model dimension	Model subcategory	Aggregate dimension Gioia	Changes to be incorporated
	Algorithmic transparency	Transparent and ethical AI deployment	• Get the public involved in moral discussions Talk about your ethical AI commitments and create KPIs based on trust • Keep an eye on fairness and perception over time
	Participatory design	Human-centric AI design and collaboration	• Encourage the use of datasets that include everyone • Check for fairness in algorithms and do long-term ethical reviews • Make sure that AI use is in line with SDG and ESG goals

Source(s) Authors' own creation

of doubting whether it is morally right to use data and AI. Leaders must be technological savvy, of course, but they must also advocate the use of guidelines founded on good ideas. There should also be checking to ensure that algorithms are not biased, prejudice is to a minimum, and that data is secure.

Individuals said that good ethical leadership requires and brings trust and openness. They claimed that at this moment, people do not trust AI much. To rectify this, businesses should be transparent on how they manage information, how their models perform, and what are the results. Leaders who understand what AI is attempting to accomplish, what it is incapable of accomplishing and what security measures are implemented are perceived as one of the main steps to ensure that employees and other people become comfortable with it. Building trust can also be supported by getting varying opinions on the development of AI and providing an avenue through which all members of the company may raise their concerns.

The data we received indicates that sustainability of things is a huge concern among the digital leaders. The respondents who participated claimed that AI plans must not merely be focused on making things work or beating the competition. They are also supposed to consider the impact of these plans in the earth and society. Ethical leaders must consider such issues as the amount of energy consumed in the data centers, e-waste, and the exclusion of particular groups of people by the algorithms. It is evident that businesses should ensure that their AI content aligns with their higher objectives of being green, social, and well-governed (ESG), and the Sustainable Development Goals (SDGs). Ethical leadership can be run not only as the right thing to do, but also a prudent step to do well in the long run.

It has been indicated by the Gioia analysis that ethical leadership must be situational, interactive, and self-ambivalent. It was claimed that people would want decisions to be made not only by IT and top managers dictating AI. They believe that we should have written ethical policies and means of holding everyone accountable to promote responsible innovation.

Although it is an improved situation, some of the problems still exist. The majority of leaders claimed that they do not fully understand AI ethics and require training on data ethics and leadership. Money issues and former habits are the barriers to ethical leadership sometimes. This is why we require further assistance of the institutions and collaboration in various spheres.

6 Conclusion

To sum up, ethical leadership is not only a theory, but also a necessity of data-driven companies that are interested in becoming successful working with AI. The results indicate that it is quite crucial that leaders develop AI systems that will be sustainable, trustful, and responsible. Ethical leadership must accompany all processes of digital change, including planning and engaging individuals, the success measurement, and establishing rules.

It is a chapter that can contribute to the emerging debate, demonstrating ethical leadership in real-life AI situations. Teams are not merely able to adhere to regulations; they can drive responsible and sustainable innovation through signifying moral administration, confidence of stakeholders and sustainability in their leadership. The AI changes that require the transformation of leadership styles must be ethical, empathetic, and long-term-oriented regarding the well-being of people.

References

Abasaheb, S. A., & Subashini, R. (2023). Maneuvering of digital transformation: Role of artificial intelligence in empowering leadership—An empirical overview. *Journal of Law and Sustainable Development, 11*(9), e1667. https://doi.org/10.55908/sdgs.v11i9.1667

Alasiri, N., & AlKubaisy, Z. (2022). Exploring the role of leadership, work environment, it alignment and company performance on the digital transformation: a study on the private sector companies in western region, Saudi Arabia. *International Journal of Professional Business Review, 7*(2), e0500. https://doi.org/10.26668/businessreview/2022.v7i2.500

Brown, M. E., & Treviño, L. K. (2006). Ethical leadership: A review and future directions. *The Leadership Quarterly, 17*(6), 595–616. https://doi.org/10.1016/j.leaqua.2006.10.004

Díaz-Rodríguez, N., Del Ser, J., Coeckelbergh, M., De Prado, M. L., Herrera-Viedma, E., & Herrera, F. (2023). Connecting the dots in trustworthy Artificial Intelligence: From AI principles, ethics, and key requirements to responsible AI systems and regulation. *Information Fusion, 99*, Article 101896. https://doi.org/10.1016/j.inffus.2023.101896

Floridi, L. (2023). *The Ethics of Artificial Intelligence: Principles, Challenges, and Opportunities* (Oxford, 2023; online edn, Oxford Academic, 24 Aug. 2023). https://doi.org/10.1093/oso/9780198883098.001.0001

Frimpong, V. (2025). The impact of AI on evolving leadership theories and practices. *Deleted Journal, 2025*(3), 188–193. https://doi.org/10.53935/jomw.v2024i4.1100

Frisch, C., & Huppenbauer, M. (2013). New insights into ethical leadership: A qualitative investigation of the experiences of executive ethical leaders. *Journal of Business Ethics, 123*(1), 23–43. https://doi.org/10.1007/s10551-013-1797-9

Heyder, T., Passlack, N., & Posegga, O. (2023). Ethical management of human-AI interaction: Theory development review. *The Journal of Strategic Information Systems, 32*(3), Article 101772. https://doi.org/10.1016/j.jsis.2023.101772

Madanchian, M., Taherdoost, H., Vincenti, M., & Mohamed, N. (2024). Transforming leadership practices through artificial intelligence. *Procedia Computer Science, 235*, 2101–2111. https://doi.org/10.1016/j.procs.2024.04.199

Mahmood, G., Khakwani, M. S., Zafar, A., & Abbas, Z. (2024). Impact of digital transformation and AI through fostering digital leadership excellence: A focus on sustainable organizational performance. *Journal of Accounting and Finance in Emerging Economies, 10*(1). https://doi.org/10.26710/jafee.v10i1.2925

Osasona, N. F., Amoo, N. O. O., Atadoga, N. A., Abrahams, N. T. O., Farayola, N. O. A., & Ayinla, N. B. S. (2024). Reviewing the ethical implications of AI in decision making processes. *International Journal of Management & Entrepreneurship Research, 6*(2), 322–335. https://doi.org/10.51594/ijmer.v6i2.773

Petrucci, T., & Rivera, M. (2018). Leading growth through the digital leader. *Journal of Leadership Studies, 12*(3), 53–56. https://doi.org/10.1002/jls.21595

Prasanth, A., Vadakkan, D. J., Surendran, P., & Thomas, B. (2023). College of Computer Studies, University of Technology Bahrain, Salmabad, Kingdom of Bahrain, Role of Artificial Intelligence and Business Decision Making, (IJACSA) International Journal of Advanced Computer Science and Applications, Vol. 14, No. 6

Samani, H. A., Koh, J. T. K. V., Saadatian, E., & Polydorou, D. (2012). Towards robotics leadership: An analysis of leadership characteristics and the roles robots will inherit in future human society. In *Lecture Notes in Computer Science* (pp. 158–165). https://doi.org/10.1007/978-3-642-28490-8_17

Smith, A. M., & Green, M. (2018). Artificial intelligence and the role of leadership. *Journal of Leadership Studies, 12*(3), 85–87. https://doi.org/10.1002/jls.21605

Srinivasan, R., & González, B. S. M. (2021). The role of empathy for artificial intelligence accountability. *Journal of Responsible Technology, 9*, Article 100021. https://doi.org/10.1016/j.jrt.2021.100021

Uddin, A. S. M. A. (2023). The era of AI: Upholding ethical leadership. *Open Journal of Leadership, 12*(04), 400–417. https://doi.org/10.4236/ojl.2023.124019

Chapter 3
AI-Powered Decision-Making and Responsible Approaches to Ethical Dilemmas

Amit Kumar, Nitendra Kumar, and Jasreen Kaur Kakar

Abstract The decision-making process has been benefitted with Artificial Intelligence (AI) integration, and it has greatly benefitted many critical areas such as governance and healthcare. AI systems are fast, efficient and accurate, and their increasing clout in decision-making raises ethical dilemma issues which challenges the established standards of moral values, transparency and accountability. This chapter evaluates the implications of AI systems on the core ethical issues of privacy of data, algorithmic bias, and human intervention on AI-powered decisions. The chapter would also study the increasing need for examining the AI-powered decision-making through the ethical lens. The chapter would study the law, management and healthcare perspectives in addressing trust and ethical issues of AI based decisions. The study would highlight the limitations of the existing regulations in addressing the ever-changing world of AI systems. The chapter would outline a framework that would suggest integrating human in the AI-driven decision-making process and stakeholder engagement. The chapter would contribute to the ever-evolving area of ethical AI and would offer insights for policymakers and practitioners in navigating the ethical dilemmas arising from the implementation of AI systems.

Keywords Artificial intelligence · Ethical dilemmas · AI systems · Algorithmic bias

---

A. Kumar (✉)
School of Business Management, Christ University, Bengaluru, India
e-mail: amitgem@gmail.com

N. Kumar
Department of Decision Sciences, Amity Business School, Amity University, Noida, Uttar Pradesh, India

J. K. Kakar
University Institute of Liberal Arts and Humanities, Chandigarh University, Mohali, India
e-mail: jasreen.e13250@cumail.in

K. Kumar et al. (eds.), *AI-Driven Circular Economy for Carbon Neutrality in Emerging Economies*, Environmental Footprints and Eco-design of Products and Processes,
https://doi.org/10.1007/978-3-032-30298-4_3

1 Introduction

Organizations are facing ethical dilemmas in the dynamic business environment of today as they may have to choose between moral principles and profit maximisation, which may lead to a compromising situation (Porkodi & Cedro, 2025). The companies are faced with the dilemma of choosing between data privacy and profit maximisation. AI is being increasingly used in different sectors which have added a different dimension to the already existing dilemma. AI systems are efficient and fast, but have their fair share of challenges as well (Buzko et al., 2016). AI has been increasingly used in content moderation, legal enforcement and healthcare diagnostics. The penetration of AI across industries has raised issues around ethics, transparency and accountability. Ethical decision-making is heavily dependent on experience and human judgements have their own sets of biases and challenges (Agrawal et al., 2024; Chen et al., 2024). AI can work on huge datasets at a high speed and it can easily identify patterns. This would help companies in better governance and maintaining the moral standards. Integration of AI in the process of decision-making is important to reinforce business ethics.

AI systems are man-made and would reflect the values, beliefs, limitations and biases of people. This raises the issues of integrating ethics in AI systems and algorithms (Atluri & Reddy, 2025; Kumar et al., 2023a, 2023b). This emphasizes that the AI systems should be ethically built and should behave like a partner to humans in decision-making. This chapter aims at examining the ethical dilemmas that exist in the application of AI in decision-making process. It would evaluate the design, development and application of AI systems in morally complex business situations across different sectors (Di Vaio et al., 2020; Singh et al., 2024). The chapter would examine the different frameworks that are being used universally and the dependence on AI for making decisions, especially in cases involving ethical principles. The world is progressing towards automation with technology and moral issues being properly aligned. Intelligent AI systems and moral reasoning would make a universe where AI systems would facilitate in making ethically sound business decisions.

2 Understanding Ethical Dilemmas in the AI Era

The penetration of AI across industries has evolved the landscape of decision-making and has also shown the way in which ethical boundaries are dealt with. AI has now been integrated into systems affecting human lives from pharma to financial services and recruitment. The society is now facing a new set of ethical dilemmas that have not been witnessed before. These dilemmas are no longer a part of philosophical discussions, but they are heavily embedded in data infrastructures (George & Thomas, 2019; Singh et al., 2025a, 2025b). The ethical challenges that are being witnessed in the AI era need a reflection on ethical principles and their application in today's environment. Ethical dilemmas in the AI era become complicated by the increase in levels of automation, transparency issues, and scale. The traditional characteristics of ethical decisions have a clear intent and focus. AI-based ethical decision-making has introduced a system where decisions are made by algorithms that have the least human interference (Jia et al., 2018; Singh et al., 2025a, 2025b). This results in a conflicting situation of functionality and fairness, outcome and intention, as well as autonomy and accountability.

In recruitment, algorithms are being used for scanning, ranking, and assessing resumes for interviews. Such algorithms might be efficient and consistent, but there are serious concerns around them being biased and discriminatory. The biases of the makers of these algorithms would reflect in the results, and the AI would enlarge these biases in the findings. For e.g., if the algorithm contains a historical bias of underrepresentation of women in leadership positions, then the AI system would magnify these biases (Awan et al., 2021). The dilemma increases when companies prioritize cost-effectiveness and efficiency over the morally crucial aspect of equal opportunity and inclusivity.

In healthcare, AI systems are being widely used in the diagnosis and treatment of patients and also in the prediction of patient outcomes based on a large set of medical data. This system operates as a 'black box' and the logic behind the decision cannot be easily interpreted by patients or professionals. In a practical situation, the recommendation of an AI system might be different from the recommendation based on the clinical judgement of a medical practitioner, and it may conflict with the values and beliefs of the patient. This raises concerns around the issue of trade-off between accuracy and ethical autonomy. This leads us towards establishing that decisions should be made on what basis—data-driven predictions or algorithmic recommendations (Dewey Bergren, 2019; Santra et al., 2026). Different countries have started using Law enforcement agencies in using AI tools in estimating the number of criminal activities in different regions for allocating patrol resources accordingly. Such a system ensures optimum resource deployment on the basis of crime data. This may result in certain areas getting unnecessarily targeted, leading to social exclusion. This gives rise to the dilemma of choosing between individual rights and social security, as well as social justice and algorithmic efficiency.

AI leads to the ethical issue of influencing public perception. All social media platforms aim at driving revenues and engagement by personalizing content. Personalization of content often leads to misleading campaigns and manipulation of behaviour patterns. Commercial agendas and more autonomy being given to users often give rise to ethical conflict (Monah et al., 2022; Porkodi & Cedro, 2025). Data privacy is the right of a consumer and personalization of content poses a threat to data privacy. Content personalization may end in losing control over data. Losing control over personal data due to AI systems is ethically incorrect. Something might be morally correct, but culturally incorrect. Biometric attendance might be considered good in some societies, but in others it might be considered as a surveillance mechanism. This becomes a challenge for organizations in implementing culturally fit AI systems across geographical boundaries (Sahoo, 2022; Shaheen et al., 2024).

Data privacy, ownership and consent are the three pillars on which the ethical issues in AI systems significantly depend on. AI systems use enormous quantities of data and use it for commercial gains without taking user consent. People are not aware that companies use their digital footprints to make models that impact job prospects, insurance premiums, or creditworthiness. The entire AI development process is consumed with ethical dilemmas at every stage—data collection, development, deployment, and feedback (Hand, 2018; Jain et al., 2025). A data analyst would find it tough to make decisions by excluding outliers that consist of marginalized groups. A product manager would face the challenge of launching a profitable and efficient AI system. Such choices enforce the requirement of having ethical considerations embedded in the design and implementation of AI systems. AI systems either influence or make the decisions of people who will get the loan first, who will get parole, or who is a national threat. When decisions are made by transparent systems, ethical violations would be less. AI systems don't have moral values, sensitivity, or empathy, as they operate on probabilities and patterns. AI is capable of handling large sets of data, but the challenge is that the ethical concerns are actively considered for development. This issue can be solved if computational logic and moral imagination can be combined to address the demands of companies and societal needs (Agrawal, 2024; Zhang, 2024). AI can be a useful tool in identifying probabilities and patterns, but the final call has to be made by people who are accountable, responsible, sensible, and culturally aware.

3 Role of AI in Ethical Decision-Making

AI is continuously evolving as it is not only used for performing routine jobs, but also helps in making ethical decisions. AI has made significant inroads in legal judgements, governance, medical field, food policy and public policy (Kumar, 2022; Liu et al., 2025; Srivastava et al., 2018). The absence of fatigue and emotions in AI systems makes it valuable. It further helps in managing huge data in healthcare industry, which was evident during COVID. This highlights the equitable impact that the usage of AI can have on the society. Companies should look beyond the speed, accuracy, and computational capacity of AI systems and look to answer the larger question of how to embed ethical principles into algorithms. Different classical theories on virtue ethics, deontology, and utilitarianism give a divergent outlook on ethical reasoning. Virtue ethics focuses on true intentions and moral character, whereas deontology focuses on duties and rules irrespective of the outcomes (Abdullah & Valentine, 2009; Graham, 2004; Srivastava et al., 2024). The utilitarian approach focuses on results and aims at achieving the best outcome for the largest number of people. Integration of these approaches in AI systems is a challenge. The compatibility of deontology and virtue-based approaches are qualitative in nature and it is difficult to quantify them, whereas utilitarian approaches are comparatively compatible as they work well with data-based models.

AI systems use machine learning algorithms that calculate probabilities and establish patterns from historical data. These systems are customised to flag preferable options, bias detection, and ethical risk. For instance, predictive analytics can be used to understand the impact of financial decisions on the below poverty line population. The sentiments of stakeholders can be assessed by using Natural Language Processing (NLP) in policy discussions (Kumar et al., 2025; Wen et al., 2025). Various algorithms can adjust and alter their decision logic on the basis of feedback and change in other external factors. These capabilities can prove to be significant if the human input is transparent and ethical. This leads to the concept of 'Value Alignment', which states that the values and objectives of AI systems should be properly aligned. Ethics should be embedded in the design and development of AI systems (Brynjolfsson & McElheran, 2016). This involves designing and developing algorithms that are easy to understand and analyse large data in a transparent manner. This is necessary to make sure that the decisions are ethically justified as well.

The application of AI in decision-making is fraught with risks as there have been many failures around the world, despite the fair intentions. For instance, the US law and order system implemented the COMPAS algorithm to give fair sentences to criminals, but COMPAS often gave biased outcomes that sentenced black people for different charges (Monte-Serrat et al., 2025; Richards & King, 2014). This happened because COMPAS was made using a historical dataset that mostly had cases where black people were sentenced for different crimes. This underscores a serious limitation of AI systems, that it can be as good as the dataset it is trained on. If there are inequities in the historical dataset, then the AI system would exaggerate the outcomes. An AI system helps in identifying patterns, but it fails to assign a fixed coefficient to each competing priority. This raises the question of bringing in human intervention, and the AI systems should help in the decision-making process, but should not act as decision-makers. A framework should be made in which the AI systems should suggest the different possible options, but the job of selecting the most appropriate option lies with the human so that the decisions are culturally suitable and ethically sound.

AI systems should be made to suit the local needs and should be flexible to accommodate different ethical frameworks. AI should complement the entire process and not act in an autonomous manner. It can clearly show the dilemmas, consequences, and biases, but it cannot take the place of the detailed logical reasoning that morally sound decisions demand. AI can actively support decision-making through interdisciplinary collaboration. Engineers, managers, legal professionals, sociologists, psychologists, and other domain experts should give their advice in the development and deployment of AI systems that make sure that such systems are technically and ethically sound (Kumar et al., 2023a, 2023b; Zhao et al., 2026). Different stakeholders should be actively engaged in the process, so that the people who are affected by AI decisions do get a voice in the entire process, from development to deployment of AI systems. A continuous evaluation process should be set up to supervise the entire process on every parameter—technological development, societal impact, and ethical correctness. AI is a useful tool in facilitating ethical decision-making, but caution should be exercised in its development and deployment. AI can significantly increase the quantum and speed, but it can overlook the other important aspects, such as empathy, wisdom, and accountability. AI ethics is more about human touch and less about technology.

4 Building Ethical AI Systems

AI systems are growing in importance globally in the context of decision-making, and appropriate care should be given to its governance mechanism and monitoring. It becomes significant to explore the institutional mechanisms and foundational needs required to build AI systems that would support ethical outcomes.

The essence of an ethical AI design is the intention with which it is built. There are a few parameters that the AI system should look out for:

- The value system with which the system has been built.
- All the communities should get a fair representation in the dataset.
- Who would be harmed by its decisions?

The ethical design should be engineered by integrating moral values and beliefs in the entire development of AI system—design, data selection, development, deployment, and continuous monitoring. Fairness should also be observed in this process, in which the ethical AI systems must not have any bias, whether it is in the dataset or in the algorithms. The data should be inclusive on every ground, whether it is age, race, gender, or social status. AI would continue to give exaggerated outcomes if the dataset is not properly calibrated (Bousdekis et al., 2021; Kaith & Sachdeva, 2024; Srivastava et al., 2021). Fairness-conscious machine learning techniques, such as adjusting model outputs or reassigning weights to training samples, are significant in addressing these concerns. Transparency should also be observed in this process, in which the ethical AI systems not only publisher datasets or code, but also explain the reasons behind a particular decision to all the stakeholders, regulators, and users. XAI can facilitate in decoding model behaviour across industries, and it assumes wider significance in healthcare, supply chain and credit ratings (Provost & Fawcett, 2013; Tripathi et al., 2022). Transparent AI systems promote responsibility and accountability, and they provide a platform through which the problems faced by affected parties can be addressed.

Oversight should also be observed in this process, in which the AI systems must have an ethical governance mechanism. It calls for establishing cross-functional monitoring committees, ethics review committees, and an independent audit system. The primary goal of these committees is to check conformance to societal norms, legal norms, organization values and beliefs, together with assessing the technical performance as well. A regular audit is important to ensure all emergent risks and fairness thresholds are revealed and properly addressed in due course (Van Maanen, 2022). Institutions and Governments around the world understand the importance of ethical and structured governance. There are many frameworks around the world that have laid down regulatory pathways and guiding principles, such as Ethically Aligned Design (IEEE), Recommendation on the Ethics of Artificial Intelligence (UNESCO), and AI Act (European Union). These frameworks focus on the different parameters that are considered important for ethical AI deployment—privacy protection, non-discrimination, and sustainability.

Implementing ethical AI is a continuous process as there may be a shift in society value system, shift in data patterns, and algorithms may behave differently with the change in these factors. Organizations should ethically align themselves by constantly monitoring their systems and constant evaluation of societal feedback. A robust AI system should not only be ethical, but it should provide good governance as well. Organizations should observe these basic pillars for contributing meaningfully in the development of a better society.

5 The Future of Ethical Decisions Through Human-AI Collaboration

Collaboration between humans and AI is happening and it is increasingly penetrating different industries placing huge emphasis on making ethically correct decisions. This indicates a significant shift where the awareness and ethical logic will be supplemented with the predictive abilities, pattern identification, efficiency, and speed of AI systems. This would aim at striking the right balance between AI and humans, especially in cases where moral issues are involved.

There is huge dependency on data for making complex ethical decisions, as it has to safeguard the interests of society, stakeholders, context, and implications. AI can identify anomalies, conduct simulations and rank different options on large data, but analysing it on ethical principles, values and beliefs becomes a challenge. These aspects are significant in the dynamically changing business environment. For instance, AI systems cannot provide compassion and empathy to a patient on a death bed. AI can supplement in scenario modelling and diagnostics, coupled with human touch can be impactful. Collaboration of AI and humans is significant as it will be accurate and ethically aligned. AI can analyse large datasets in removing bottlenecks and humans will decide on the best possible action. Humans will decide the backward classes which will benefit from these actions, as it would help these communities to survive and recover from the different issues which they have been facing since generations.

Humans have many challenges of less memory, bias, and inconsistency, and it also suffers from accountability and fairness issues. Proper collaboration between AI and humans can solve many issues systematically. Analysing various perspectives in the light of moral obligations, the formation of strategy, and stakeholder expectations involves a deep human interference. AI may highlight that the vendor has violated the labour laws, but the consequences of continuing or cancelling the contract would need human judgement. The collaboration of ethical reasoning and algorithmic vigilance is a strategic requirement as organizations are increasingly becoming responsible, transparent, and accountable.

AI-human collaboration has its fair share of challenges as well. Decision makers may become over-reliant on AI and its recommendations, especially if the results are objectivity and precision in it. The algorithmic results might conflict with stakeholder beliefs and ethical intuition. Humans should develop the ability of critically analyse the AI results and its limitations and question its assumptions as well. Ethical AI decision-making needs a huge investment in developing ethical leadership by conducting training and developing a conducive culture. Managers, educators, public officials, and healthcare professionals need trainings and development programmes to analyse these AI findings, evaluating the ethical consequences, and integrate the stakeholder opinions in the decision-making process. Moral reasoning is an important qualitative skill and acts as a prerequisite for effective leadership in a technology-oriented environment.

The users and the affected communities should have a voice in the design, development, deployment, and monitoring of the AI systems. This would help in refining the blind spots and making the AI systems better, inclusive, and democratic. For instance, the application of AI models in urban planning may help in energy usage estimation and traffic flow estimation. Involving the citizens in urban planning would help in developing heritage preservation sites and bringing equity in development. Institutional safeguard is a critical element in the AI-human collaboration. European Union ensures human interventions in AI systems.

The future is about striking the right balance between AI and humans. AI systems can help humans in understanding the cognitive behaviour of data, based on which the humans will check for its ethical correctness. Finding the right balance is important for ensuring that the societal interest is safeguarded.

6 Limitations and Risks Involved

AI can facilitate in ethical decision-making, but it also has many limitations. This gives rise to complications around ethics, transparency and accountability in an AI-dominated era. The alignment of ethical principles and technology can be achieved by understanding the limitations of AI as it is increasingly being deployed in education, healthcare, justice, and governance. The transparency in AI systems is an area of concern because many AI systems operate as a 'black box'—the system makes a decision without explaining the process of reaching the outcomes. It creates a serious concern, especially in those companies where there is a lot at stake and the accountability is high. The involvement of AI in every field without human involvement can be dangerous, as the AI might flag somebody to be a national threat, or a patient could be denied medical care based on the recommendations of AI.

AI systems are built on historical data, and any bias in the dataset would reflect the inequalities of gender disparities, etc. The results of such models would highlight the flaw of the data, as the results would amplify the biases. The objective of every technology is to be transparent and fair, but it could become a carrier of systemic injustice. The algorithmic outputs should be objective so that the decisions based on it appear to be more scientific and legitimate. The usage of AI in decision-making has increased significantly universally leading accountability and transparency concerns. Many situations in organizations require care and understanding and over dependence on AI may lead to wrong decisions.

AI may compromise on some aspect and ignore some other important aspect. Improving efficiency in a public service delivery model may reduce the flexibility of the employee. Excessive monitoring may compromise on the privacy of people. Such issues should be dealt with utmost care, as it may affect the morale of employees. Ethics should be there in every organization that wants to make a long-term impact on the society. A framework should be developed to counter the risks arising from AI implementation. The future prospects of AI not only depend on its effectiveness, but also on how people are using it.

7 Conclusion and Future Directions

AI affects in the decision-making process of different sectors. Businesses take decisions by taking into account the rights, justice and well-being of people. Ethical reasoning and empathy are lacking in AI systems, but its relevance in the today's data-centric world cannot be ignored. AI should collaborate with humans and create a framework which is ethical and can handle huge data. The strength of AI in finding inaccuracies, identifying risks and patterns can be used by humans in the decision-making process. A judge cannot fully depend on the algorithmic criminal score given by the AI for passing his judgment, as the ethical responsibility of the final judgment rests on him. Similarly, the final decision to accept or reject the AI recommendation for a treatment depends on the doctor himself, as the ethical responsibility of the patient's final outcome depends on him only. The collaboration of human and AI would help in enriching the outcome.

AI systems that are accountable and transparent will be integral in generating trust amongst the different stakeholders. This involves standardizing the ethical principles, regulator clarity, regular AI audits, and providing a framework that would address the grievances. When international bodies and governments start adopting AI frameworks, it would symbolize that the AI systems have started to focus more on fairness, accountability, and dignity in the entire AI lifecycle. People using AI in ethical decision-making in different sectors, such as healthcare, law enforcement, and education, must obtain training on not only its functioning, but also its limitations as well. Professionals using AI in decision-making should also critique its recommendations and also integrate the different stakeholder perspectives. Leadership can be established today by combining ethical and data literacy. AI applications should be adaptable to ethical standards. The important issues of social impact and justice should be addressed by the AI systems. It is not about how good AI applications would become, but it will be about how are they placed ethically. Human race should do innovations in AI, but machines should follow in this process.

References

Abdullah, H., & Valentine, B. (2009). Fundamental and ethics theories of corporate governance. *Middle Eastern Finance and Economics, 4*(4), 88–96.

Agrawal, D. (2024). Corporate environmental, social and governance (ESG) practices. In *Reference Module in Social Sciences*. Elsevier. https://doi.org/10.1016/B978-0-443-13701-3.00141-9

Agrawal, D., Kumar, A., & Kumar, K. (2024). Ethical sourcing. In *Reference Module in Social Sciences* (p. B9780443137013000165). Elsevier. https://doi.org/10.1016/B978-0-443-13701-3.00016-5

Atluri, A., & Reddy, V. (2025). Cognitive HR management: How oracle HCM is reinventing talent acquisition through AI. *International Journal of Artificial Intelligence, Data Science, and Machine Learning, 6*(1), 85–94.

Awan, U., Shamim, S., Khan, Z., Zia, N. U., Shariq, S. M., & Khan, M. N. (2021). Big data analytics capability and decision-making: The role of data-driven insight on circular economy performance. *Technological Forecasting and Social Change, 168*, Article 120766.

Bousdekis, A., Lepenioti, K., Apostolou, D., & Mentzas, G. (2021). A review of data-driven decision-making methods for industry 4.0 maintenance applications. *Electronics, 10*(7), 828.

Brynjolfsson, E., & McElheran, K. (2016). The rapid adoption of data-driven decision-making. *American Economic Review, 106*(5), 133–139.

Buzko, I., Dyachenko, Y., Petrova, M., Nenkov, N., Tuleninova, D., & Koeva, K. (2016). Artificial Intelligence technologies in human resource development. *Computer Modelling and New Technologies, 20*(2), 26–29.

Chen, P., Chu, Z., & Zhao, M. (2024). The road to corporate sustainability: The importance of artificial intelligence. *Technology in Society, 76*. Scopus. https://doi.org/10.1016/j.techsoc.2023.102440

Dewey Bergren, M. (2019). Data governance and stewardship. *NASN School Nurse, 34*(3), 149–151. https://doi.org/10.1177/1942602X19838798

Di Vaio, A., Palladino, R., Hassan, R., & Escobar, O. (2020). Artificial intelligence and business models in the sustainable development goals perspective: A systematic literature review. *Journal of Business Research, 121*, 283–314.

George, G., & Thomas, M. R. (2019). Integration of artificial intelligence in human resource. *International Journal of Innovative Technology and Exploring Engineering, 9*(2), 5069–5073.

Graham, G. (2004). *Eight theories of ethics*. Routledge. https://www.taylorfrancis.com/books/mono/10.4324/9780203486511/eight-theories-ethics-gordon-graham

Hand, D. J. (2018). Aspects of data ethics in a changing world: Where are we now? *Big Data, 6*(3), 176–190. https://doi.org/10.1089/big.2018.0083

Jain, M. N., Maheshwari, M. M., Singhal, S., & Vishnoi, M. D. (2025). Cybersecurity in HR Tech: A review of data privacy challenges in the digital HR ecosystem. *Advances in Consumer Research, 2*(2), 856.

Jia, Q., Guo, Y., Li, R., Li, Y., & Chen, Y. (2018). *A conceptual artificial intelligence application framework in human resource management*. https://aisel.aisnet.org/iceb2018/91/

Kaith, A., & Sachdeva, G. (2024). Bibliometric insights into green human resource management: A connection with business ethics and values. *PURUSHARTHA-A Journal of Management, Ethics and Spirituality, 17*(1), 51–63. https://doi.org/10.21844/16202117104

Kumar, K. (2022). Emerging phenomenon of corporate sustainability reporting: Evidence from top 100 NSE listed companies in India. *Journal of Public Affairs, 22*(1), Article e2368. https://doi.org/10.1002/pa.2368

Kumar, K., Singh, R., Mishra, P., Lehri, B. P. S., & Kumar, A. (2025). What influences companies to go beyond mandatory corporate social responsibility rule? Empirical evidence from India. In P. Singh, S. Daga, & K. Yadav (Eds.), *Integrating Environmental Sustainability and Corporate Social Responsibility: A Move Beyond Profit* (pp. 45–63). Springer Nature Switzerland. https://doi.org/10.1007/978-3-031-81469-3_2

Kumar, R., Kaushal, S. K., & Kumar, K. (2023). Does source credibility matter in promoting sustainable consumption? Developing an integrated model. *Social Responsibility Journal, 19*(7), 1320–1347.

Kumar, R., Kumar, K., Singh, R., Sá, J. C., Carvalho, S., & Santos, G. (2023). Modeling environmentally conscious purchase behavior: Examining the role of ethical obligation and green self-identity. *Sustainability, 15*(8), 6426.

Liu, B., Cifuentes-Faura, J., Liu, X., & Ding, C. J. (2025). Towards green governance and real economy development: How digital technology affects health human resource allocation? *Applied Economics, 57*(21), 2696–2710. https://doi.org/10.1080/00036846.2024.2331033

Monah, S. R., Wagner, M. W., Biswas, A., Khalvati, F., Erdman, L. E., Amirabadi, A., Vidarsson, L., McCradden, M. D., & Ertl-Wagner, B. B. (2022). Data governance functions to support responsible data stewardship in pediatric radiology research studies using artificial intelligence. *Pediatric Radiology, 52*(11), 2111–2119. https://doi.org/10.1007/s00247-022-05427-2

Monte-Serrat, D. M., Cabella, D. M. S., & Cattani, C. (2025). Ethics and sustainability in education that makes use of artificial intelligence and robots. In S. Papadakis & G. Lampropoulos (Eds.), *Social Robots and Artificial Intelligence in Education* (pp. 121–152). Springer Nature Switzerland. https://doi.org/10.1007/978-3-032-03977-4_6

Porkodi, S., & Cedro, T. L. (2025). The ethical role of generative artificial intelligence in modern HR decision-making: A systematic literature review. *European Journal of Business and Management Research, 10*(1), 44–55.

Provost, F., & Fawcett, T. (2013). Data science and its relationship to big data and data-driven decision making. *Big Data, 1*(1), 51–59. https://doi.org/10.1089/big.2013.1508

Richards, N. M., & King, J. H. (2014). Big data ethics. *Wake Forest l. Rev., 49*, 393.

Sahoo, S. (2022). Big data analytics in manufacturing: A bibliometric analysis of research in the field of business management. *International Journal of Production Research, 60*(22), 6793–6821. https://doi.org/10.1080/00207543.2021.1919333

Santra, S., Kumar, K., & Chattopadhyay, N. (2026). Developing a theory of gendered digital inclusion (GDI) for sustainable outcomes: A multi-method qualitative study on gender digital divide. *Gender Issues, 43*(1), 12. https://doi.org/10.1007/s12147-025-09396-2

Shaheen, N., Jaiswal, S., Chinta, D. U., Singh, N., Goel, O., & Chhapola, A. (2024). Data privacy in hr: Securing employee information in us enterprises using oracle hcm cloud. *International Journal of Research Radicals in Multidisciplinary Fields, ISSN*, 319–341.

Singh, R., Kumar, K., & Khan, S. (2024). A comprehensive view of artificial intelligence (AI)–Based technologies for sustainable development goals (SDGs). *Artificial Intelligence Enabled Management: An Emerging Economy Perspective, 183.*

Singh, R., Kumar, K., Khan, S., & Sharma, A. (2025). An empirical examination of factors influencing adoption of innovation for sustainable economy: Evidence from MSMEs' perspective. In R. Singh, V. Corvello, I. M. Coniglio, & S. Khan (Eds.), *Innovation Ecosystems and Sustainable Technologies* (pp. 271–291). Emerald Publishing Limited. https://doi.org/10.1108/978-1-83662-368-720251013

Singh, R., Sharma, A., & Kumar, K. (2025). Navigating carbon neutrality: Strategies for sustainable future and environmental challenges. In R. Singh & W. L. Filho (Eds.), *Climate Neutrality Through Smart Eco-Innovation and Environmental Sustainability* (pp. 69–80). Springer Nature Switzerland. https://doi.org/10.1007/978-3-031-83250-5_5

Srivastava, S. K., Sinha, A. K., & Kumar, S. (2024). Workplace spirituality empowered by spiritual leadership to promote organizational effectiveness: A bibliometric analysis. *PURUSHARTHA—A Journal of Management, Ethics and Spirituality, 17*(1), 1–22. https://doi.org/10.21844/162021 17101

Srivastava, S. K., Singh, S., & Srivastava, M. (2021). Food waste at household and social gatherings: Drivers and possible remedies. *PURUSHARTHA-A Journal of Management, Ethics and Spirituality, 14*(2), 117–133.

Srivastava, S. K., Anshul, A., Pathak, P., & Bansal, J. (2018). Responsible consumption for curbing food wastage: An exploratory enquiry. *PURUSHARTHA-A Journal of Management, Ethics and Spirituality, 11*(2), 104–116.

Tripathi, V., Singh, K., Parihar, S., & Srivastava, S. K. (2022). A bibliometric analysis on supply chain collaboration. *PURUSHARTHA-A Journal of Management, Ethics and Spirituality, 15*(2), 66–103. https://doi.org/10.21844/16202115205

Van Maanen, G. (2022). AI ethics, ethics washing, and the need to politicize data ethics. *Digital Society, 1*(2), 9. https://doi.org/10.1007/s44206-022-00013-3

Wen, S.-F., Shukla, A., & Katt, B. (2025). Artificial intelligence for system security assurance: A systematic literature review. *International Journal of Information Security, 24*(1), 43. https://doi.org/10.1007/s10207-024-00959-0

Zhang, D. (2024). The pathway to curb greenwashing in sustainable growth: The role of artificial intelligence. *Energy Economics, 133*. Scopus. https://doi.org/10.1016/j.eneco.2024.107562

Zhao, P., He, G., & Guan, J. (2026). The ethical costs of artificial intelligence: Investigating how and when workplace artificial intelligence usage promotes employee unethical outcomes. *Journal of Business Ethics, 203*(3).

Chapter 4
Sustainability and Ethics in the Digital Economy: Exploring Corporate Responsibility Among Nifty 50 Companies

Akancha Srivastava, Shefali Mishra, and Dipanjan Bagchi

Abstract The digital economy has transformed the corporate world. Companies must now connect innovation with ethics and sustainability. As AI and data-driven services become more common, firms need to manage their environmental and social impacts. The Nifty 50, which includes India's largest corporations, plays a key role in this shift. Owing to the background, this chapter aims to discuss the integration of ethical and sustainable practices into companies by Nifty 50 organizations. The current chapter discusses the approaches and activities that these companies undertake to align themselves with international conventions and initiatives like the United Nations Sustainable Development Goals (SDGs). For this the textual analysis was conducted through a mixed approach (quantitative and qualitative) in 3 steps. The initial stage of text mining was carried out using R extracting the keywords about ethics and SDGs. The second step involved the manual exploration of the most commonly used keywords (=> 0.5). In the third phase Sentiment analysis was done in R software to attain more objectivity to the analysis as compared to the key word analysis. It served to cognize the way the discussion of ethics and SDGs is being relayed. The chapter offers a basis for exploring how companies construct digital ethics and sustainability practices that produce inclusive, transparent, and environmentally responsible business ecosystems in India & globally.

Keywords Digital economy · Nifty50 · Ethics · Sustainability · Governance

A. Srivastava (✉)
Jindal School of Psychology & Counselling, O.P. Jindal Global University, Sonipat, India
e-mail: akancha.srivastava@jgu.edu.in

S. Mishra
Department of Psychology, Christ University, Bengaluru, India
e-mail: drshefali1709@gmail.com

D. Bagchi
School of Liberal Studies and Humanities, UPES, Dehradun, India
e-mail: dipanjan.bagchi@ddn.upes.ac.in

K. Kumar et al. (eds.), *AI-Driven Circular Economy for Carbon Neutrality in Emerging Economies*, Environmental Footprints and Eco-design of Products and Processes,
https://doi.org/10.1007/978-3-032-30298-4_4

1 Introduction

Corporate responsibility is taking a new form in the digital economy, whereby more focus is being put on how organizations strike a balance between innovation, sustainability and ethics. The colossal pressures that the digital economy in a fast-growing economy puts on companies, has raised the responsibility of the corporation to a higher standard, which is beyond the conventional indicators of economic performance, but the ethical values and the environment-friendly practices. Recent research also emphasized the high degree to which the Environmental, Social, Governance (ESG) performance has a strong correlation with firm valuation and investor confidence, particularly in emerging markets like India (Deb et al., 2023). Sustainability and ethics in this context cannot be a choice since it is a key ingredient to survival, growth and development of the corporate world.

The rising digital economy has provided both opportunities and challenges, ranging from data privacy and algorithmic transparency to digital inclusion and responsible innovation. Ethical responsibilities and sustainable development have emerged as a most prominent issue for global as well as India's prominent corporations listed in the Nifty 50 index (Rajan & Sivakumar, 2025; Rao et al., 2023). The Nifty 50 represents the weighted average of the fifty largest Indian companies listed on the National Stock Exchange. Since these companies play a significant role in making the economy robust, their sustainable development initiatives are analyzed to have an in-depth understanding of their contribution. The Nifty 50 companies have a significant financial and ecological effect due to their carbon emissions, reserve consumption, and waste management practices. Their sustainability inputs aid in resolving environmental deterioration and safeguarding the social community's trust (Babu et al., 2024). Companies that incorporate sustainability into their business practices gain an advantage from augmented brand reputation, operational efficiency, and threat reduction. Evaluating the sustainability performance of Nifty 50 firms allows us to track progress toward global targets and identify areas for improvement.

This chapter discusses Nifty 50 organizations and their methods of incorporating ethical and sustainable practices in their companies. In particular, the paper (i) examines the ethical issues and questions that such companies face in the digital economy, and (ii) discusses how such companies address the global standards including the United Nations Sustainable Development Goals (SDGs). The chapter offers reflections to investigate how business ecosystems built by corporate digital ethics and sustainability activities in India and globally would be inclusive, transparent, and environmentally responsible. The evaluation of the digital responsibility of Nifty 50 companies reveals valuable insights to create ethical and responsible digital innovation and help achieve sustainable development on the global level.

According to the State of India's Digital Economy Report 2024, India ranks as the third most digitalized nation worldwide in terms of economy-wide digitalization and holds the 12th position among G20 nations with respect to the digitalization of individual users. According to the report India's digital footprint is rapidly increasing. China accounts for 1.78 billion of the world's total 8.36 billion mobile subscriptions, with India coming in second at 1.14 billion. India tops the world in 5G deployment speed and has an average monthly cellular data consumption of 18.39 GB per user. During the fiscal year 2022–23, the country issued over 1.3 billion biometric identification numbers and processed the most real-time digital payments, mostly through UPI transactions. Furthermore, India is the world's second-largest exporter of ICT services, accounting for 23% of AI-related projects on GitHub.

Advanced technologies such as AI, big data analytics, cloud computing and Internet of Things are profoundly transforming how businesses operate, interact and expand within the digital economy. The most original feature about the digital economy is that it redesigns corporate responsibility. Firms present financial performance in maintaining ethical practices, operational transparency and sustainability to the environment. As the use of digital tools and analytics grows, such crucial concerns as data privacy, algorithmic bias, cyber crimes, digital equity, and ecological sustainability have become major challenges. The cross-section of innovation and responsibility means that the drastic shift has taken place—ethics and sustainability are now at the center of the way of achieving the lasting trust and resiliency in the digital era (Bansal, 2025; Shashwat et al., 2025). The digital economy has steadily been integrated by organizations across different fields, leading to significant advancements in innovation, development and growth, strategic planning, and community development, through serving the weak. This change has made profit maximization a secondary agenda, and long-term sustainability and inclusive growth have gradually become prioritized.

This change is especially significant to Nifty 50 firms—the most influential firms that make a decisive change in influencing the economy and environment in the Indian context. These companies are major emitters of carbon gas, depletion of resources and industrial wastes are taking up more roles in ensuring their sustainability. Moreover, they can also establish the standard of the ethical integration of digital activities and conformity to international programs like the United Nations Sustainable Development Goals (UNSDG).

1.1 Change in Corporate Responsibility

The past decade has changed corporate responsibility drastically where it was viewed as a narrow profit-oriented concept to a broader and value-based approach through the introduction of models like Environmental, Social and Governance (ESG) and the so-called triple bottom line that concentrates on economic, social and sustainability. The change in India has been enhanced by the increased regulations, investor expectations, and increased stakeholder scrutiny.

In May 2021, one of the major achievements related to the subject was the introduction of the Business Responsibility and Sustainability Reporting (BRSR) framework by the Securities and Exchange Board of India (SEBI). The BRSR are obligatory of the top-listed companies to report data related to the ESG in compliance with the international standards like Global Reporting Initiative (GRI) and Sustainability

Accounting Standards Board (SASB). The progressive introduction of the reasonable assurance requirements of the top 150 companies in the 23–24 financial year and the 1000 companies by 26–27 financial year represent the strategic nature of how SEBI has introduced the concept of ESG into the mainstream corporate reporting framework. Moreover, sustainability disclosures also showed substantial annual growth in the 2021–2022 financial year, including energy consumption, greenhouse emissions, human rights and supply chain governance, and average reporting rate has grown by 30–50.

1.2 Nifty 50 Companies, Role, Reach, and Responsibility

The Nifty 50 index is a collection of 50 most successful and most powerful publicly traded businesses in India that represent a wide range of industries like energy, finance, IT and consumer goods. These companies have high market value and a wide range of operation resulting in a collective impact on areas of essential economic development, labor market and environmental accountability. Even though the regulatory requirements such as the Business Responsibility and Sustainability Reporting (BRSR) framework initially influenced the decisions undertaken by many 50 companies, the ESIR programs of many companies have been redesigned in a bid to achieve greater alignment to the strategic and value-based goals. Sustainability is also becoming one of the most important elements of corporate governance, and different organizations tie incentives to ESG performances and redefine the goals to ensure climate resilience and sustainable development. Internal ESG tracking systems were adopted by the key companies, and they invested in climate innovations, supplier interactions, and initiatives aimed at the ecosystem (Doreswamy, 2025; IiAS, 2025; PwC India, 2024) by the conclusion of the 2023 financial year.

In spite of these prevailing issues, these companies are transforming the CSR environment. The expected growth in corporate contributions to the corporation giving between Rs 35,000 crore in the 23–24 financial year to Rs. 120,000 crore by FY 2034–2035 is a strong emphasis on strategic and high effective intervention that will be in line with the national development agenda. To put it briefly, Nifty 50 companies have transformed into a sustainable role that is beyond the usual economic position, and they actively institute ESG principles and social and environmental objectives (Give Grants, 2025; CSRBOX 2024).

2 Method

The sample used in this study was the Nifty 50 companies. To investigate the sustainability development practices, secondary data was gathered on Nifty 50 companies. The Nifty 50 was an Indian (National Stock Exchange) stock market index, which consisted of the top 50 companies in the market capitalization of the National Stock

Exchange. An annual report was a wide-ranging report which contained details of company activities over the previous year. The analysis was done based on the annual reports of the latest year available. This study used the annual reports of all the Nifty50 companies to gather data on the Sustainable Development Goals (SDGs). Automated text mining methods and manual review of the selected key terms and targeted SDGs were done. The methodological frame blended quantitative examination alongside qualitative scrutiny to combine breadth coverage with context accuracy.

2.1 Data Source and Strategy

The first phase consists of downloading the latest annual reports of all the Nifty 50 companies. Table 1 presents the details of the companies used in the study.

Note: Table 1 presents the sorting of Nifty 50 companies across their industry categories and merged with different sectors. The companies were divided according to the industries. These include fast moving consumer goods, automobile and auto components, capital goods, construction, construction materials, consumer durables, consumer services, financial services, healthtech, information technology, metals and mining, oil gas and consumable fuels, power services and telecommunications.

Data Source Latest available Annual reports of Nifty 50 companies FY 2023-24.

2.2 Planning the Review

These predefined criteria were established at the commencement of the investigation.

2.2.1 Context

To explore the Nifty50 companies in attainment of the SDGs.

2.2.2 Study Objectives

The objectives of this study are:

- To identify the strategies and policies adopted by Nifty 50 companies to align with corporate ethics and the United Nations Sustainable Development Goals (SDGs),
- To highlight their role in promoting economic, social, and environmental sustainability.

Table 1 Details of the companies used in the present study

Company name	Industry	Merged sectors
Bajaj Auto Ltd.	Automobile and Auto Components	Automobile
Eicher Motors Ltd.	Automobile and Auto Components	
Hero MotoCorp Ltd.	Automobile and Auto Components	
Mahindra & Mahindra Ltd.	Automobile and Auto Components	
Maruti Suzuki India Ltd.	Automobile and Auto Components	
Tata Motors Ltd.	Automobile and Auto Components	
Bharat Electronics Ltd.	Capital Goods	Capital goods
Grasim Industries Ltd.	Construction Materials	
Larsen & Toubro Ltd.	Construction	
UltraTech Cement Ltd.	Construction Materials	
Asian Paints Ltd.	Consumer Durables	
Eternal Ltd.	Consumer Services	
Titan Company Ltd.	Consumer Durables	
Trent Ltd.	Consumer Services	
Hindustan Unilever Ltd.	Fast Moving Consumer Goods	FMCG
ITC Ltd.	Fast Moving Consumer Goods	
Nestle India Ltd.	Fast Moving Consumer Goods	
Tata Consumer Products Ltd.	Fast Moving Consumer Goods	
Axis Bank Ltd.	Financial Services	Financial
Bajaj Finance Ltd.	Financial Services	
Bajaj Finserv Ltd.	Financial Services	
HDFC Bank Ltd.	Financial Services	
HDFC Life Insurance Company Ltd.	Financial Services	
ICICI Bank Ltd.	Financial Services	
IndusInd Bank Ltd.	Financial Services	
Jio Financial Services Ltd.	Financial Services	
Kotak Mahindra Bank Ltd.	Financial Services	
SBI Life Insurance Company Ltd.	Financial Services	
Shriram Finance Ltd.	Financial Services	
State Bank of India	Financial Services	
Apollo Hospitals Enterprise Ltd.	Healthcare	Healthcare
Cipla Ltd.	Healthcare	
Dr. Reddy's Laboratories Ltd.	Healthcare	
Sun Pharmaceutical Industries Ltd.	Healthcare	
HCL Technologies Ltd.	Information Technology	IT
Infosys Ltd.	Information Technology	

(continued)

Table 1 (continued)

Company name	Industry	Merged sectors
Tata Consultancy Services Ltd.	Information Technology	
Tech Mahindra Ltd.	Information Technology	
Wipro Ltd.	Information Technology	
Adani Enterprises Ltd.	Metals & Mining	Extractive Industries*
Hindalco Industries Ltd.	Metals & Mining	
JSW Steel Ltd.	Metals & Mining	
Tata Steel Ltd.	Metals & Mining	
Coal India Ltd.	Oil Gas & Consumable Fuels	
Oil & Natural Gas Corporation Ltd.	Oil Gas & Consumable Fuels	
Reliance Industries Ltd.	Oil Gas & Consumable Fuels	
NTPC Ltd.	Power	Power
Power Grid Corporation of India Ltd	Power	
Adani Ports and Special Economic Zone Ltd.	Services	Services
Bharti Airtel Ltd.	Telecommunication	Telecommunication

Note * 'metal and mining' and oil gas and consumable fuels were merged into extractive industries

2.2.3 Applications

It is important to understand the impact of the initiatives undertaken by the Nifty50 companies in addressing the pressing issues of contemporary times.

2.2.4 Inclusion and Exclusion Criteria

Inclusion Criterion:

- Companies which are listed in Nifty 50 list.
- Companies which have uploaded the latest available FY report.
- Companies which have included strategies to practice environmental sustainability and digital ethics.

Exclusion Criterion:

- The companies that will not be part of this study are those that are not listed under Nifty 50 list.
- Business firms that have failed to incorporate strategies to exercise environmental sustainability.

Table 2 Steps involved in the present study

Phase	Steps
Selection of database	Nifty50 annual reports
Text mining	Creation of corpus Keyword analysis Qualitative analysis of keywords Lemmatization Sentiment analysis

Table 2 represents the phases and procedural steps followed in this study. The study was carried out using secondary data sources. Annual reports of Nifty 50 companies served as database. For the text mining phase corpus creation, keyword analysis lemmatization followed by qualitative analysis of keywords. This was followed by sentimental analysis.

2.2.5 Organizing and Screening of the Reports

Data screening and arrangement of the collected information was done through a two-stage process. The initial stage involved data extraction and compilation. Subsequently, the second stage encompassed the screening process, which was undertaken through collaborative consultation with fellow authors via Google Meet.

1. Phase 1: Data Extraction and Corpus Creation: This phase entailed the conversion of the annual reports in pdf format to text format. Text mining was done by using the R package "tm" to process the corpus. The processing includes stripping white space, converting the corpus to lowercase, and removing punctuations, stopwords and numbers. Moreover, the corpus was divided into sectors to distinguish the strategies used in corporate ethics and SDGs.

Table 3 illustrates the primary analytical activities that have taken place in the Text Mining Workflow along with the R Packages used for each activity.

Table 3 Packages used in this study for specific tasks

Task	Key packages used
PDF text extraction	pdftools
Text cleaning and preprocessing	tm, tidyverse
Corpus creation and transformation	tm
Lemmatization	UDPipe
Keyword frequency analysis	tm, dplyr, tidyr
Data export to Excel	writexl
Heatmap and bar plots	ggplot2, viridis scales
Topic modeling (LDA)	topicmodels, tm, tidytext
Sentiment analysis	syuzhet, dplyr, ggplot2

Table 4 Keywords related to corporate ethics and SDG's used in this study

Topic	Keywords
Corporate ethics	"corporate ethics", "moral accountability", "data privacy", "business integrity", "ethical governance", "compliance", "transparency"
SDGs	"poverty", "hunger", "good health", "well being", "quality education", "gender quality", "clean water", "sanitation", "clean energy", "inequalities", "responsible consumption", "climate action"

2. Phase 2: Keyword Analysis: The keywords that had a value of 0.50 or above were utilized in selecting the strategies applied in each sector.
3. Phase 3: Qualitative analysis of keywords: A manual search of keywords was done on the processed text. The keywords were chosen depending on their relevance to the corporate ethics and SDGs goals. The sentencing and paragraphs with each target keyword were also read carefully to gain not only a deeper insight into the context, tone, and purpose of the use of the target keywords, but also to conduct an extensive analysis of the strategies that companies use.
4. Phase 4: Lemmatization is a very key step in improving the language modelling procedures which in the end results into high document retrieval, with the lemmatization application producing the best outcome (Balakrishnan & Lloyd-Yemoh, 2014). UDPipe R package was lemmatized with the aid of the current study.
5. Phase 5: Sentiment analysis: The sentiment analysis enables a further in-depth analysis of the keywords frequency by understanding the contextual feeling of how companies convey ethical ideas. It shows if industries only discuss ethics or negotiate their position in a positive, defensive, or negative way - enabling stakeholders evaluate real action of ethical and SDG values.

Table 4 includes the keywords terms that were identified to be analyzed through a keyword analysis during the text-mining process, and have been grouped into two categories: Corporate Ethics, and Sustainable Development Goals (SDG's).

3 Result and Discussion

3.1 Keyword Frequency Analysis

There are clear trends in the use of ethics-related keywords across all sectors based on the number of times they are used per document. The use of "compliance" (0.82 frequency/document) and "transparency" (0.71 frequency/document) are the most frequent in the financial sector. This is consistent with what we know about the financial sector's regulatory environment. Similarly, the use of certain terms by the healthcare sector, including "data privacy" (0.78 frequency/document) and "compliance" (0.69 frequency/document) also points to the fact that it deals with sensitive information and therefore is required to do so in an ethical manner.

Figure 1 is a graphical representation of the relative frequencies of selected keywords associated with corporate ethics (e.g., Business Integrity; Compliance; Data Privacy; Ethical Governance; Moral Accountability; Transparency) within each of the merged sectors utilized for the purpose of this research. The color intensities are indicative of the normalized frequencies per document. A darker purple indicates a low frequency while a lighter yellow indicates a high frequency. "ggplot" package was used and "Viridis" pallette was used for the colour scheme.

The extractive industry, which includes mining, oil, and gas, appears to be somewhat moderate in the use of the terms "corporate ethics" (0.58 frequency/document) and "transparency" (0.63 frequency/document) and may be influenced by pressures from ESG (environmental, social, and governance) issues. There is also a moderate level of use of the term "ethical governance" (0.52 frequency/document) in the capital goods sector. On the other hand, there are much lower levels of use of the term "corporate ethics" (0.31 frequency/document) in the automobile sector and (0.28 frequency/document) in the consumer sector.

The construction sector appears to have little or no interest in ethics. In this sector, most of the terms that would be considered relevant to ethics appear to be used at frequencies of less than 0.25 frequency/document. However, the frequency of the term "compliance" (0.34 frequency/document) in this sector may suggest that the regulatory environment of this sector may lead to increased use of some of these terms.

Overall, the results of this research illustrate how the nature of a sector can affect the language that is used to discuss corporate ethics. For example, the use of highly regulated sectors such as finance ("compliance" 0.82) and health care ("compliance" 0.69) is vastly different than the lower rates of use of the same terms in sectors that are subject to fewer regulatory environments. These differences could represent opportunities for improved ethical communications. They may also point to the absence of certain terms, such as "media accountability" which is used at levels below 0.1 frequency/document in most sectors indicative of further investigation.

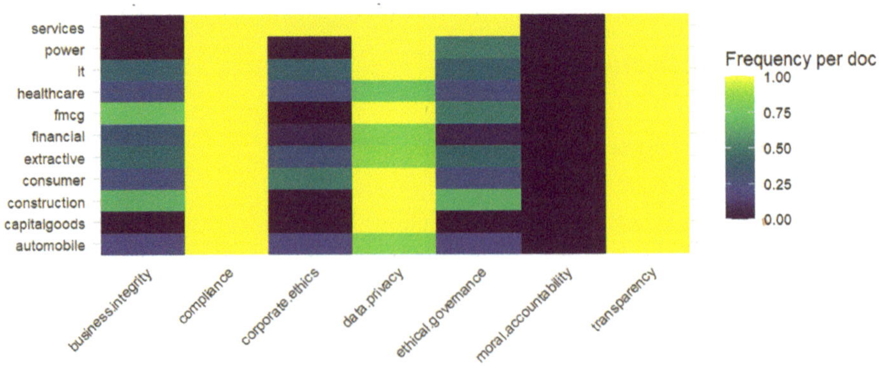

Fig. 1 "Corporate ethics" related keywords frequency

Figure 2 represents the frequency of SDG-related keyword usage. It depicts the frequency of clean energy, clean water, climate action, gender equality, good health, quality education, responsible consumption and well-being across the industries covered in the study; light coloration represents a high frequency of usage (yellow), while dark coloration indicates low frequency (purple). "ggplot" package was used and "Viridis" pallette was used for the colour scheme.

The analysis of SDG-related keywords shows different levels of engagement of different sectors that can be attributed to factors like industry operations, regulatory exposure and stakeholder expectations. The most frequently occurring keywords were identified to be "quality education," "climate action," and "well-being." On the other hand, the keywords with low frequency were "sanitation," "poverty," and "clean water" across different sectors. On further analysis, it can be found that the service sector has a strong relation with the keywords "quality education" (0.85 frequency/document). This can be attributed to its dependence on skilled workers and focus on employee welfare. A high association can be seen in power sectors with keywords like "responsible consumption" (0.78 frequency/document) and "climate action" (0.65 frequency/document). This further highlights its focus on the environment and impetus to shift to renewable energy. The IT sector shows association with SDG keywords of "quality education" (0.68 frequency/document) and "well-being" (0.80 frequency/document) which highlights its role in digital skills training and "tech for good" projects.

The healthcare sector was found to have a balanced approach to meet the SDGs. The keywords with high frequency are "clean water" (0.58 frequency/document), "climate action" (0.75 frequency/document), "poverty" (0.55 frequency/document), "quality education" (0.80 frequency/document), "responsible consumption" (0.68 frequency/document) and "wellbeing" (0.95 frequency/document). This underlines its effectiveness in improving health outcomes, equal access to healthcare and understanding environmental responsibilities. On the other hand, the FMCF sector has

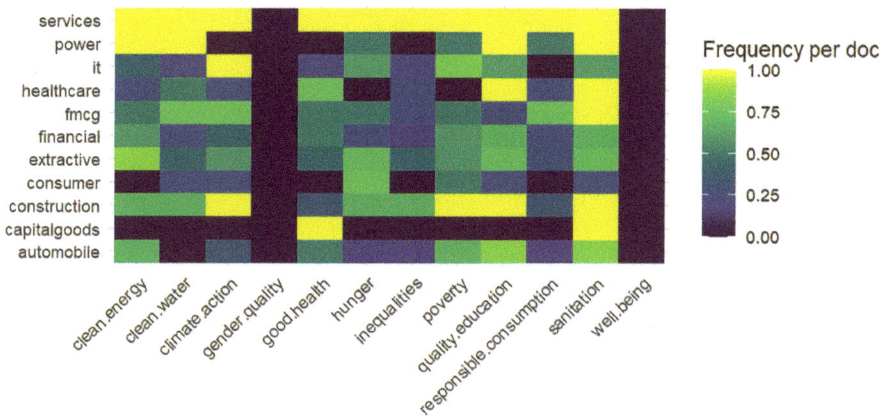

Fig. 2 "SDG" keyword prevalence across sectors

more frequently occurring keywords like "clean water" (0.70 frequency/document), "sanitation" (0.65 frequency/document) and "quality education" (0.85 frequency/document). This can be interpreted in the context of its focus on consumer health and Corporate Social Responsibility (CSR) outreach.

The frequency analysis reveals that the financial and consumer goods sectors have average levels of engagement with the SDG keywords. The most common keywords in this sector are "poverty" (0.58 frequency/document), "education" (0.75 frequency/document) and "well-being" (0.80 frequency/document) that highlights its involvement in inclusive finance and community development. Similar trends could be observed in the consumer goods sector with reference to the keywords like "sanitation" (0.65 frequency/document) and "climate action" (0.50 frequency/document) that might be attributed to pressure from consumers demanding sustainable investment. The extractive sectors which are often criticized for their impact on the environment, in this analysis show frequently occurring SDG keywords like "climate action" (0.78 frequency/document), "poverty" (0.58 frequency/document), and "clean water" (0.60 frequency/document). This can be attributed to increasing pressure on ESG measures being adopted to secure community support for operations. Moderate SDG frequency of keywords like "sanitation" (0.52 frequency/document) and "quality education" (0.55 frequency/document) were found in spite of having low ethics-related disclosures. As opposed to other sectors the capital goods and automobile sectors have low frequency of SDG keywords. For example, "climate action" has 0.40 frequency/document in the capital goods sector and 0.38 frequency/document in the automobile sector. This non-alignment with SDG keywords may be related to its importance in economic infrastructure and traditional focus on product rather than CSR integration.

The findings reveal a clear pattern linking the sectors with direct human and environmental interaction like healthcare, FMCG, services and extractive sectors show more alignment with SDG keywords. Low frequency of keywords like "sanitation," "poverty," and "clean water" are crucial to achieving SDG goals but have not been addressed consistently.

3.2 Qualitative Analysis of Keywords

The qualitative analysis of the keywords with values above 0.5 was done. This was done to understand the context of the frequent terms and identify the strategies that are being implemented.

Ethical oversight and sustainable development within a company are no longer merely dispositional and regulatory "check boxes" to be ticked off, rather they have become pillars of focus. A closer look at the Nifty 50 companies indicates the top companies in India are gradually accepting the principles of corporate ethics and the Sustainable Development Goals (SDGs) in the day-to-day functioning and governance of the companies.

Table 5 shows the major ethical governance aspects which were reported by all of the organizations in each of the main industries and how they are similar (such as privacy compliance, cyber security, and whistleblower protection) and also compliance with national and international regulations. Data Source Annual Report of Nifty 50 companies.

Sustainable Development Goals (SDGs)

Table 6 above represents a qualitative mapping of sustainability related keywords to relevant SDGs by major sector for disclosures contained within Nifty 50 Annual Reports. The Nifty 50 companies have exhibited ethical behavior at best as an industry shift from compliance attitudinal behavior to a culture of compliance acceptance. The automotive industry is one of the leaders in this change, driven by Tata Motors, Maruti Suzuki, and Mahindra & Mahindra. This sector demonstrated strength in compliance with data privacy and cybersecurity governance. Their adherence to data privacy safeguards and cybersecurity breach policies coupled with a perfect data breach track record and compliance with benchmarks like GDPR and ISO 27001 is a sign of emerging digital ethical norms. This indicates an understanding of data and privacy safeguards that transcend compliance providing psychological safety and

Table 5 Key qualitative findings for ethical keywords

Sector	Companies	Key ethics dimensions
Automotive	Tata Motors, Maruti Suzuki, Hero MotoCorp, Mahindra & Mahindra, Eicher Motors, Bajaj Auto	GDPR/DPDPA compliance, Zero data breaches, ISO 27001, Governance committees, Cybersecurity controls
Consumer/ FMCG	HUL, Nestlé India, ITC, Tata Consumer, Titan, Asian Paints, Trent	Whistleblower policies, Fair trade compliance, Responsible sourcing, Privacy-by-design apps
Financial	HDFC Bank, Axis Bank, Kotak, SBI, ICICI Bank, Bajaj Finserv, HDFC Life, Shriram Finance	Data encryption, privacy frameworks, RBI/SEBI compliance, Fair lending, Transparent disclosures
Healthcare	Apollo Hospitals, Cipla, Sun Pharma, Dr. Reddy's	Patient data security, WHO/FDA compliance, Drug safety, Public health transparency
IT & Tech	Infosys, TCS, Wipro, HCLTech, Tech Mahindra	GDPR, AI governance, Stakeholder engagement, ISO standards, Whistleblower frameworks
Steel/Energy	Tata Steel, JSW Steel, Hindalco, ONGC, NTPC, Coal India, Adani Enterprises	ISO 27001, SEBI compliance, ESG-linked audits, Anti-corruption policies
Capital Goods/ Infra	L&T, BEL, Grasim, UltraTech Cement	Cyber risk governance, Ethics audits, Board-led compliance, Data privacy (selective)
Ports & Services	Adani Ports & SEZ	Anti-bribery compliance, Whistleblower protections, Stakeholder dialogue

trust which is critical for sustaining brand equity in a digitally interconnected global landscape.

Ethics is applied in the banking industry both in customer interaction and the computer systems in the background. HDFC Bank, SBI, and ICICI Bank exhibit basic level practices of ethics like encryption and privacy-by-design, equitable lending systems, and transparent disclosures of upfront products. In addition to being in line with the regulatory requirements as dictated by RBI and SEBI, such practices enhance business resiliency because they instill trust in consumers, particularly the disadvantaged who are new entrants in the financial system. In this case, ethics plays the role of a social and economic trigger. HUL, ITC, and Tata Consumer Products are taking the lead over the market through ethical advertising, fair trade sourcing, and being transparent in the packaging and claiming sustainability. Capital goods and infrastructure is another sector in which Larsen and Toubro are dominant and are yet to embrace ethics. These areas are more compliance oriented as opposed to being culture oriented. Simply put, ethical governance is no longer a checklist, but a formal corporate culture driven by investor pressure, changing laws and increased outside pressure. Those companies that are successful with this aspect demonstrate that information confidentiality, openness, and system of internal governance are not the obstacles in the way of innovation, reputation, and factor of endurance; instead, they are the stimulus.

Table 6 Key qualitative findings for SDG keywords

Sector	Companies	Primary SDGs aligned
Automotive	Tata Motors, Maruti Suzuki, Hero, M&M, Eicher, Bajaj Auto	SDG 7 (Clean Energy), SDG 13 (Climate Action), SDG 4 (Education), SDG 6 (Sanitation)
Consumer/ FMCG	HUL, Nestlé India, ITC, Tata Consumer, Titan, Asian Paints	SDG 12 (Responsible Consumption), SDG 2 (Hunger), SDG 1 (Poverty), SDG 6 (Water)
Financial	SBI, ICICI Bank, HDFC, Bajaj Finance, Axis Bank	SDG 1 (Poverty), SDG 4 (Education), SDG 6 (Sanitation), SDG 7 (Clean Energy)
Healthcare	Apollo, Cipla, Dr. Reddy's, Sun Pharma	SDG 3 (Good Health), SDG 4 (Education), SDG 6 (Sanitation), SDG 6 (Water Access)
Steel/Energy	NTPC, Coal India, Hindalco, Adani, Tata Steel	SDG 13 (Climate), SDG 7 (Clean Energy), SDG 1 (Poverty), SDG 6 (Sanitation), SDG 12 (Consumption)
IT & Tech	Infosys, TCS, HCLTech, TechM, Wipro	Indirect support through education skilling (SDG 4), environmental management (SDG 13)
Capital Goods/ Infra	L&T, UltraTech, Grasim, BEL	SDG 9 (Infrastructure), SDG 4 (Education), SDG 13 (Climate), SDG 6 (Sanitation), SDG 5 (Gender Equality)
Ports & Services	Adani Ports & SEZ	SDG 10 (Inequality), SDG 13 (Climate), SDG 6 (Sanitation), SDG 12 (Consumption)

SDG Integration: Sector Strategic Dimensions

The contribution of Nifty 50 companies to the sector in which they operate cannot be overestimated in relation to the United Nations Sustainable Development Goals (SDGs). As an example, players in the automotive sector are developing climate action (SDG 13) and clean energy (SDG 7) by making a significant move to electric mobility (with carbon neutrality and zero-waste production goals). The amount of change the sector can bring about in society is seen in examples of water positivity of Hero MotoCorp of 455% and investments in various green energy sources by Maruti Suzuki. In this case, the intensity of the impact is transferred socially and ecologically in climatic change and resource constraints.

The priority under the responsibility of consumption (SDG 12) and zero hunger (SDG 2) is given to the FMCG companies such as HUL, Nestle India, and Asian Paints. Their extensive changes inside supply chains, their waste minimization programs and their community nutrition programs reveal their obsession not only with the manufacture of responsible products, but also with the communication and education of the responsible products to their consumers.

The financial services sector is a special intermediary to facilitate the SDGs across the sectors. Financial service providers such as Axis Bank, Kotak Mahindra, and Bajaj FinServ in their microfinance activities, digitization and financial inclusion, education-related Corporate Responsibility activities and sanitation-related Corporate Responsibility activities are tackling the principal development challenges of poverty (SDG 1), education (SDG 4), and sanitation (SDG 6).

Health care companies such as Apollo Hospitals, Cipla and Sun Pharma have been committed in supporting SDG 3 (Good Health) and SDG 6 (Sanitation) by associating affordable health care with the community wellness programs that the companies are offering. Their activities in rural, tele-health, and community-level sanitation outreach represent the move toward the provision of products and services as opposed to preventive, inclusive rehabilitative care systems.

Surprisingly, the steel, energy and infrastructure industry, which has always been termed as hard-to-abate, have come up as some of the brightest spots in sustainability today. The companies like NTPC, Hindalco and Adani Enterprises are putting huge investments in clean energy transitions, afforestation of locals population in addition to community development projects in the education and water arena.

Although there has been a breakthrough, there are still gaps. As an example, automotive still has a talk-under-sanitation (SDG 6) that has always been poorly represented in the larger CSR narratives with no definite success metrics or impact data that should measure the problem by innovation gating. Similarly, where firms invest in education (SDG 4) via scholarships and training opportunities there are virtually none that quantifies the results or contribution so that the nation meets its national educational goals—scalability is lower and accountability is decreased.

Those discrepancies suggest that even though SDG rhetoric is omnipresent, its strategic nature differs significantly. Others consider the SDGs as key to their business model (e.g. electric vehicle producers or green financial services) whereas others view them as a mere addition to their CSRs which they must pay lip service to, and more robust models need to be built to integrate them and put a serious emphasis on serious measurement is possible.

The chapter is a one-of-a-kind attempt to assess the primary business strategies of Nifty 50 companies, which refer to the Sustainable Development Goals (SDGs) and corporate ethics on a sectoral basis. It manages to go beyond mere compliance checklist strategies by showing how ethics and sustainability are becoming the main focus of innovation, trust and long-term value. Another distinct feature brought by the report is an analysis of sectoral comparisons of ethical maturity and SDG alignment and depicts clear images of sectoral gaps, and emerging trends.

3.3 *Sentiment Analysis*

3.3.1 Sentiment Analysis of Keywords Related to Ethics

The trends of emotional tones in sentiment analysis of the ethical keywords in different industries are predominantly positive or neutral. No strong negative feeling when it comes to any ethical term was found in the documents under review. This has been captured in a color scheme in which the color green reflects a high level of sentiment and the absence of red shows that there is little negative conversation. Sentiment scores are usually approximately 5 or above 20, which shows various degrees of excitement or intensity in ethical communication in different sectors and keywords.

Figure 3 illustrates the Mean Sentiment Scores of Ethical Keywords per Industry Category are shown in this graph. Positive sentiment for ethics-related keywords is significantly stronger in both extractive and consumer categories than for all other categories. All other sectors have a moderate but consistent positive engagement with sustainability, transparency, etc. Sentiment analysis was done using "syuzhet" package and the graph was done using "ggplot2".

A mining activity as well as oil and gas extractive industry record one of the best positive sentiments particularly in the area of ethical governance and business integrity. The sentiment scores of these keywords are on average at the level of 20. This indicates a definite attempt to portray the ethical reforms favorably, which may be in response to the criticism of ESG in the past. Equally, the services industry expresses favorable feelings towards sustainable development, which indicates that there is a good perception towards growth, innovation and benefits in sustainability to stakeholders.

Conversely, the financial sector has moderate levels of sentiment, albeit with a considerable level of discussion in the ethical arena as revealed in the analysis of the keywords in the first section. This can be the way these words are put in a matter of fact or necessity, as opposed to the use of a passionate language.

FMCG, healthcare and other consumer-facing industries appear to be less strong on such keywords as data privacy and corporate ethics. That may point to a more restrained and institutional form of communication, as opposed to involving stories.

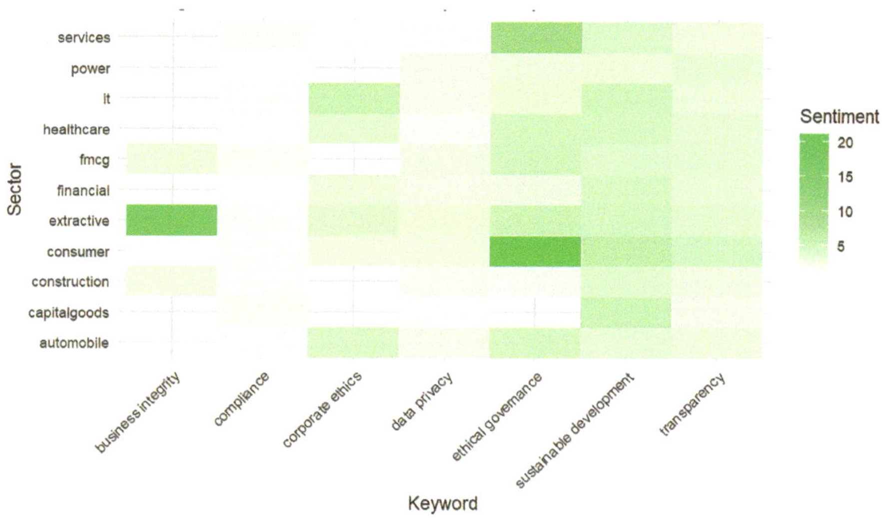

Fig. 3 Average sentiment of ethical keywords by sectors

It may be due to the cautious language taken when discussing privacy or regulatory concerns.

3.3.2 Sentiment Analysis of Keywords Involving Ethics.

The SDG related keywords sentiment analysis by sector demonstrates that the sentiment is predominantly positive as there are no negative sentiment cells including red or negative ones. Discussion of SDG topics in sectors tends to be positive, supportive, but the level of sentiment differs greatly depending on sector and keyword.

Figure 4 illustrates average sentiment ratings for SDG-related words by industry segments. A heat map is used to display a mean sentiment rating on an SDG word (such as clean energy, climate action, inequality, quality education) across various industries, like services, information technology, health care, finance, consumers, and construction. The darker the green color of a cell represents the highest positive sentiment rating, while the lightest the green color represents the lowest sentiment rating level. This image illustrates the variation among the sectors in terms of frequency and positivity of references to SDG concepts. Sentiment analysis was done using "syuzhet" package and the graph was done using "ggplot2".

The consumer segment has the best overall feeling, especially over quality education with a score of above 30 in sentiment with a high emotional intensity. It implies that consumer-oriented companies, which are probably motivated by brand awareness and customer demands, are encouraged to invest in education in their sustainability strategies and position it as positive. It is likely that in their message, they will focus on empowerment, readiness for the future, and change within communities.

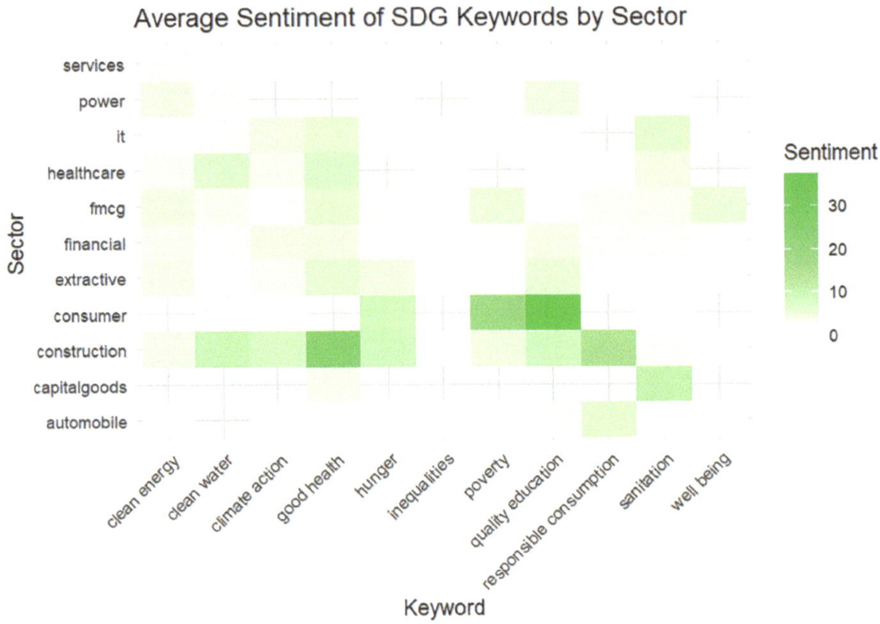

Fig. 4 Average sentiment of SDG keywords by sectors

Other sectors such as healthcare, FMCG and financial services have moderate values of sentiment of 5–15 to different SDG terms especially clean water, good health, sanitation, and well-being. In the case of healthcare and FMCG, these keywords correspond to important business goals such as the influence of products on hygiene or nutrition hence the positive framing is not only strategic but also anticipated. By contrast, the extractive and capital goods industries and automobile industries exhibit less sentiment intensity with respect to most of the SDG terms. Interestingly, extractives have a little more sentimentality on poverty and climate action perhaps in impact mitigation and CSR programs. Conversely, capital goods and IT contain a number of blank or close to zero cells, which means that these keywords have not been widely used or that they did not bear enough emotional background to demonstrate any significant sentiment. However, what is interesting to note is that such keywords as sanitation, hunger, and inequalities are ranked as having the least sentiment in general. It implies that companies can talk about them in a more factual or technical manner without using emotional terms, even though the issues are critical. However, the industrial sector presents SDGs in a more technical or functional way resulting in less sentiment scores.

4 Conclusion

The case of Nifty 50 companies is a good indication that corporate ethicality and responsible SDG alignment are in a convergent course of becoming the pillars of core strategy. Ethical governance offers the platform of trust of innovation and business development, and scrutinized SDG pledges solidify inclusive, equal, and value-added creation that is future-proof. In the same light, firms effective in the two areas are not only meeting the expectations of the society, but they are also setting the story of responsible capitalism in India. In order to maintain this momentum, companies have to move beyond output reporting, outcome measurements, compliance, purpose-driven leadership.

There are several implications that are brought out in this chapter. The chapter explains that to be more credible, stronger, and innovative, organizations need to go beyond regulatory compliance and include ethics and sustainability as the central part of their strategy. The implications determine the necessity of further consolidated reporting requirement and sector-based advice to the policymakers in the process of even-handed integration of SDGs and ethical maturity by sector. In the case of the academic researcher, the research process allows the researcher to engage secondary data to explore business sustainability even more. The analysis is valuable in helping the investors and ESG professionals make ethical investment decisions when evaluating the sustainable performance of organizations. Finally, the chapter demonstrates how the existence of organizations with an ethical and sustainable orientation contributes to the inclusive growth and influences the future of responsible capitalism in India.

4.1 Limitations

The chapter has certain limitations. The data used for this study is secondary data and hence only snapshots could be captured and not real-time generalization or interpretation is possible. In quantitative keywords text mining extraction cannot fully account for context dependence and other intermingled issues. The findings represent only large-sized companies which are contributing immensely to running the economy. The chapter does not provide any information regarding small sized companies or local-based companies. Since the entire chapter is based on secondary data, it lacks the depth of primary data. Stakeholder and employee's perspective if utilized could have provided value to the interpretation.

References

Babu, E. N., Prakash, G. S., & Sandhya, G. (2024). Sustainability practices and ESG scores: A sectoral study of Nifty 50 firms. *Research Review International Journal of Multidisciplinary, 9*(7), 36–46.

Balakrishnan, V., & Lloyd-Yemoh, E. (2014). Stemming and lemmatization: A comparison of retrieval performances. *Lecture Notes on Software Engineering, 2*(3), 262–267. https://doi.org/10.7763/LNSE.2014.V2.134

Bansal, C. (2025). AI ethics and sustainability: Accelerating paradigm shifts toward sustainable development. *Journal of Strategy & Innovation, 36*(1), Article 200537.

CSRBOX Research Desk. (2024). *India CSR Outlook Report 2024*. CSRBOX. https://csrboximpact.in/storage/publications/India-CSR-Outlook-Report-2024.pdf

Deb, R., Behra, A., & Dusmanta, K. (2023). ESG score and firm performance: Evidence from Indian-listed firms. *South Asian Journal of Management, 30*(4).

Doreswamy, H. (2025). *India's path to net zero: A journey through the Nifty Fifty companies. Journal of Management Studies and Reporting*. (Forthcoming).

Give Grants. (2025, June 28). India Inc's annual CSR spend to touch ₹1.2 lakh crore by 2035. *Fortune India*. https://www.fortuneindia.com/business-news/india-incs-annual-csr-spend-to-touch-12-lakh-crore-by-2035-says-give-report/124448

Indian Council for Research on International Economic Relations (ICRIER), & Ministry of Electronics & Information Technology. (2025). *State of India's Digital Economy Report 2024*. ICRIER. https://icrier.org/publications/state-of-indias-digital-economy-report-2024

Institutional Investor Advisory Services. (2025). *Greening of the NIFTY50: Net zero & carbon neutral commitments by 30 September 2024*. https://esgsummit.ficci.in/Compendium-ESG.pdf

PwC India. (2024). *Navigating India's transition to sustainability reporting: BRSR analysis of top 100 listed companies*. https://www.pwc.in/assets/pdfs/navigating-indias-transition-to-sustainability-reporting.pdf

Rajan, R., & Sivakumar, N. (2025). *Effect of ESG performance on portfolio performance—A study of portfolios of Nifty-indexed stocks in India*. (Forthcoming).

Rao, A., Dagar, V., Sohag, K., Dagher, L., & Tanin, T. I. (2023). Good for the planet, good for the wallet: The ESG impact on financial performance in India. *Finance Research Letters, 56*, Article 104093. https://doi.org/10.1016/j.frl.2023.104093

Salwan, P., & Bisht, B. S. (2023). Digital governance: Government of India initiatives in the public sector. In *Administration in India* (pp. 214–234).

Securities and Exchange Board of India. (2023). *BRSR Core—Framework for assurance and ESG disclosures for value chain* (Circular No. SEBI/HO/CFD/CFD-SEC-2/P/CIR/2023/122). https://www.sebi.gov.in/legal/circulars/jul-2023/brsr-core-framework-for-assurance-and-esg-disclosures-for-value-chain_73854.html

Shashwat, S., Chhabra, M., & Giri, A. K. (2025). Environmental externalities of digital financial inclusion: A green growth perspective. *Discover Sustainability, 6*(1), 81. https://doi.org/10.1007/s43621-025-00862-1

Chapter 5
Corporate Responsibility: Sustainability and Ethics in the Digital Economy

Anil Kumar, Neha Nupoor, and Mukesh Babu Gupta

Abstract In the fast-changing digital economy, corporate responsibility is changing in a big way, and it is becoming more and more influenced by the need for sustainability and moral imperatives. This chapter utilized the bibliometric analysis and topic modelling techniques of content analysis. This chapter looks at the connections between corporate ethics, digital transformation, and sustainable development. The study explores the new dimensions of discussion and common themes in the dialog around corporate responsibility. The analysis is conducted on metadata from 241 peer-reviewed documents indexed in Scopus. We used a term co-occurrence analysis to find conceptual clusters and then latent Dirichlet allocation (LDA) for topic modelling to systematically find knowledge domains. The findings of chapter indicate that technological disruption is redefining corporate responsibility, emphasizing transparency, inclusivity, and adherence to regulations. The novelty of this chapter lies on the contribution made towards academic and policy discussions by offering a data-driven foundation for comprehending how enterprises address ethical and environmental challenges in the digital economy.

Keywords Content analysis · Corporate responsibility · Digital economy · Ethics · Sustainability

A. Kumar (✉)
School of Business, UPES, Dehradun, Uttarakhand, India
e-mail: anilbhu03@gmail.com

N. Nupoor · M. B. Gupta
Sarala Birla University, Ranchi, Jharkhand, India
e-mail: nehanupoor1617@gmail.com

M. B. Gupta
e-mail: mukesh.gupt@outlook.com

K. Kumar et al. (eds.), *AI-Driven Circular Economy for Carbon Neutrality in Emerging Economies*, Environmental Footprints and Eco-design of Products and Processes,
https://doi.org/10.1007/978-3-032-30298-4_5

1 Introduction

Corporate responsibility (CR) in the twenty-first century is how businesses behave in terms of morality and tactics towards society, the environment, and the economy. The contemporary CR does not focus on profit maximization. Rather, it looks at the way a company ought to behave in a manner that is clear, long-lasting, and to the greater good of all the stakeholders, not only the shareholders. This will include employees, societies, and the future (Carroll & Brown, 2018). This definition is being redefined by global issues such as climate change, inequality, and scarcity of resources. Such problems are compelling companies to take more inclusive and responsible business models (World Economic Forum, 2020). In modern times, CR implies taking care of the environment, making sure your workers are treated fairly, assisting in your community, conducting business ethically, and contributing to charity. It is not an independent activity but an important aspect of the objectives of a company (Porter & Kramer, 2011). CR is currently employed globally to assist in sustaining the existence and validity of companies due to the preferences of people, regulators, and investors (Eccles et al., 2020). Now that companies use digital means to communicate worldwide, their responsibility is expected to be both ethically right and strategically smart. Communities expect companies to openly share miles they cover, use non-financial goals, and work internationally with the SDGs (United Nations, 2015). This makes twenty-first-century CR more than just following the rules or building a good reputation. It is an integrated, dynamic approach that shapes a company's identity, builds stakeholder trust, and spurs innovation for long-term growth (Brammer et al., 2012).

The digital economy includes billions of everyday Internet interactions between people, businesses, gadgets, data, and processes. Making decisions, customizing services, and creating ideas all rely on data, so data-driven models are necessary (OECD, 2020). To understand how people shop, improve how they operate, and provide what the customers require, businesses examine large amounts of data. Thanks to companies like Amazon, Uber, and Airbnb, single individuals are now able to take part in the market, transactions are cheaper, and the enterprise grows quickly and soon dominates through network effects (Parker et al., 2016). AI and automation are significant elements driving today's digital economy. These tools make things more efficient by doing things that used to need human help, including chatbots for customer care and predictive maintenance in manufacturing (Brynjolfsson & McAfee, 2014). Automation changes not simply the way work is done, but also the way people work and the job market. All of these things together make for a rapidly changing economy where innovation, flexibility, and connectedness are the keys to growth and a competitive edge.

Digital economy is also rendering the world smaller and friendlier through enabling individuals and companies to conduct business in markets that are not necessarily in their respective geographical locations. Small businesses and start-ups can now reach their customers and resources around the globe more easily due to e-commerce Websites, online payment systems, and work-from-home technologies (UNCTAD, 2021). However, with this rapid transition into the digital age, other issues such as concerns with data privacy, cybersecurity risks, and the digital divide may become more severe unless addressed efficiently (World Bank, 2021a, 2021b). Since the digital economy is expanding, businesses and regulations should collaborate to ensure that all people have equal access to it, the data is applied ethically, and the digital growth is sustainable. Corporate responsibility (CR) has become increasingly significant in the digital era as business works in a highly connected, open, and fast-paced world. Digital technologies such as social media, big data, and AI have transformed the requirements of stakeholders to the business, and they are now more responsible, ethical, and socially responsible (Schwab, 2017). Due to the rapid flow of information, every company is under constant scrutiny, and the inability to behave responsibly can ruin the brand instantly. Therefore, CR as part of the digital strategy can assist businesses to build trust, remain legitimate, and ensure that their operations remain aligned with the evolving values of the customers, investors, and authorities. The digital revolution also provides us with tremendous means to enhance and monitor CR projects. Data analytics are able to assess sustainability metrics in real time, which assists in the decision-making and reporting (EY, 2021). On the other hand, the supply chain can be opened with the help of blockchain. Digital platforms enable an enterprise to reach communities, crowdsource solutions, and positively impact more as environmental and social issues become more severe. In other words, it is not only a strategic necessity to look into CR in the digital world that would enable the creation of long-term value and the pursuance of global objectives such as the sustainable development goals (SDGs); however, it is also about the necessity to comply with the rules and develop a brand (UNGC & Accenture, 2022).

The digital age has made stakeholders reach out to one another and participate, which helps businesses unite employees, customers, and people in the community to assist in creating responsible solutions. Companies can get feedback quickly, follow open governance processes, and demonstrate that they use the Internet to help protect social and environmental causes (Ismail et al., 2017). With a participative approach, the business becomes more legit and ensures its core values are respected in keeping with the beliefs of people globally. Companies can adapt and are capable of overcoming fast changes in technology, society, and rules, and regular digital tools are helping them keep their corporate responsibility (CR) plans current. That makes their role more important as business leaders who care about the future.

2 Review of Literature

2.1 Corporate Sustainability in the Digital Economy

In today's digital world, companies are working hard to fulfil new expectations such as the ESG criteria and SDGs, when it comes to sustainability (ICMA, 2023a, 2023b; World Economic Forum, 2022). Because they rise in some of the most inventive environments, digital-first companies can ensure sustainability is part of their very foundations. Recent publications indicate that these companies are now treating sustainability as a key strategy, not just a duty or charitable act, and it guides how they build their products, interact with stakeholders, and build their worth (Bocken et al., 2019; OECD, 2023). Studies concluded that organizations that use ESG values in their technology gain more ethical investors and remain better recognized by customers (Deloitte, 2022; KPMG, 2021). Technology is the key that facilitates and magnifies sustainability initiatives. As an example, blockchain has enhanced the transparency and traceability of the supply chain, making it possible to reduce unethical actions and waste considerably (Accenture, 2020; IBM, 2021). Simultaneously, artificial intelligence (AI) and the Internet of things (IoT) are being leveraged to optimize energy use in smart cities and manufacturing, contributing directly to SDG 7 (Affordable and Clean Energy) and SDG 11 (Sustainable Cities and Communities) (UNDP, 2021; World Bank, 2021a, 2021b).

However, the environmental costs of digitalization are non-trivial. The increasing reliance on data centres and cloud computing has expanded the digital carbon footprint, due to high energy consumption and cooling needs (Nature, 2023a, 2023b; IEA, 2022). Additionally, frequent upgrades of electronic devices contribute significantly to global e-waste, complicating efforts to achieve SDG 12 (Responsible Consumption and Production) (UNEP, 2022). While digital tools can support environmental goals, their deployment must be guided by conscious efforts to mitigate their ecological impacts (Gartner, 2022). As a result, many scholars and policymakers are now suggesting strategies that combine attention to ethics with the development of sustainable technology. Technology-makers should add life cycle assessments to their development, use circular economy methods for electronics, and motivate consumers to limit their digital usage (Ellen MacArthur Foundation, 2021; Geissdoerfer et al., 2020). Corporate sustainability benefits greatly from digital technology, though it is essential to know both the positives and negatives (Schwab, 2022).

The findings of recent research point to the value of working with partners from various sectors and involving key stakeholders when growing resilient digital ecosystems. Innovations for social and environmental progress are often possible today because governments, NGOs, and academic institutions are becoming partners through public–private partnerships (McKinsey & Company, 2022; World Bank, 2021a, 2021b). For example, activities meant to improve technology education and help all people use it address SDG 4 (Education) and SDG 10 (Inequality). With the growth of instant data and new ways for citizens to participate in government, stakeholder theory is being adjusted and recognized in ways that match the digital world (Freeman et al., 2021).

Digital platforms make it easier for marginalized groups to access information, education, and markets in society and government (ITU, 2021; World Bank, 2021a, 2021b). The use of fintech solutions, mHealth apps, and edtech platforms is helping close the gaps facing many people. However, the digital divide, especially in rural and underserved regions, remains a persistent challenge that exacerbates social inequality (UNCTAD, 2022; World Bank, 2021a, 2021b).

From a governance perspective, digitalization has introduced new mechanisms for transparency and accountability. Companies are increasingly disclosing non-financial metrics such as ESG performance through automated dashboards and AI-based reporting systems (PwC, 2023). These tools empower stakeholders to demand ethical and sustainable corporate conduct. Businesses with effective digital governance practices usually achieve better environmental, social, and governance results and fewer breaches of rules (EY, 2023). Yet, because data ethics, privacy, and exploitation of information are key issues, there is a strong demand for robust governance frameworks (OECD, 2023). Adopting circular economy ideas in digital industry business models shows increasing popularity. For example, Dell, Apple, and Fairphone have taken measures such as collection programs and making products out of parts that can be changed or reset, to keep e-waste away and promote material use (Ellen MacArthur Foundation, 2021; Circular Economy Action Plan, EU, 2020). Thanks to Digital PLM systems, manufacturers can make better choices about using and discarding resources across their value chains (Geissdoerfer et al., 2020).

Sustainability is now integrated into how companies do Research & Development (R&D). Simulation tools and predictive analytics are used at the beginning of the design process to see what effects a project might cause (MIT Sloan, 2022). This practice minimizes risks and makes it possible to come up with new green inventions. Areas such as energy-saving algorithms and naturally decomposable electronic components become increasingly important (Deloitte, 2022). Lastly, a multi-stakeholder model is becoming more necessary in the digital transformation and sustainability intersection point. The collaboration structures play a significant role in the setting of standards, mobilization of resources, and fair access to innovation (UN Global Compact, 2023). Such examples as green data centres, e-waste policies, and digital equities are provided. With the digital economy getting mature, future studies should concentrate on cross-sector and interdisciplinary solutions to balance the long-term sustainability focus with the technological progress (OECD, 2023; Schwab, 2022).

2.2 *Ethical Challenges in the Digital Economy*

Information has emerged as one of the most important possessions of the dynamically evolving digital economy. It promotes innovation, personalization, and decision-making. However, with this change comes a large moral issue: data ethics. Business organizations and Websites are gathering increasing amounts of personal information, and in many cases, the user is not even aware of how much data is being gathered or why. The issues of privacy, surveillance capitalism, and the lack of informed permission are of concern to people (Keogh, 2021). The digital trails of individuals are being auctioned, and more than ever, there are concerns over the ownership of the information, how transparent these are, and how they can be managed in an ethical manner (Digital Regulation, 2024).

Worryingly, further is the introduction of AI and machine learning systems that are currently applied to hiring technologies, as well as algorithms that determine whether to grant a loan. Such algorithms will be effective and unbiased, yet they tend to inherit or even increase the prejudices that are already present in the data they are trained on (Nature, 2023a, 2023b). This increases the importance of the question of algorithmic fairness. Unless AI programs are well scrutinized, they are likely to discriminate against marginalized groups. This may cause disparity in key areas such as housing, employment, and even in the criminal justice system. People who are harmed by the algorithmic decision-making have an even more difficult time holding anyone responsible and seeking assistance since it is extremely difficult to visualize how it functions (ICMA, 2023a, 2023b).

Meanwhile, the so-called gig economy of flexible, platform-based employment has generated its own moral problem. Gig workers do not necessarily enjoy the same protection as ordinary workers, such as the security of jobs, medical care, and the ability to negotiate as a union. Another concern that people have is the health and liberation of workers due to algorithmic management, i.e. the working schedule, performance rating, and wages are determined by applications and systems (Human Rights Watch, 2025). An increasing demand exists regarding the ethical labour standards that consider the peculiarities of digital work. It is increasingly becoming hard to balance productivity with human dignity.

Another issue in the digital economy is ethical marketing and targeted advertising. When firms are able to access specific consumer information, they can tailor adverts to meet the habits and preferences of individual people. But it is also easier to control people like this. Targeted marketing can capitalize on the mental vulnerability of people and either get them to purchase excessively or spread misinformation. It creates issues to the liberty of individuals and the well-being of the society at risk (Singh, 2022). With the continued advancement in technology, there is a need to ensure that ethical concerns lie at the core of digital transformation in ensuring that the future is sustainable and accessible to all people.

Increasingly, the population is realizing that ethical governance in the digital economy should be based on digital literacy and citizen engagement. The only way to ensure real consent and accountability is to make people aware of how their data is used, the influence of algorithms on their lives, and their rights (ICMA, 2023a, 2023b). This will include introducing digital ethics in the schools, promoting open dialogue among tech companies, authorities, and the civil community, and promoting participatory decision-making. Ultimately, a digitally just society cannot rely solely on top-down regulation; it must also be built through an informed and active citizenry that can critically engage with the technologies shaping their world.

3 Martial

Data is base of any study. For getting data, we constructed three search strings exhibited in Appendix 1, with the "OR" and related keywords. During the search, "AND" has been used between these search strings. Further, during the search, we have applied three filters related to subject area, types of documents, and language. First, we filtered out documents other than business management, accounting, economics, econometrics, and finance. Second, we restrict our search results to article who are in the English language. After extracting data, we removed all irrelevant documents, documents with less or missing information. After this data cleaning process, we have a corpus of 241 documents. Bibliometric and LDA were performed on the metadata of these 241 documents.

4 Discussion

Figure 1 exhibits a keyword network built on authors' keywords. VOSviewer was utilized to create this keywords co-occurrence network. This network presents a comprehensive view of the co-occurrence relationships between keywords. Keywords co-occurrence network serves as a valuable tool for understanding the intellectual structure and evolving trends in the discipline. This network is made of nodes and edges. Different colours exhibited different clusters. These clusters are generated from the co-occurrence of keywords in the titles, abstracts, and keyword sections of academic articles. The proximity of nodes indicates the strength of their association, while node size reflects the frequency of occurrence.

First, the Blue cluster is found in the middle of the network. This is based on the interaction of corporate social responsibility and artificial intelligence, sustainable development, big data, digitalization, and corporate digital responsibility. This cluster is an indication of how digital CSR looks, where companies use digital tools such as artificial intelligence, machine learning, and legal machine learning to increase transparency, social performance, and sustainability promises. The fact that we can find keywords like ChatGPT, AI, big data, and digitization in the context of CSR is a signal of the growing use of AI-based tools in corporate morality. Besides, this group of people had the tendency to shift conventional CSR into tech-enabled, measurable, and scalable activities using AI and data technologies.

The second cluster is identified as related to sustainability and ESG governance in yellow colour. This cluster is constructed on the sustainability, CSR, ESG, and sustainable development goals. form another dense cluster, showing a strong conceptual overlap between sustainability reporting, stakeholder engagement, and ESG performance. The connection among fintech, COVID-19, and the banking sector suggests an increasing emphasis on the financial sector's role in sustainable recovery and ESG-aligned investing. The presence of social sustainability and performance

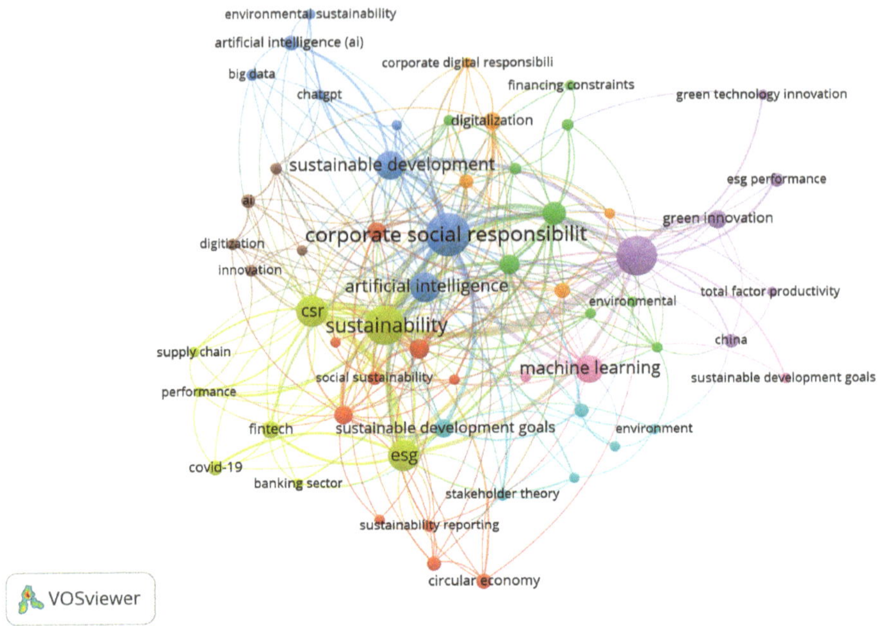

Fig. 1 Keywords co-occurrence network

shows that stakeholder theory is still influential, indicating the evolving measurement of firm success beyond financial terms to include social and environmental outcomes. Thus, this cluster emphasizes that corporate responsibility in the digital age must align with the SDGs, especially as ESG reporting becomes standardized across sectors.

Third, cluster, purple, exhibited a relationship between green innovation and technology performance. This cluster consists of keywords such as green innovation, green technology innovation, ESG performance, and total factor productivity. The presence of China signifies a growing body of literature from environmental economics and innovation in the country. In addition, the existence of China implies the dominance in the academic production in the region, which may indicate the strategic focus of the country on the transitions to green and the compliance with ESG. The availability of such keywords as productivity and green innovation suggests that the researchers are looking into the ways in which sustainability-based technological upgrades can propel economic productivity in the long term. In short, this purple cluster factor in the economic performance and innovativeness with respect to the environment, responsible digital technologies.

Forth cluster is depicted in red colour, and it is oriented on circular economy and stakeholder engagement. The cluster is built on terms such as circular economy, sustainability reporting, stakeholder theory, and sustainable development goals. These ideas are modelling a more prescriptive and governance-oriented perspective of corporate responsibility, which focuses on ethical business models and the inclusive management of stakeholders. The relationship between sustainability reporting and ESG and stakeholder theory, showing how corporate reporting is being re-packaged in the digital economy. The red circle points to the fact that the ethical aspects of corporate conduct in the digital era should also involve circular value chains and open communication with all stakeholders.

The fifth cluster is related to machine learning and environmental modelling. This cyan cluster framed with keywords such as machine learning, environmental, environment, and sustainable development goals, and this cluster focuses on the technical dimension of AI in sustainability. This cluster indicates the use of AI and ML for environmental monitoring, climate modelling, carbon accounting, and issues related to SDG progress tracking. This cluster connects technical AI capabilities to real-world environmental outcomes, supporting the role of data science in achieving SDGs. The importance of this cluster lies in its contribution to evidence-based sustainability strategies, assisting firms to track and optimize their environmental impact.

LDA:

We have applied LDA analysis to identify the potential research theme. Firstly, we identify the top 10 topics exhibited in Appendix 2. Thereafter, we calculated the coherence score for each topic (see Appendix 3). Then we decided on the number of themes for review with the help of the elbow method (see Appendix 4). After deciding on the number of themes, we selected topics on the coherence score.

Topic 2: Blockchain and corporate social responsibility in the digital age

Blockchain, corporate, social, CSR, research, sustainability, responsibility, business, and technology are some of the words that describe Topic 2. This group shows how more and more people are studying and using new technologies like blockchain to change how businesses are responsible and sustainable. Blockchain technology makes supply chains and corporate governance more open, accountable, and traceable than ever before, which are all important for corporate social responsibility (CSR). The words "corporate," "social," and "CSR" combined with "technology" and "blockchain" show that there is a growing conversation on how to connect corporate moral duties with technical frameworks. The problem demonstrates that digital technologies not only help to make things more efficient, but they are becoming increasingly significant concerning the ethical corporate practices in the context of the topic of a sustainable economy, Corporate Responsibility: Sustainability and Ethics in the digital economy. Indicatively, blockchain will be able to ensure that companies do not purchase raw materials that are sourced by illegal traders or that violate environmental regulations. It is able to make ESG reporting systems more effective and transparent, and they can be reviewed the moment in real time by the regulators and stakeholders. It is a massive shift from passive reporting to active and verifiable accountability.

Topic 3: Digital services, platform economies, and sustainable infrastructure

Topic 3: The third theme is based on such words as new, service, information, development, social, services, digital, infrastructure, sustainable, and platforms. Collectively, the concepts narrate a story of the creation of digital service platforms (gig economy applications, SaaS, and financial platforms) and their implications on social welfare and sustainable development. This issue is very related to such concepts as digital capitalism and platform governance that are both growing closer to corporate accountability. With the digital economy, services are becoming different due to data-driven models, which require new forms of ethical regulation. The fusion of social, sustainable infrastructure, and platforms implies that companies should not only develop infrastructure but also ensure that these platforms are beneficial to society and the environment. This brings up pertinent concerns about what is supposed to be the responsibility of the businesses. Are platforms easy to visit by everybody? Are employees paid enough? Should we deal with user data? Due to this fact, Topic 3 is extremely significant since it indicates that it is the responsibility of companies to ensure their online services are not only ethical but also sustainable. It implies that in addition to their profits, platform companies must also be accountable regarding their impacts on society, which is one of the most significant problems of corporate responsibility in the digital era.

Topic 5: Digital transformation and green innovation

The subjects of this section are digital transformation and sustainable innovation. Topic 5 comprises digital, corporate, transformation, ESG, innovation, sustainability, performance, CSR, social, and environmental. In this topic, it is demonstrated that organizations using digital tools not only have more efficient operations but also make their environmental impact better. It demonstrates the convergence of digital change and ecological innovation. This issue is highly relevant to the ESG (environmental, social, and governance) frameworks employing the so-called digital technologies, including IoT, AI, and big data analytics, to monitor, report on, and reduce the environmental impact. This theme structure demonstrates the possibility of digital and environmental operations that can work together in a manner that can be good business. As an example, the optimization of the use of energy, organization of the trash, and measurements of carbon footprints are now optimized with smart technologies. All of them are tools that could aid in sustainability efforts at the moment. By including the word performance to the list of keywords, it is evident that digital transformation should not only generate actual ESG outcomes but should also enable businesses to be more efficient. Moreover, words such as corporate and CSR demonstrate that these attempts are included in the strategies of the company as its moral responsibilities. The topic provides a comprehensive insight into the need to have digital infrastructure to achieve long-term environmental objectives, which is why it is a crucial aspect of the issue that should be brought up when discussing ethics and sustainability in the digital economy.

Topic 6: Corporate ESG and ethical AI integration

Topic 6 discusses such issues such as ESG, corporate governance, learning, businesses, trade, machine learning models, ethics, CSR, and reporting. The topic is very detailed on the corporate governance procedures, including the emergence of machine learning and AI-based ESG evaluations and strategy reporting. Ethics, machine learning, and learning are terms that frequently occur, so it will be safe to assume that artificial intelligence is both an opportunity and a challenge to contemporary corporate liabilities. In cases where companies apply AI to assist them in decision-making on various issues, including recruitment, investments, and credit rating, they encounter numerous ethical issues, including algorithmic bias, lack of transparency, and data privacy concerns. Consequently, an increasing number of people anticipate that corporate ESG models will incorporate ethical AI governance. By the term reports, we are likely referring to ESG or CSR reports, which now have AI-assisted data-driven appraisals. This not only makes them more objective but also makes it possible to check them in real time and brings up issues of transparency and human control. This problem demonstrates the significance of ethical principles in the application of AI in companies, which is viewed through the prism of the article, "Corporate Responsibility: Sustainability and Ethics in the Digital Economy." Although AI has much potential, the technology must adhere to ethical standards, such as fairness, inclusion, and environmental protection, even in the process of automating significant business.

Topic 7: AI-driven corporate responsibility and social innovation

The Topic 7 keywords will primarily be social, AI, responsibility, corporate, CSR, intelligence, artificial, sustainability, study, and development. This subject matter has close connections to artificial intelligence as well as corporate social responsibility. It indicates that the area of research regarding the application of AI to social innovation and business ethics is expanding. Topic 7 appears to focus on the emerging and new applications of AI in CSR initiatives, whereas Topic 6 is more of governance and ESG. The term social responsibility and AI describes an initiative that involves applying AI to address societal issues, including applying AI to predict how to alleviate poverty, applying AI to assist in health issues diagnosis in underserved communities, or even improving educational resources. The level of accountability in commercial environments with such applications is quite high as AI is applied in making things improved for everyone, rather than the money of a single individual. This is a topic that has a lot of significance to your primary theme. Businesses in the digital economy are obligated not merely to abide by the rules or to donate money to charity. They must bring social innovation to their conventional business models. Artificial intelligence, as a versatile technology, presents both isolated risks and opportunities. When applied ethically, it has the potential to transform the manner in which businesses relate to the communities, facilitate a developmental agenda, and promote purposeful innovation. These five topics present a synthesis and thematic implications in that the entire spectrum of how the digital economy transforms the corporate responsibility and sustainability plans is represented. All the topics, such as the role of blockchain in making supply chains more open (Topic 2), the use of AI to establish social innovation and responsible governance (Topics 6 and 7), platform accountability (Topic 3), and the green digital transformation (Topic 5) all indicate the same trend: with the development of technology, businesses will have more responsibilities.

5 Conclusion

These five topics taken together form the range of ways that the digital economy is transforming corporate responsibility and sustainability practices. Blockchain and supply chain transparency (Topic 2), AI and social innovation and responsible governance (Topics 6, 7), platform accountability (Topic 3), and green digital transformation (Topic 5) are the themes that manifest a definite pattern: the corporate responsibility needs to be realigned with technological innovation. The chosen themes of the LDA model are wealthy in providing the academic and practical knowledge when it comes to corporate responsibility in the digital economy. They present us with a mixed up and still consistent landscape where the technological advances, sustainability vows, and ethical regulation lie deep-rooted. This thematic integration is not just another conceptual mandate but a strategic benefit to companies that desire to enjoy long-term legitimacy and value creation in the twenty-first century. The analysis offers a solid foundation of an essay or policy flaw that needs the redefinition of corporate ethics in a digital-focused age, an age where accountability is no longer a minor and somewhat accidental complement, but a key catalyst of creativity, trust, and expansion. The ethical infrastructure is necessary because the digital technologies gain greater importance to the business models, it has to be structured and applied ethically. This entails data security, algorithmic justice, and easy access by everyone. The digital technology can make the CSR operations more efficient, expansive, and less monitorable. Blockchain, artificial intelligence, and the Internet of things are used in new ways to monitor the sustainability indicators automation, the ESG compliance, and the stakeholder engagement. The ESG objectives of the company should be connected with the digital innovation. AI to good and sustainability platforms and green innovation should no longer be isolated in a single experiment, but it should be in the long-term business goals. They present us with a mixed up and still consistent landscape where the technological advances, sustainability vows, and ethical regulation lie deep-rooted. This thematic integration is not just another conceptual mandate but a strategic benefit to companies that desire to enjoy long-term legitimacy and value creation in the twenty-first century. The analysis offers a solid foundation of an essay or policy flaw that needs the redefinition of corporate ethics in a digital-focused age, an age where accountability is no longer a minor and somewhat accidental complement, but a key catalyst of creativity, trust, and expansion. The digital technology can make the CSR operations more efficient, expansive, and less monitorable. Blockchain, artificial intelligence, and the Internet of things are used in new ways to monitor the sustainability indicators automation, the ESG compliance, and the stakeholder engagement. The ESG objectives of the company should be connected with the digital innovation. AI to good and sustainability platforms and green innovation should no longer be isolated in a single experiment, but it should be in the long-term business goals.

Appendix 1: Search String

String 1	"corporate responsibility" OR "corporate social responsibility" OR "CSR" OR "corporate accountability" OR "business ethics" OR "corporate ethics"
String 2	"digital economy" OR "digital transformation" OR "artificial intelligence" OR "machine learning" OR "blockchain" OR "platform economy" OR "gig economy" OR "FinTech" OR "digital platform"
String 3	sustainability OR "sustainable development" OR "ESG" OR "SDGs" OR "sustainable practices" OR "green practices"

Appendix 2: Token and Probabilities

Topic 1		Topic 2		Topic 3		Topic 4		Topic 5	
research	0.1849	blockchain	0.1601	new	0.1585	sustainability	0.1693	digital	0.1706
supply	0.1254	corporate	0.1198	service	0.1296	social	0.1237	corporate	0.1307
sustainability	0.1192	social	0.1141	information	0.1248	corporate	0.1174	transformation	0.122
digital	0.0972	csr	0.1058	development	0.1128	environmental	0.1031	esg	0.1005
review	0.0881	research	0.0912	social	0.0971	fintech	0.1021	innovation	0.0939
chain	0.0869	sustainability	0.0845	services	0.0962	esg	0.0821	green	0.0895
corporate	0.0802	sustainable	0.0835	digital	0.0764	study	0.0774	performance	0.0855
sustainable	0.0755	responsibility	0.082	infrastructure	0.0758	business	0.0773	csr	0.076
analysis	0.0721	business	0.0796	sustainable	0.0677	responsibility	0.0753	social	0.0686
literature	0.0704	technology	0.0795	platforms	0.061	learning	0.0723	environmental	0.0626
Discard	Blockchain and CSR	Digital services	Discard	Digital transformation and green innovation					
Topic 6		Topic 7		Topic 8		Topic 9		Topic 10	
esg	0.1543	social	0.1979	environmental	0.1617	policy	0.1257	information	0.1345
corporate	0.1158	ai	0.1694	social	0.148	climate	0.1079	strategic	0.131
learning	0.1087	responsibility	0.1147	development	0.1094	energy	0.1075	sustainability	0.1168
companies	0.0994	corporate	0.1141	sustainable	0.1008	innovation	0.1055	research	0.1135
business	0.0977	csr	0.1014	esg	0.0979	sustainable	0.102	food	0.0988
machine	0.0958	intelligence	0.0719	study	0.0939	development	0.0964	data	0.0893
model	0.089	artificial	0.0707	digital	0.0847	corporate	0.0927	leadership	0.0842

(continued)

(continued)

ethics	0.0814	sustainability	0.055	sustainability	0.0713	economy	0.0913	csr	0.0839
csr	0.0799	study	0.0527	responsibility	0.0704	flexible	0.0885	organizations	0.0746
reports	0.0779	development	0.0521	governance	0.0619	mining	0.0826	study	0.0735
Corporate ESG and AI integration	AI-driven corporation responsibility and social innovation	Discard	Discard	Discard					

Appendix 3: Topic Coherence Score

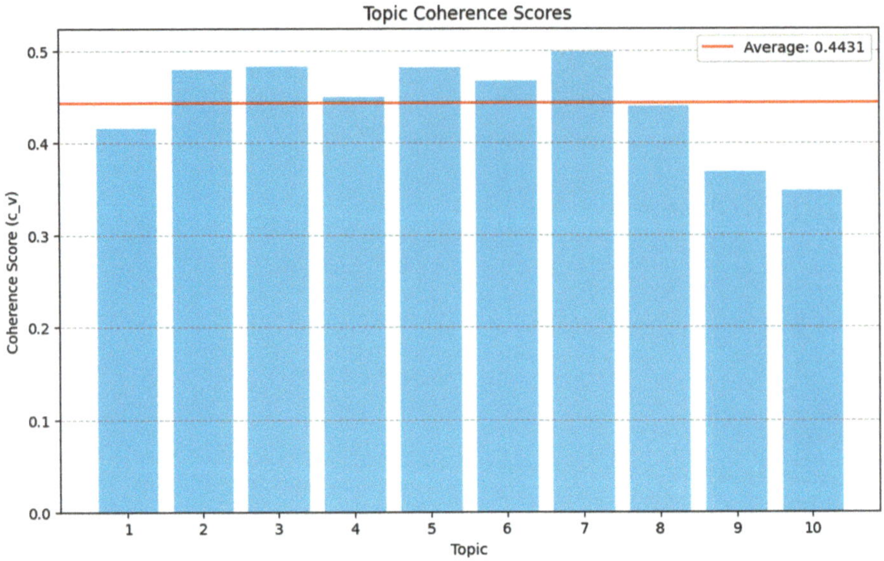

Appendix 4: Elbow Method: Coherence Versus Number of Topics

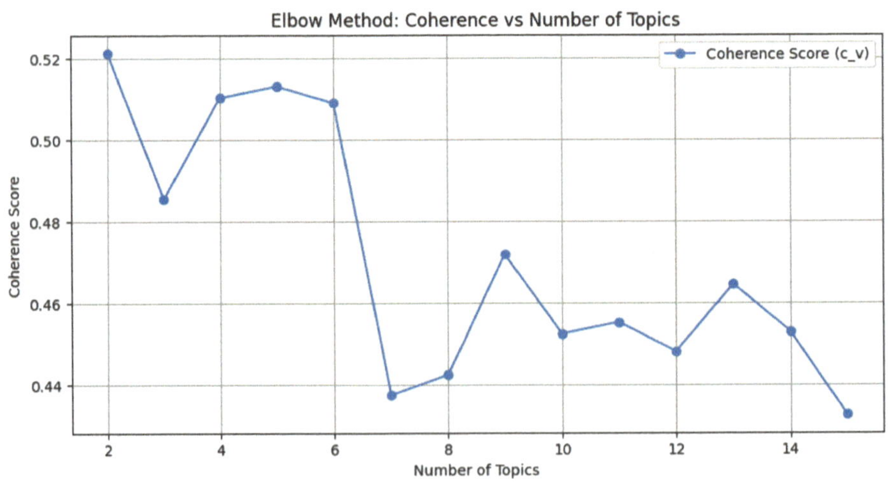

Appendix 5: Visualization of LDA Topics

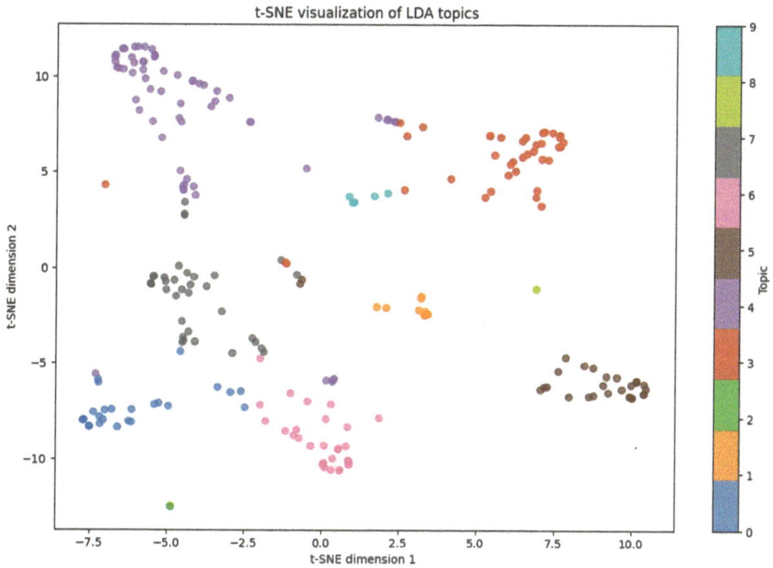

References

Accenture. (2020). Blockchain for sustainable and inclusive value chains.

Bocken, N. M. P., Short, S. W., Rana, P., & Evans, S. (2019). A literature and practice review to develop sustainable business model archetypes. *Journal of Cleaner Production.*

Brammer, S., Jackson, G., & Matten, D. (2012). Corporate social responsibility and institutional theory: New perspectives on private governance. *Socio-Economic Review, 10*(1), 3–28. https://doi.org/10.1093/ser/mwr030

Brynjolfsson, E., & McAfee, A. (2014). *The Second Machine Age: Work, Progress, and Prosperity in a Time of Brilliant Technologies.* W.W. Norton & Company.

Capriotti, P., & Moreno, Á. (2017). Corporate citizenship and public relations: The importance and interactivity of social responsibility issues on corporate websites. *Public Relations Review, 43*(3), 412–422.

Carroll, A. B., & Brown, J. A. (2018). Corporate social responsibility: A review of current concepts, research, and issues. In A. Farazmand (Ed.), *Global Encyclopedia of Public Administration, Public Policy, and Governance.* Springer. https://doi.org/10.1007/978-3-319-31816-5_1191-1

Circular Economy Action Plan, EU. (2020). European Commission Circular Economy Strategy.

Deloitte. (2022). *Digital ESG Reporting and Sustainable Transformation.*

Digital Regulation. (2024). *Navigating data governance: A guiding tool for regulators.* https://digitalregulation.org/navigating-data-governance-a-guiding-tool-for-regulators/

Eccles, R. G., Klimenko, S., & Stroehle, J. (2020). The investor revolution. *Harvard Business Review, 98*(3), 106–116.

Ellen MacArthur Foundation. (2021). *Circular Economy in the Digital Age.*

EY. (2021). *How analytics is driving ESG performance.* Retrieved from https://www.ey.com/

EY. (2023). ESG Reporting Through Automation.

Freeman, R. E., Harrison, J. S., & Wicks, A. C. (2021). *Stakeholder Theory: The State of the Art.*

Gartner. (2022). *Green IT and Sustainable Digital Trends.*

Geissdoerfer, M., Savaget, P., Bocken, N. M. P., & Hultink, E. J. (2020). *The circular economy—A new sustainability paradigm?*
Human Rights Watch. (2025, May 12). *The gig trap: Algorithmic, wage and labor exploitation in platform work in the US*. https://www.hrw.org/report/2025/05/12/gig-trap/algorithmic-wage-and-labor-exploitation-platform-work-us
IBM. (2021). *Blockchain for Ethical Supply Chains.*
ICMA. (2023). The four pillars of ethical digital engagement. https://icma.org/articles/pm-magazine/four-pillars-ethical-digital-engagement
ICMA. (2023). The Role of ESG in Digital Finance.
International Energy Agency (IEA). (2022). Data Centres and Energy Demand.
Ismail, M. H., Khater, M., & Zaki, M. (2017). Digital business transformation and strategy: What do we know so far. *Cambridge Service Alliance, 10*(1), 1–35.
ITU. (2021). *Digital Inclusion and Access.*
Keogh, B. (2021). 50 attorneys general investigate Google: Surveillance capitalism and legal privacy frameworks. *Transnat'l l. & Contemp. Probs., 30*, 267.
KPMG. (2021). *The Future of ESG Integration.*
McKinsey & Company. (2022). *Tech for Good: Sustainability in the Digital Economy.*
MIT Sloan Management Review. (2022). Green Innovation in Digital R&D.
Nature. (2023). Algorithmic bias: The unseen danger in machine learning. *Nature, 615*(7951), 4–5. https://doi.org/10.1038/d41586-023-00412-z
Nature. (2023). Ethics and discrimination in artificial intelligence-enabled recruitment. *Humanities and Social Sciences Communications, 10*(1), 1–9. https://www.nature.com/articles/s41599-023-02079-x
OECD. (2020). Measuring the digital transformation: A roadmap for the future. https://www.oecd.org/publications/measuring-the-digital-transformation-9789264311992-en.htm
OECD. (2023). *Digital transformation and the environment.*
Parker, G., Van Alstyne, M., & Choudary, S. (2016). *Platform revolution: How networked markets are transforming the economy*. W. W. Norton & Company.
Porter, M. E., & Kramer, M. R. (2011). Creating shared value. *Harvard Business Review, 89*(1–2), 62–77.
PwC. (2023). *Smart ESG reporting platforms.*
Schwab, K. (2017). *The fourth industrial revolution.* Penguin Random House.
Schwab, K. (2022). *Shaping the future of the fourth industrial revolution.* World Economic Forum.
Singh, R. (2022). The ethics of data collection and targeted advertising in digital marketing. LinkedIn. https://www.linkedin.com/pulse/ethics-data-collection-targeted-advertising-digital-marketing-singh
UN Global Compact. (2023). *Accelerating digital inclusion and sustainable innovation.*
UNCTAD. (2021). *Digital Economy Report 2021. United Nations Conference on Trade and Development*. https://unctad.org/
UNCTAD. (2022). *Digital Economy Report: Bridging the Digital Divide.*
UNDP. (2021). *Technology and the SDGs.*
UNEP. (2022*). E-waste and circular economy solutions.*
UNGC & Accenture. (2022). *CEO study on sustainability 2022. United Nations global compact.* https://www.unglobalcompact.org/
United Nations. (2015). *Transforming our world: The 2030 agenda for sustainable development.* https://sdgs.un.org/2030agenda
World Bank. (2021). *World development report 2021: Data for better lives.* https://www.worldbank.org/
World Bank. (2021). *World development report: Data for better lives.*
World Economic Forum. (2020). *Toward common metrics and consistent reporting of sustainable value creation.* https://www.weforum.org/
World Economic Forum. (2022). *Global risks report: ESG in the digital era.*

Chapter 6
Artificial Intelligence, Sustainability, and the Future of Human Resources in the Organisation

Devika Agrawal, Teena Bagga, and Mahanish Panda

Abstract Artificial intelligence (AI) technologies strive to transform the landscape of human evolution. The integration of AI technologies in the organisation raises complex, ethically centred challenges, and unprecedented opportunities. Human involvement may possess some biases, a lack of transparency, and inefficiency, whereas AI promises to tackle these challenges with greater accuracy in various HR processes such as recruitment, appraisal, and employee engagement in an organisation. This chapter will explore the emerging landscape of human-centric AI—a balanced approach that ensures fairness, transparency, empathy, and dignifies the deployment of human resources within an organisation, coinciding with the aspects of sustainability. This chapter will examine the integration of human-centric AI and sustainability within the value chain system, encompassing various processes such as recruitment, training and development, performance appraisal, and succession planning. It will draw some insights from the real-world cases that will reinforce the strategic advantages and dilemmas associated with artificial intelligence in HR processes. The chapter also outlines the framework for a responsible adoption of AI along with its principles, ranging from inclusive data practices, continuous algorithmic audits, employee consent, and human oversight to ensure the robustness of the social aspect of sustainability. Further, we will also discuss the challenges, such as technological barriers, organisational resistance, and regulatory bottlenecks in the implementation of human-centric AI within the organisation, while offering practical insights to overcome these issues. The book chapter offers both theoretical narratives and actionable strategies to bridge the gap between human values and technological

D. Agrawal (✉)
Institute of Business Management, GLA University, Mathura, India
e-mail: devikaagrawal.phd@gmail.com

T. Bagga
Department of Management Studies, Jamia Millia Islamia, Delhi, India
e-mail: tbagga@jmi.ac.in

M. Panda
International Management Institute, Kolkata, West Bengal, India
e-mail: mahanishresearch@gmail.com

K. Kumar et al. (eds.), *AI-Driven Circular Economy for Carbon Neutrality in Emerging Economies*, Environmental Footprints and Eco-design of Products and Processes,
https://doi.org/10.1007/978-3-032-30298-4_6

advancement. With the blend of human experience within AI design and application, the chapter advocates a balanced approach to assist humans rather than replace them. We will also conclude a forward-looking perspective to embrace responsible AI to foster ethics-driven performances and an inclusive workplace in the digital era.

Keywords Artificial Intelligence · Generative AI · Human resource management · Human-centric Artificial Intelligence · Human resource sustainability

1 Introduction

The landscape of artificial intelligence has greatly influenced the operation across an organisation's various sectors. It is now being deployed to facilitate strategic decision-making processes at various levels of management, streamlining the flow of activities that operate within it (Gong et al., 2025; Srivastava et al., 2024). AI has surely reflected a ground-breaking potential to achieve a competitive advantage; however, its rapid transition put forth both challenges and opportunities in human resource management (HRM). Human resource management is the most important function of an organisation, grounded in empathy, ethical stewardship, and interpersonal relations within the organisation (Sharma et al., 2025; Zolak Poljašević et al., 2025). Many corporations are constantly deploying various AI tools to entirely automate their recruitment process, employee learning system, performance appraisal system, and salary system, but the question remains: how can we ensure that the integration of AI into human resource management will preserve the sacrosanctity of the presence of the human element rather than spoiling it?

The last decade has seen the evolution of HCAI (human-centric AI) as a response to impersonal application of AI and efficiency-driven organisations. Human-centric AI is a system designed with a clear strategy to assist in augmenting opportunities for human resources rather than replacing them (Alwali & Alwali, 2025). The systems are governed through a set of rules and relations, but mainly by principles of fairness, inclusivity, well-being, transparency, and accountability. Within HRM, a human-centric AI framework presents how algorithms come in contact with humans and make them retain their decisions, how data is being analysed, and how employees can get due recognition while maintaining their dignity in every phase of the employee life cycle (Kumar & Prakash, 2019). Organisations globally are navigating their operations towards sustainability (Agrawal, 2024). With special uphold on societal and regulatory pressure, it is imperative that business practices are constantly being aligned with UN-SDGs (United Nations Sustainable Development Goals), particularly SDG 8, SDG 9, and SDG 10, which work for decent work opportunities and economic growth, innovation, and infrastructure, and aims at reducing the inequalities among the workforce (Kandpal, 2020; Kumar & Tarkar, 2024). HRM is expected to play an important role in being an equitable, resilient, and inclusive workplace. The interaction of HCAI and sustainability provides a compelling opportunity to build an intelligence system that will foster ethical labour practices, celebrate diversity, increase employee engagement, while maintain organisational productivity (Agrawal et al., 2024a). The study aims to address the intersection of artificial intelligence throughout HR activities, strategically crafted to achieve any organisation's sustainability goal. The study does not view AI as a mere automation tool that can delve into a transformational force and creates are responsible, fair, and inclusive workspace (Singh et al., 2026).

A major mindset shift is needed from the narrative of '*AI replacing humans*' to '*AI supporting humans*' in their various decision-making processes while maintaining their emotional intelligence. This will build an organisation into a more resilient and adaptive entity. Additionally, the urgency to adopt a human-centric approach has now been highlighted as, after COVID-19, hybrid working models, safety, and mental health of all the employees were prioritised over long working hours and having a workaholic mindset. The use of very heavy AI equipment while the employees work remotely raises many ethical concerns regarding employee psychological safety and autonomy at work (Boon et al., 2025; Kumar et al., 2023a). The increased usage of artificial intelligence during the entire process of human resources, from recruitment to performance appraisal, pose certain risks regarding algorithmic biases. These biases can have severe consequences if they are not addressed and audited. A well-crafted plan is crucial, which includes ethical intent and human values and considers technology as a well-embedded tool (Srivastava et al., 2021; Torelli et al., 2025).

HCAI (human-centric artificial intelligence) is definitely enhancing long-term value creation, and this fact cannot be denied. Organisations prioritising employee empowerment, transparency, engagement, and empathy can retain their human resources to a great extent, which further improves their innovation ability and builds a strong brand reputation. This can further be empirically proved through ESG disclosure and sustainability reports, as per the format presented by SEBI in India (Kumar, 2022). The SEBI has also launched initiatives like business responsibility and sustainability reporting for the top 1000 listed companies (based on market capitalisation), which mandate the integration of HRM and ethical AI into measurable quantitative indicators ('sustainability Reporting for Boosting National Commitment and Overcoming Challenges: A Hierarchical Model', 2024).

Moving ahead and framing the central argument of this particular study, the major objective of this study is to present an in-depth understanding of human-centric AI frameworks, the principles governing them, and their importance in the modern era. We will discuss how the convergence of HCAI can ensure sustainability in an organisation and can provide a road map for a resilient, ethical, and forward-looking HRM. We will also discuss the convergence of HCAI and ensuring sustainability to provide a blueprint for an ethical, resilient and forward-looking HRM. The study also provided a few case studies from global and Indian scenarios, which tell us about the major organisational practices and regulatory developments, providing us with scholarly insights. By putting human resources at the centre of an organisation and making sure that all the other activities can revolve around it, we can say that AI can bring transformation and an ecosystem that will be efficient while being humane and sustainable.

2 Understanding Human-Centric AI

Human-centric artificial intelligence is a setup, or we can say a framework that revolves around human values, employee well-being, and their needs and desires while fulfilling its core operations of design, development, and deployment of artificial intelligence through the various levels of management (Nti et al., 2022; Tripathi et al., 2022). It focuses on being a strategic collaborator that can enhance human capacities while maintaining the dignity of social, ethical, and psychological aspects of human personality. We can also say that AI is a tool to ensure efficient automation within an organisational system, leading to outcomes that are inclusive, transparent, fair, and aligned with human dignity.

The traditional AI systems are often designed with a single motive of optimising resources; they either work for productivity, speed, or cost, whereas human-centric artificial intelligence comes without a very broad perspective, as it seems to maintain the trust, safeguarding the interest and human oversight (Mabrouk, 2025). Within the realms of HRM, where the decisions revolve around career aspirations, maintaining livelihood, ensuring mental health, and equity, it becomes a crucial force for implementation in an organisation.

Discerning HCAI requires a deep understanding of traditional AI systems to differentiate it from human-centric AI, even though both are considered computer algorithms, data analytics, and machine learning. Their approaches differ significantly, as their intent, ethical orientation, and design policies differ. For instance, now the company is using ATS for recruitment, which is an AI tool that filters candidates solely based on the keywords that their resume has, instead of going through the entire resume of the potential candidate, which can amplify the risk of algorithmic and historical biases (Kaith & Sachdeva, 2024). Whereas human-centric artificial intelligence will also offer certain transparency during the evolution process, because as with the help of artificial intelligence, human involvement will still be there while making important decisions regarding recruitment performance evaluation and other processes of human resource management, thereby ensuring the contextual judgement and proper fairness.

The baseline of HCAI lies in the set of technical governance and ethical principles that serve as a guiding light for the responsible allocation and deployment of these AI tools in the organisational structure, particularly in human resource management.

AI should not control humans; rather, humans should keep control of the AI system, and therefore, it is said that AI should augment the opportunities are alternatives in human decision-making rather than overriding it (Kumar et al., 2023b; Sharma et al., 2025). This is particularly important in critical HR functions, such as the entire recruitment, selection, performance appraisal, and exit interview process, where it is essential to maintain empathy and ethical judgement.

HCAI systems are designed to recognise and minimise the algorithmic biases related to cast, creed, race, gender, age, and background. It is crucial in AI outcomes that it should ensure equity and equality in fostering inclusivity at the workplace, which also supports the attainment the SDG-8, which talks volumes about reducing inequalities (Kaith & Sachdeva, 2024). The most important principle, an AI must ensure, is transparency in its entire function. Employers should be informed well in advance when an AI is involved in the decision-making process, and that could influence them in a certain way, so that they can comprehend beforehand as to how and why certain outcomes were derived (Agrawal et al., 2024b). The foundation of artificial intelligence is privacy and data governance, where HCAI should ensure that employee data is protected in every circumstance under a clear governance structure that can protect against unauthorised usage and misuse of any information about employees. A clear line of accountability must be established which are related to AI-related decisions, and these boundaries should ensure ethical oversight boards and frameworks that should be regularly audited (Batra, 1996).

On global platforms, human well-being as part of overall sustainability performance has now been talked about for volumes; therefore, beyond operational efficiency, it is important to promote mental well-being, work–life balance, psychological safety, and long-term human development at the workplace (Kandpal et al., 2024). This directly helps us to attain SDGs 3 and SDG 8, which focus on good health, well-being, decent work, and economic growth. HR revolves around making decisions at any given moment; decision-making is not just a technical function; it should also integrate social, ethical, and emotional aspects (Santra et al., 2026). The traditional decision-making models were mainly dependent on the sensitivity of the current situation or the experience of the person involved in the decision-making. Whereas AI offers various merits of making decisions by deriving insights based on data analytics; however, we can say that it cannot work in isolation when the HR scenarios are quite ambiguous and emotionally driven (Yildirim, 2022).

HCAI ensures that we are hiring through inclusive practices and by removing biases in our historical data, and providing a wide range of shortlisted candidates. It also supports transparency by making the candidate understand how I can assess his or her work, and it is important to note that final decisions would be made after human intervention. Traditional AI evaluation systems prioritised the quantitative information like attendance or the number of hours worked in a premise, whereas HCAI focuses on holistic development and holistic performance assessment by involving the qualitative inputs as well.

3 Application of Human-Centric AI Across HR Functions

About a decade ago, we all talked about artificial intelligence as something very futuristic, but now it has transitioned from just a concept to a practical tool that is transforming human resource management in very practical and applied ways. Traditionally, AI has focused only on automating the entire HR process, reducing costs, and gaining efficiency. Today, as companies are getting more inclined towards ethical, social, and sustainability norms within their framework, they have understood the importance of moving towards human-centric artificial intelligence, which will do the traditional tasks along with augmenting human capabilities while ensuring transparency and inclusivity (Sahan et al., 2025).

3.1 Talent Acquisition and Recruitment

The first stage of an HR process is recruitment. It is one of the very earliest activities of HR that has integrated AI, but now the shift has been made towards HCAI, which has transformed its potential to a great extent. The traditional AI tools used to navigate thousands of resumes very quickly and identify the candidate who can be matched with the current job description (Pattanayak, 2025). HCAI is now used for bias detection algorithms, so one should not ignore the underrepresented candidates due to the bias in the algorithm. For instance, an HCAI-enabled screening tool can find potential biases in keyboards and encourage the recruiters to review the profiles manually. Earlier, communication used to happen through telephone or mail; however, the AI-powered chatbots are now bringing in the element of empathy, and they are providing the candidate with quick updates, various interview tips, and FAQs, which can increase their chances of selection and the overall experience of a candidate throughout the recruitment process. Earlier, the transactional bots were programmed with a more robotic voice, but now HCAI chat bots try to maintain the respective tone and provide clear explanations and avoid responding to sensitive queries, and escalate to humans (Atluri & Reddy, 2025). HCAI platforms can quickly assess the adaptive skill and tailor the question based on the candidates' earlier responses. The responses can highlight if there is a growth potential in a candidate rather than making a quick judgement based on the current skill, the focusing on long-term talent development.

3.2 Onboarding and Orientation

Once the candidate is selected, it has to go through an onboarding process, which is very crucial for the engagement and retention of the employees. AI is applied in this area to be inclusive, supportive, and give a personalised experience (Chen et al., 2024). AI can analyse the role, skill set, and learning style required for a job and can chalk out a personalised onboarding plan, and the system ensures that there is a due cultural integration to maintain the work–life balance. It also gives advice to mentors and social groups that how they can support each other to smooth out the social adaptation (Di Vaio et al., 2020). Some organisations are also using AI-simulation to make their employees familiar with the entire workplace and safety protocols, so these tools ensure accessibility features and can be used by differently abled employees from the first day of their employment. These tools can also incorporate feedback and detect agitation and frustration in the survey responses and trigger the HR managers for quick interventions.

3.3 Training and Development

Earlier, the AI model used for learning and development was very generic, but now it has transformed itself into a highly personalised ecosystem that is driven by career growth. HCAI can create, aid programs after analysing the employees' skill gap, performance feedback, and their career aspirations (Adegoke et al., 2024). It is not purely a data-driven tool as it also considers the preference of the employee's work–life balance and accessibility requirements. AI also enables a virtual assistant that can give constant and instant feedback during the employee's work hours and help them to solve problems without affecting their productivity. Human-centric systems make sure that these recommendations are not intrusion to privacy and sensitive to the context. After analysing market trends and organisational needs, these tools can help employees to visualise their career opportunities and role, and upskilling programs, which help the employees to take ownership of their career development. The next step in the HR processes is performance appraisal, and this entire system is often criticised for being biased, very inconsistent, and focusing on short-term performance and the current relationship shared between the employee and managers. HCAI can overcome these limitations.

3.4 Performance Management

The system incorporates inputs from self-assessment, peers, supervisors, and customer feedback. This multi-dimensional approach makes sure that the evaluation has both qualitative and quantitative nature. These tools can also identify patterns and can quickly generate flags if there is a bias in ratings, and alert HR managers to take corrective actions. Instead of giving the reviews annually, this platform focuses on continuous performance evaluation. It ensures that feedback is constructive, balanced, and actionable and can give due time for the employee to make changes regarding his or her development.

3.5 Employee Engagement

HCAI tools take into consideration email language, collaboration platform activity, and feedback to identify and detect certain trends that give an idea about employee morale. Unlike a surveillance-based approach, the modern system anonymises the data while working on collective patterns and ensuring individual privacy (Shaheen et al., 2024). AI platforms also encouraged the employees to submit their ideas, which can help in improving the process, and these ideas can be categorised in priorities, based on the management review. HCAI recognises and rewards the people who are coming forward to submit their ideas and promotes a culture that recognises and appreciates the employees with a proactive approach. These tools also recommend mental health resources, flexible working options, and breaks in between the workload, fostering a healthy and conducive work environment.

3.6 Compensation Management

HCAI tools can analyse time-tire structure related to compensation and generate flags if they identify disparities due to gender or other demographics, or race, caste, creed, and ethnicity and ensure timely corrections and due compliance with the HR laws prevailing in a country (Murphy, 2018). HCAI can customise the benefits, changing from wellness program health plans and stipends suitable for employee needs. Human-centric approach is necessary to make sure that these packages are inclusive of different family structures and stages of an employee's career. Considering factors like pay competitiveness, industry standard workloads, and various scores, the HCAI systems can flag employees on the borderline and can recommend timely interventions. A strategic plan for the workforce benefits is greatly driven by AI's predictive capabilities, especially when it is blended with a human-centred approach (Agrawal, 2025). The system can project talent needs based on industry and market trends, adoption of the current technology, allowing the HR department to prepare the upskilling initiatives beforehand. AI can simulate working scenarios on-site, off-models, and hybrid models to assess the impact on diversity, productivity, and various sustainability outcomes. HCAI system provides data-driven, actionable insights into hiring rates, retention of underrepresented groups, promotion patterns, and sets up the organisation with measurable goals.

3.7 Employees' Mental Well-Being and Exit Management

AI's role in safety and health has expanded beyond just physical safety, and it includes psychological safety as well. HCAI tools can identify stress through patterns in communication, signs of disengagement, and absenteeism while maintaining individual identity confidentiality unless it is important to intervene. AI-powered cameras and sensors follow the safety protocol specifically in logistics and manufacturing environments, so these designs can be used for safety and not just as a measure to create punitive vigilance (Jain et al., 2025). AI chatbots, which are designed empathetically, can provide mental health support while connecting the employees with a desired human counsellor. Even after an employee decides to leave an organisation, HCAI can give them a respectful exit, ensuring a long-lasting relationship. AI can analyse patterns to uncover systematic issues in the data generated and an exit interview. Human-centric interpretation is used for improvement, not just for retribution (Yadav & Dabhade, 2014). AI-enabled platforms can help maintain the relationship with the former employees for brand advocacy and to open opportunities, in case they want to rejoin (Srivastava et al., 2018).

The usage of AI across human resource functions is extensive, but the real transformation lies in the human element, which is designed to augment human judgement, boosting ethical values, and contributing to a sustainable organisation. The recruitment performance management, learning and development, and well-being initiatives (Walker & Larson, 2025). HCAI can help HR leaders with various tools and platforms to create workplaces that are efficient, resilient, and inclusive. By integrating HCAI into HR, corporations can align their business objective, simultaneously advocating for governance and social dimensions of sustainability, creating a win–win situation for both people and performance.

4 Challenges and Ethical Dilemmas Concerning Sustainability

Human-centric AI offers the potential to change the operation and decision-making process by integrating human values, oversight, and ethics in artificial intelligence. However, the integration is in a very sensitive domain, so it brings certain dilemmas along with a set of ethical challenges. These challenges stem from the tension between human-centric values such as transparency, fairness, and trust, and technology upgradation. The following are some most challenging issues.

4.1 Bias in Data and Algorithms

One of the fundamental and ethical challenges in HCAI is biases in data and algorithms. An AI system is driven through historical datasets, which reflect on the organisation's cultural biases, societal inequities, and discriminatory practices. If these biases are not recognised and minimised, they can become an integral part of AI models, leading to inefficient in biased decision-making (Atluri & Reddy, 2025). In HR functions, these biases can turn out to be a major recruitment algorithmic problem that will unintentionally favour certain demographic groups and will be skewed towards a particular behaviour trend and disproportionately benefit one ethnicity, gender, or race over another. A classic example would be an AI model favouring male applicants, as the historical hiring data was dominated by males in a certain industry, such as transport or logistics (Agrawal et al., 2024a). HCAI demands that the AI system should be designed in such a way that deliberate effort should be put to eliminate the biased variables, integrate fairness proportions, and timely audit the model design. This process might become a bit complex because biases are not very explicit and can be deeply embedded in the entire model, making decision-making a pressing issue.

4.2 Lack of Transparency

Another critical challenge in HCAI is the lack of transparency during the decision-making process, especially in the black box model. A black box model is like a deep learning system, and the reasoning behind a particular decision can be so integrated that developers still find it difficult to explain and explore (Buzko et al., 2016). In human resources, the lack of interpretability factor is always there between the employee and management. For example, if an employee is rated poorly without a clear explanation, AI will fetch the data, as it is. This leads to disengagement, perception of unfairness, and dispute between management and employees (Agrawal et al., 2025; Vrontis et al., 2023). From a human standpoint, decisions affecting people's careers must be given due attention and should be held accountable. HCAI requires integration of certain principles where a reasoning should be made to enable non-technical users to understand. This not only improves accountability but ensures that employees can appeal and challenge the decision made by AI. However, explanation comes at the cost of accuracy and model complexity (Wen et al., 2025). Thus, we need to trade-off between the two challenges.

4.3 Employees' Privacy Concerns

HCAI involves data collection and analysis of a large volume of employee data from communication patterns, wellness information, and performance metrics. While such data is used for decision-making, there are certain privacy concerns also. Employees are aware of data collection, and they may feel burdened to consent as they have a fair idea of data monitoring, which can lead to an imbalance. An AI-enabled system can monitor and track physical movements if there is a change in body language, which can lead to a perception of surveillance, and the employees might not provide the desired support. Respecting privacy is very important when the data is collected, as it can give us unbiased and transparent information to achieve the purpose for which the data is being collected (Agrawal & Sharma, 2024). Employees should have an understanding of what data to collect, what is necessary, and how it will be used. Additionally, consent should not be forced; it should be genuine, and proper laws regarding data protection, like GDPR and DPDPA, must be followed to balance the individual rights in data-driven insights.

The successful implementation of HCAI depends more on people and less on technology, as humans might be resistant to AI adoption. HR professionals may also fear that AI will come and take up their job, replace them, and hamper their professional identity and expertise. Employees should trust that the AI system is there to augment the opportunities for them by building more fairness and reliability. In some cases, employees might take an AI-based HR system as a superficial thing and might think that only human judgement is capable of evaluating a unique situation or circumstance; however, this perception can affect the adaptability of AI. To introduce HCAI, we need a planned change management strategy. A proper plan, clear communication, and capacity-building program are important to help them address their resistance to adopting the system. Here, the HR professional should be in a pivotal role and collaborate rather than passive recipient. AI-driven decisions should be integrated with human judgement to ensure that the employees' individuality will always be preserved. The governance of HCAI is constantly changing as there are changes in data protection regulations, labour laws, and the anti-discrimination framework. As a result, there are grey areas where accountability is a concerning issue for AI-driven decisions.

5 Responsible and Human-Centric AI Integration in HR—A Framework

The congruence of HR activities of human resource management along with AI is going through a transformation and will face major transformation in the next few decades, driven by the dynamic business environment, changes in organisational need, technology, and societal expectations. One of the most important aspects of AI is the presence of emotional quotient or emotional intelligence. AI is helping us in data-driven decision-making, but it falls short of understanding the emotional dynamics and stimuli present behind any decision. The future of AI will go through major phases that will recognise employee settlement by various behavioural cues, kinesics, and tone of the voice and will provide empathetic and contextual interaction. Areas like mental health support, employee engagement, and conflict resolution will be solved through emotionally intelligent AI and will make the HR process smoother. Another emerging trend surrounding AI these days is generative AI, which is a tool to help in highly personalised learning and development programs and can generate a large set of data based on a single prompt. Generative AI can be optimised for creating real-life scenarios and building inclusivity in training programs. These skills can reduce the burdens from admin roles' monotonous work and enhance the learning experience, but require careful consideration to prevent bias to a certain extent.

As AI gets deeply integrated into HR processes, it is crucial to establish an AI Ethics Board comprising HRD leaders, legal experts, technology experts, and employee representatives, which will help standardise the practices to be followed throughout the organisation and ensure that the application of AI is in line with corporate values, ethical standards, and legal requirements. This board shall serve as an oversight body to review algorithms for monitoring data usage, framing accountability and establishing fairness. Considering that the challenges posed by the use of AI are not only limited to technical glitches, but impacting social, logical, and psychological domains, there is a strong need for interdisciplinary collaboration within this landscape. This shall entail coming together of researchers and practitioners from the field of labour laws, ethical, and organisational behaviour and technology to create a system which is not only technically sound but also culturally sensitive and socially responsible. A comprehensive and multidisciplinary partnership can help in designing HR to lay the foundation for strong trust and acceptance among the stakeholders.

Another aspect of AI being integrated into the HR process which needs attention is the study of its long-term impact. While there is an increasing body of literature on the short-term benefits of AI over an employee's improved recruitment process, there is limited evidence on the long-term impact on organisational culture and employee satisfaction. To obtain valuable insights on the contribution of AI-driver human resource system, longitudinal studies can help to understand the relationship of trust and human connection over a period of time and the development of a sustainable workforce. In conclusion, it can be said that the future of AI in HR systems lies in balancing deep commitment to human values with technological soundness. To understand and manage the long-term implications, emerging trends such as generative AI applications in collaboration and blending AI with emotional intelligence will be useful to shape the next phase of innovation.

6 Conclusion and Future Outlook

In this era of AI, organisations are standing at the intersection of human connections and operational efficiency. AI offers unparalleled opportunities in processing large volumes of data sets, identifying patterns in that data, and predicting its outcomes, but its value in HR will only be measured depending on how it serves the people and not first the data processing. The challenge is to ensure that AI becomes an empowering force for HR professionals and employees, and not come as a fear for them that it can replace them.

Staying human-centric in this entire transformation process requires deliberate actions in strategies. Organisations must design systems that enhance human judgement rather than overpassing it. It means positioning of AI in the decision-making process based on data-driven insights that professionals and HR can easily interpret and tailor as per the organisation's culture and values. As earlier mentioned, AI is not about replacing humans but augmenting the opportunities for them. The presence of human elements, such as moral reasoning, empathy, and contextual judgement, would be essential, and the human experts can be complemented by AI. Human-centric AI holds the future of making data-driven decisions that can be efficient and fair. However, a proper attention to transparency, privacy, and governance is important to reinforce equality. Addressing these issues is a deep commitment to preserving human values, ethical foresight, and data value creation.

References

Adegoke, A. A., Oyindamola, K. A., & Offonabo, N. (2024). The role of HR in sustainability initiatives: A strategic review. *International Journal of Research in Engineering, Science and Management, 7*(5), 31–37.

Agrawal, D. (2024). Corporate environmental, social and governance (ESG) practices. In *Reference module in social sciences*. Elsevier. https://doi.org/10.1016/B978-0-443-13701-3.00141-9

Agrawal, D. (2025). Tech-driven progress: Leveraging technology in advancing sustainable development goals in emerging economies. In P. Singh, S. Daga, K. Yadav, & P. M. Nguyen (Eds.), *Advancing the SDGs in developing and emerging economies* (pp. 75–89). Springer Nature Switzerland. https://doi.org/10.1007/978-3-031-86500-8_6

Agrawal, D., & Sharma, H. P. (2024). Green investing: Decoding investor psyche towards high-scored ESG companies. In *Green transition impacts on the economy, society, and environment* (pp. 50–67). IGI Global Scientific Publishing.

Agrawal, D., Kumar, A., & Kumar, K. (2024a). Ethical sourcing. In *Reference module in social sciences* (p. B9780443137013000165). Elsevier. https://doi.org/10.1016/B978-0-443-13701-3.00016-5

Agrawal, D., Kumar, K., & Rahman, O. (2024b). ESG investment strategy. In *Reference module in social sciences*. Elsevier. https://doi.org/10.1016/B978-0-443-13701-3.00300-5

Agrawal, D., Maurya, S., & Kumar, K. (2025). A comprehensive view of FinTech as a catalyst for achieving sustainable development goals (SDGs): Policy insights for developing economies. In *Investing for a Greener Planet*(pp. 137–150). CRC Press.

Alwali, J., & Alwali, W. (2025). Transformational leadership and moral norms: Green human resource management and behaviour. *Management Decision, 63*(5), 1417–1442.

Atluri, A., & Reddy, V. (2025). Cognitive HR management: How Oracle HCM is reinventing talent acquisition through AI. *International Journal of Artificial Intelligence, Data Science, and Machine Learning, 6*(1), 85–94.

Batra, G. S. (1996). Human resource auditing as a tool of human resource valuation: Interface and emerging practices. *Managerial Auditing Journal, 11*(8), 23–30.

Boon, C., Jiang, K., & Eckardt, R. (2025). The role of time in strategic human resource management research: A review and research agenda. *Journal of Management, 51*(1), 172–211. https://doi.org/10.1177/01492063241264250

Buzko, I., Dyachenko, Y., Petrova, M., Nenkov, N., Tuleninova, D., & Koeva, K. (2016). Artificial Intelligence technologies in human resource development. *Computer Modelling and New Technologies, 20*(2), 26–29.

Chen, P., Chu, Z., & Zhao, M. (2024). The Road to corporate sustainability: The importance of artificial intelligence. *Technology in Society, 76*. Scopus. https://doi.org/10.1016/j.techsoc.2023.102440

Di Vaio, A., Palladino, R., Hassan, R., & Escobar, O. (2020). Artificial intelligence and business models in the sustainable development goals perspective: A systematic literature review. *Journal of Business Research, 121*, 283–314.

Gong, Q., Fan, D., & Bartram, T. (2025). Integrating artificial intelligence and human resource management: A review and future research agenda. *The International Journal of Human Resource Management, 36*(1), 103–141.

Jain, M. N., Maheshwari, M. M., Singhal, S., & Vishnoi, M. D. (2025). Cybersecurity in HR tech: A review of data privacy challenges in the digital HR ecosystem. *Advances in Consumer Research, 2*(2), 856.

Kaith, A., & Sachdeva, G. (2024). Bibliometric insights into green human resource management: A connection with business ethics and values. *PURUSHARTHA-A Journal of Management, Ethics and Spirituality, 17*(1), 51–63.

Kandpal, V. (2020). Reaching sustainable development goals: Bringing financial inclusion to reality in India. *Journal of Public Affairs*. https://doi.org/10.1002/pa.2277

Kandpal, V., Jaswal, A., Santibanez Gonzalez, E. D. R., & Agarwal, N. (2024). Circular economy principles: Shifting towards sustainable prosperity. In V. Kandpal, A. Jaswal, E. D. R. Santibanez Gonzalez, & N. Agarwal (Eds.), *Sustainable energy transition* (pp. 125–165). Springer Nature Switzerland. https://doi.org/10.1007/978-3-031-52943-6_4

Kumar, K. (2022). Emerging phenomenon of corporate sustainability reporting: Evidence from top 100 NSE listed companies in India. *Journal of Public Affairs, 22*(1), Article e2368. https://doi.org/10.1002/pa.2368

Kumar, K., & Prakash, A. (2019). Examination of sustainability reporting practices in Indian banking sector. *Asian Journal of Sustainability and Social Responsibility, 4*(1), 2. https://doi.org/10.1186/s41180-018-0022-2

Kumar, K., & Tarkar, P. (2024). Influence of sustainable human resource management on sustainability performance of the organization: Exploring the mediating role of organizational citizenship behaviour towards sustainability. *Journal of the Knowledge Economy*. https://doi.org/10.1007/s13132-024-02182-7

Kumar, R., Kaushal, S. K., & Kumar, K. (2023a). Does source credibility matter in promoting sustainable consumption? Developing an integrated model. *Social Responsibility Journal, 19*(7), 1320–1347.

Kumar, R., Kumar, K., Singh, R., Sá, J. C., Carvalho, S., & Santos, G. (2023b). Modeling environmentally conscious purchase behavior: Examining the role of ethical obligation and green self-identity. *Sustainability, 15*(8), 6426.

Mabrouk, A. (2025). Human resouce management in the age of artificial intelligence: Concepts tools and steps. *Solo International Collaboration and Publication of Social Sciences and Humanities, 3*(01), 1–12.

Murphy, K. R. (2018). The legal context of the management of human resources. *Annual Review of Organizational Psychology and Organizational Behavior, 5*(1), 157–182. https://doi.org/10.1146/annurev-orgpsych-032117-104435

Nti, E. K., Cobbina, S. J., Attafuah, E. E., Opoku, E., & Gyan, M. A. (2022). Environmental sustainability technologies in biodiversity, energy, transportation and water management using artificial intelligence: A systematic review. *Sustainable Futures, 4*, Article 100068.

Pattanayak, B. (2025). *Human resource management*. PHI Learning Pvt. Ltd. https://books.google.com/books?hl=en&lr=&id=2GpsEQAAQBAJ&oi=fnd&pg=PP1&dq=Human+Resource&ots=ThsgdI44DS&sig=-Hu-4mdI9Apk2O9hsqzgzJ2S-xQ

Sahan, U. M. H., Jaaffar, A. H. H., & Osabohien, R. (2025). Green human resource management, energy saving behavior and environmental performance: A systematic literature review. *International Journal of Energy Sector Management, 19*(1), 220–237.

Santra, S., Kumar, K., & Chattopadhyay, N. (2026). Developing a theory of gendered digital inclusion (GDI) for sustainable outcomes: A multi-method qualitative study on gender digital divide. *Gender Issues, 43*(1), 12. https://doi.org/10.1007/s12147-025-09396-2

Shaheen, N., Jaiswal, S., Chinta, D. U., Singh, N., Goel, O., & Chhapola, A. (2024). Data privacy in HR: Securing employee information in us enterprises using oracle hcm cloud. *International Journal of Research Radicals in Multidisciplinary Fields*, 319–341.

Sharma, A., Kumar, K., & Dwesar, R. (2025). "Keenly aware of environmental issues, yet wavering in conviction": Understanding the factors influencing the consumers' willingness to pay for green energy. *Sustainable Futures, 9*, Article 100427.

Singh, R., Khan, S., Joshi, A., Katragadda, R., & Kumar, V. (2026). Exploring the role of artificial intelligence on business innovation and entrepreneurship. *Strategy & Leadership, 54*(3), 302–322.

Srivastava, S. K., Singh, S., & Srivastava, M. (2021). Food waste at household and social gatherings: Drivers and possible remedies. *PURUSHARTHA-A Journal of Management, Ethics and Spirituality, 14*(2), 117–133.

Srivastava, S. K., Sinha, A. K., & Kumar, S. (2024). Workplace spirituality empowered by spiritual leadership to promote organizational effectiveness: A bibliometric analysis. *PURUSHARTHA-A Journal of Management, Ethics and Spirituality, 17*(1), 1–22.

Srivastava, S. K., Anshul, A., Pathak, P., & Bansal, J. (2018). Responsible consumption for curbing food wastage: An exploratory enquiry. *PURUSHARTHA-A Journal of Management, Ethics and Spirituality, 11*(2), 104–116.

Paridhi & Ritika. (2024). Sustainability reporting for boosting national commitment and overcoming challenges: A hierarchical model. *Business Strategy and Development, 7*(1). Scopus. https://doi.org/10.1002/bsd2.334

Torelli, R., Fiandrino, S., & Scarpa, F. (2025). Value-enhancing drivers of corporate governance in improving human rights due diligence: Worldwide evidence. *Corporate Social Responsibility and Environmental Management, 32*(1), 1279–1290.

Tripathi, V., Singh, K., Parihar, S., & Srivastava, S. K. (2022). A bibliometric analysis on supply chain collaboration. *PURUSHARTHA-A Journal of Management, Ethics and Spirituality, 15*(2), 66–103.

Vrontis, D., Christofi, M., Pereira, V., Tarba, S., Makrides, A., & Trichina, E. (2023). Artificial intelligence, robotics, advanced technologies and human resource management: A systematic review. In *Artificial intelligence and international HRM* (pp. 172–201).

Walker, D. O. H., & Larson, M. (2025). Leveraging generative artificial intelligence (AI) for human resource management: The AI job description assignment. *Journal of Management Education, 49*(1), 113–141. https://doi.org/10.1177/10525629241294075

Wen, S.-F., Shukla, A., & Katt, B. (2025). Artificial intelligence for system security assurance: A systematic literature review. *International Journal of Information Security, 24*(1), 43. https://doi.org/10.1007/s10207-024-00959-0

Yadav, R. K., & Dabhade, N. (2014). Human resource planning and audit–a case study of HEG limited. *Human Resource Management, 4*(5), 44–62.

Yildirim, S. (2022). The consumer role for sustainable development: How consumers contribute sustainable development goals. In *Research anthology on measuring and achieving sustainable development goals* (pp. 872–888). IGI Global Scientific Publishing. https://www.igi-global.com/chapter/the-consumer-role-for-sustainable-development/290948

Zolak Poljašević, B., Gričnik, A. M., & Šarotar Žižek, S. (2025). Human resource management in public administration: The ongoing tension between reform requirements and resistance to change. *Administrative Sciences, 15*(3), 94.

Part II
Digital Transformation and Green Growth

Chapter 7
Dynamic Capabilities for Socially Sustainable Digital Circular Transformation: A Four-Pillar Framework for Emerging Economies

Hosam Alden Riyadh, Shahid Latif, Maher Ali Al-Shmam, Ali Malallah Al-Sendy, and Sinan Zuhair Mohammed Jameel

Abstract This chapter argues how digital transformation in emerging economies can be harnessed to support the social sustainability and circular economy goals. Based on dynamic capabilities theory, a four-pillar framework has been proposed that echoes socially sustainable digitalization. This framework explains (i) organizational readiness for digital circular transformation, (ii) ethical and participatory governance, (iii) multi-stakeholder collaboration platforms, and (iv) policy instruments tailored to low- and middle-income contexts. Each pillar is interwoven with a continuous learning cycle that assists emerging economies in aligning the technological innovation with social equity and environmental objectives. By reflecting how this approach can enable inclusive services and efficient resource cycles, we assert that digital transformation can reduce the inequalities sand foster circular economy paths when managed dynamically. This integrated model provides a roadmap for policymakers, practitioners, and scholars to share digital agendas that are socially and environmentally sustainable.

H. A. Riyadh (✉)
Accounting Department, School of Economics and Business, Telkom University, Bandung, Indonesia
e-mail: Hosamalden@curtin.edu.my

H. A. Riyadh · S. Latif
Department of Accounting and Finance, Faculty of Business, Curtin University Malaysia, Miri, Malaysia
e-mail: shahidlatif@curtin.edu.my

M. A. Al-Shmam · A. M. Al-Sendy · S. Z. M. Jameel
Department of Accounting, College of Administration and Economics, University of Mosul, Mosul, Iraq
e-mail: aliaudit1970@uomosul.edu.iq

S. Z. M. Jameel
e-mail: sanan_zuhair@uomosul.edu.iq

K. Kumar et al. (eds.), *AI-Driven Circular Economy for Carbon Neutrality in Emerging Economies*, Environmental Footprints and Eco-design of Products and Processes,
https://doi.org/10.1007/978-3-032-30298-4_7

Keywords Capabilities · Transformation · Social sustainability · Ethical and governance

1 Introduction

Digital transformation has captured the significant attention of scholars due to its tendency to reshape the emerging economies (Jawad et al., 2021; Qadri et al., 2025). It has become the significant tool to address the global concerns and sustainable development goals, particularly related to social sustainability, circular economies, and carbon neutrality (Qadri et al., 2025; Varriale et al., 2025; Zhang et al., 2025). Digital transformation can be described as the integration of digital technologies and tools with diverse functionalities of business to augment the customer value and enterprise performance (Leão & da Silva, 2021) particularly in the context of emerging economies (David et al., 2025). Therefore, digital technologies can play a critical role in emerging economies success. In recent years, different scholars (e.g., Kanwal et al., 2025; Kant et al., 2025; Li & Wei, 2025; Uwuigbe et al., 2025) have asserted that emerging economies are entering into a pivotal era in which digital infrastructure coincides with increasing demands to address and mitigate the greenhouse gas emissions to attain the meaningful social progress. This global shift toward carbon–neutral economic systems has placed significance emphases on emerging economies to divert their attention on such agenda (Behera et al., 2023; Chen et al., 2024). These economies are induced to leverage digital technologies to augment the efficiency of production systems and public services, while participating in the integration of data-intensive global value chains (Nyagadza et al., 2022).

In addition to the significance of digital infrastructure, its acceleration accelerating without explicit safeguards for social welfare may aggravate inequality, erode decent work conditions, and undermine community resilience (Rushambwa, 2024). Economies with a digital transformation agenda can effectively implement sustainability programs (Guandalini, 2022). From the organizational perspective, the absence of digital transformation within social aspects puts organizations struggling to address the circular economy and sustainability concerns (e.g., Qadri et al., 2025; Toşa et al., 2024). New emerging technologies (e.g., cloud platforms, social media, artificial intelligence, and big data) can all initiate social innovation, but only when a firm can deploy and reconfigure them in changing environments (Gupta et al., 2020). Dynamic capabilities theory defines those higher-order abilities as the capacity to sense and shape opportunities and threats, seize opportunities, and reconfigure assets to stay competitive (Teece, 2007). Although the business literature recognizes dynamic capabilities as vital, far less is known about how they underpin socially sustainable digital transformation (Grego et al., 2025; Qadri et al., 2025). From the perspective of organizations in emerging economies, the digital circular pathway via digital transformation and with the balance of social mandate is also of critical nature that warrants scholarly attention (e.g., Ateeq, 2024).

This chapter responds to this research gap by proposing an integrated, four-pillar framework tailored to the realities of emerging economies. Grounded in dynamic capabilities theory, the framework links organizational readiness, ethical and participatory governance, multi-stakeholder collaboration, and adaptive policy instruments into a single learning cycle. By doing so, it clarifies what socially sustainable digital transformation means in practice and identifies the capabilities that firms, communities, and governments must build to advance a circular, carbon–neutral, and inclusive development agenda.

2 Digital Transformation and Social Sustainability in Emerging Economies

Digital transformation and social sustainability are significantly associated with each other as former is considered as a catalyst for later (Robertsone & Lapiņa, 2023). Social sustainability can be illustrated as the pursuit of inclusive, equitable, and community-oriented development, where firms manage their influence on employees, customers, and local communities in a proactive manner (Olanrewaju, 2025). In context of emerging economies, the tendency of social challenges—for instance, unequal educational access, poor healthcare infrastructure, and limited opportunities—can be high and addressed with the digital schemes. In this perspective, digital technologies may offer exclusive opportunities to build social conscious business models or adaptation in the existing models to incorporate socially sustainable practices to foster the effects of social sustainability (Qadri et al., 2025). For example, digital technology infrastructure can improve the educational accessibility and health systems in disadvantaged societies (Maia & Marin, 2021), work-life balance, and employees well-being (Ahmad & Zulkifli, 2022; Guan et al., 2022). These digital progressions strengthen the productivity while warranting the social inclusion (Papa et al., 2020; Qadri et al., 2025).

Recently, scholars have also highlighted the association of digitalization with socioeconomic development. David et al. (2025) have argued in perspective of emerging economies that internet of things (IoT) and mobile penetration are significantly associated with higher economic growth and human development improvement, particularly where effective governance mechanism exists. Digital innovation in areas such as e-government and fintech has enhanced the access to public services and financial inclusion (Gibson et al., 2024), thereby reflecting the equality for marginalized populations. Nevertheless, these digital transformation benefits for social sustainability are not automatic, as the digital divide and skill gaps concerns may exacerbate the exclusion if not addressed appropriately (e.g., Jamil, 2021). While digital transformation and technologies can reduce the income inequality, its impact is contingent on digital literacy and inclusive access (Rydzewski, 2025). Consequently, policymakers are required to prioritize investment in digital infrastructure, education, and affordable access, alongside the robust digital governance, to warrant that such transformation supports the sustainable and inclusive development (David et al., 2025). The alignment of digital transformation with social sustainability goals requires the strategic intent. While organizations often incorporate the digital technologies and tools for efficiency gains, their social impact remained underexploited (Qadri et al., 2025). Alongside the commercial aims of the organizations, associating digital capabilities to sustainable goals can enable such technologies (e.g., data analytics and AI) to address social challenges (e.g., Anshari et al., 2025; Li et al., 2023). In this regard, the role of dynamic capabilities (Teece, 2007) can be influential in addressing the social sustainability concerns under the digital transformation context of emerging economies.

3 Digital Transformation and the Circular Economy in Emerging Countries

Emerging economies are not increasingly on the path toward exploring the circular economy models to contribute to sustainable development goals. The circular economy can be described as the industrial system that seeks to reduce waste and promote the continuous reuse of resources in contrast to traditional models (Merli et al., 2018; Möller et al., 2025). For developing countries, encountering resource scarcity, pollution problems, and improper waste management systems, circular practices may offer appropriate remedies for environmental degradation issues. However, for implementation of effective circular economy strategies, high level coordination and information sharing mechanism can be required. Advance digital technologies and tools associated with fourth industrial revolution can address these requirements by enhancing the visibility, traceability, and resources efficiency (e.g., Duc & Dinh, 2024; Hariyani et al., 2024). In particular, emerging digital technologies (e.g., big data analytics, artificial intelligence, cloud computing, IoT, and blockchain) enable organizations to achieve their circular economy goals (Möller et al., 2025). By harnessing these digital technologies, firms can optimize the material flows, predict needs for maintenance, and design proper waste management mechanism.

Recent empirical studies have revealed the acceleration of circular economy due to digital transformations. For instance, Al-Khatib (2024) has asserted that Industry 4.0 readiness significantly influences the circular economy outcomes in Jordan's manufacturing firms. Similarly, Neri et al. (2023) revealed that digital tools assist organizations to develop dynamic capabilities that are essential to implement the circular economy initiatives. These studies reflect the role of digital transformations in enhancing the circular economy outcomes. Despite its significance and potential, emerging economies confront barriers, including increased financial costs for its implementation, limited infrastructure availability, skill gaps, and weak regulatory support, creating hindrances in adopting the digital solutions and transformations of circularity (e.g., Pittri et al., 2025). Nevertheless, digital transformation through digital technologies acts as the driver of circularity, enabling the tracking, analyzing, and optimizing the resource use at scale (Neligan et al., 2023). This digital enablement may assist emerging economies leapfrog to more sustainable production and consumption patterns.

4 Dynamic Capabilities Theory and Balancing Digital Circular Transformations with Social Sustainability

Managing the twin challenges of digital transformation and circular economy may require more than mere adopting new technologies; it is based on the organizational capabilities to adapt over time. Dynamic capabilities theory can explain this process. This theory refers to the capability of the firm to identify opportunities and threats *(sensing capability),* act on them by resource mobilization *(seizing capability),* and reorganize its resources to maintain the competitiveness *(reconfiguration capability)* (Teece, 2007). To achieve the twin transformation is a challenging task, and therefore, dynamic capabilities are essential that allow firms to adjust their business models and align strategies with technological advancement and sustainability requirements, supporting the long-run competitiveness (Breiter et al., 2024).

However, emerging economies encounter notable social sustainability concerns alongside environmental problems (e.g., Cezarino et al., 2022). Social sustainability concerns the protection of individual and community well-being through fair distribution of benefits including labor rights, community impact, and inclusive growth (e.g., Kalkanci et al., 2019; Molderez, 2020). Although circular economy transition in emerging countries can generate value, it may also reinforce exploitative practices if social aspects are neglected. Many circular initiatives pay limited attention to social sustainability, warning that without fair and ethical labor practices, circular transitions may harm workers and undermine community support (Redaputri & Suhartanti, 2024). Therefore, the balancing digital circular innovation with social sustainability concerns mandates firms to embed social sustainability aspects into their strategies, alongside innovating technologically.

Dynamic capabilities provide a valuable lens for this dual challenge (Neri et al., 2023; Xu et al., 2025). Firms must *sense* technological opportunities (e.g., AI or IoT for tracking waste) and social signals, such as community needs, regulatory pressures for decent work, or consumer demand for ethical products. For instance, an emerging-market manufacturer might sense growing concern for job security and see an opportunity to train workers on new remanufacturing skills. In the *seizing* phase, they may invest in workforce development, community partnerships, or inclusive business models that turn social needs into strategic advantages (e.g., collaborating with local suppliers to ensure living wages while also sourcing recycled inputs). In the *transforming* phase, firms reconfigure their operations and supply chains to institutionalize these practices—redesigning processes to be circular and socially inclusive and aligning organizational structures (e.g., HR policies, incentive systems) with these goals. This could involve establishing a cross-functional team to manage circular projects that include community liaison roles or reorienting R&D toward products that are easy to disassemble and repair by local workers. Therefore, dynamic capabilities enable firms in emerging economies to navigate the complex interplay of digital circular innovation and social sustainability. Considering these assertions, a four-pillar framework comprising of *(i)* organizational readiness for circular digital transformation, *(ii)* ethical and participatory governance, *(iii)* multi-stakeholder collaboration platforms, and *(iv)* policy instruments tailored to low- and middle-income contexts has been suggested to address the need for digital circular transformations and societal balance.

5 Four-Pillar Framework

5.1 Pillar 1: Organizational Readiness for Digital Circular Transformation

Dynamic capabilities theory begins with sensing: Organizations must notice new opportunities and threats before they can act (Teece et al., 1997). In emerging economies, this task can be multifaceted (Olaleye et al., 2024) because managers are required to analyze two simultaneous tasks. On one side, digitalization creates possibilities to monitor assets in real time, open new service channels, and lower transaction costs; on the other, the circular economy agenda demands that material loops be closed, and social equity protected (e.g., Provenzano & Seminara, 2024). If firms lack the projections to perceive both signals together, they risk adopting technology that deepens inequity or locks them into linear business models. The concept of *absorptive capacity* can elucidate what those projections look like. Absorptive capacity is critical for acquiring, assimilating, transforming, and exploiting knowledge to develop organizational dynamic capability (Roberts, 2015; Zahra & George, 2002). Organizations that already store relevant knowledge (e.g., life-cycle analysis methods, data-privacy standards, local community needs) can recognize new information and give it meaning (e.g., Cohen & Levinthal, 1990). In this regard, organizations with a strong digital culture and well-developed learning routines convert incoming data into supply chain resilience and sustainability gains more rapidly (Rodríguez-González et al., 2023). Therefore, absorptive capacity can provide the cognitive ground on which circular digital opportunities first register.

However, knowledge alone is insufficient; tangible and intangible resources must also be in place. The resource-based view argues that advantage flows from assets that are valuable, rare, and hard to copy (Ferreira & Ferreira, 2025). A study by Martínez-Caro et al. (2020) indicated that digital organizational culture, business digitalization, and digital technologies augment organizational performance. If these digital resources are coupled with circular economy-based models, they can enhance organizational performance and customer value (Chaudhuri et al., 2022). The plausible reason for this is that digitalization is an integral part of the environmental strategy, and it can assist in developing circular models and products (Agrawal et al., 2022). van Eechoud and Ganzaroli (2023) have asserted that dynamic capabilities provide the structure for sensing and seizing the digital circular model innovation (e.g., lean methodologies, project management, and supply chain collaboration. Recently, Sang and Anh (2025) have argued that digital technology adoption enhances circular economy practices, which have a social impact. Therefore, organizational readiness is critical in fostering the digital circular transformation to ensure human values and natural resources (Trad, 2024).

5.2 Pillar 2: Ethical and Participatory Governance

The seizing phase of dynamic capabilities theory stresses that organizations capture value only when they can mobilize resources in ways that other actors accept and support (Teece et al., 1997). In low- and middle-income economies, this mobilization is fraught with social risk (Kaplinsky & Kraemer-Mbula, 2022). Digital platforms have the tendency to generate the digital divide and digital inequality (Heeks, 2022). Therefore, ethical and participatory governance mechanisms are not optional add-ons; they are the capability that turns promising technology into a legitimate, socially sustainable practice.

5.2.1 Why Participation Anchors Legitimacy

Classic work on citizen involvement shows that the more power communities hold over decisions, the greater the eventual trust and policy compliance (Arnstein, 1969). When firms or governments introduce e-waste take-back schemes without such power sharing (Atasu et al., 2013), citizens may read the move as surveillance or green-washing and resist accordingly. Institutional theory helps explain this outcome: Initiatives survive only when they gain *regulative*, *normative*, and *cognitive* legitimacy within their environment (Ranta et al., 2018)_ Embedding participatory procedures can build that legitimacy by translating opaque technical choices (Offenhuber, 2020), such as algorithmic sorting of recyclables (Kim et al., 2025), into decisions that affected groups can scrutinize and amend.

5.2.2 Design Levers for Ethical Governance

Recent scholarship provides concrete design tools that embed ethics upstream. For instance, *a data stewardship* mechanism that spotlights the rules for all ecosystem participants agrees on ex-ante rules on data access, reuse, and deletion; the model relies on discourse ethics to balance power between platform owners and users (Lee, 2024). Koskinen et al. (2023) proposed the ethical governance model for data economy ecosystems by a discourse ethical approach. Mustafa et al. (2025) proposed an ethically imbued blockchain governance model. Arguably, these mechanisms turn abstract ethical principles (e.g., fairness, transparency, accountability) into routine checkpoints that managers can budget for and regulators can verify.

5.2.3 Channels for Meaningful Voice

Ethics statements alone do not guarantee voice; platforms must incorporate *interaction spaces* where ordinary users can flag exclusionary designs and propose alternatives. Linders (2012) labels such architectures "*we-government*," noting that social media interfaces, budget allocation apps, and public blockchains let citizens co-produce rather than merely consume public services. Lember et al. (2019) has reasoned that digital co-creation and co-production are essential to enhance interaction with people. Arguably, meaningful voice mechanisms operationalize the social license that digital circular initiatives depend on.

5.2.4 Dynamic Capabilities in Action

Participatory governance is a critical element that supports the involvement of community actors apart from those in command and control (Haesevoets et al., 2024). From a dynamic capabilities perspective, participatory governance can be treated as the *seizing* capability as it can mobilize the social and political resources to act on sensed opportunities. It can support the transforming phase of dynamic capabilities by producing data on equity impacts that regulators can embed in policy standards, tightening the feedback loop between Pillar 2 and Pillar 4. Organizations that adopt these routines can realize two key advantages: faster diffusion of their circular digital solution and lower risk of reputational damage due to exclusion or oversight.

5.3 Pillar 3—Multi-stakeholder Collaboration Platforms

Firms endowed with robust dynamic capabilities excel in making product and process decisions and deciding *which collaborative ties are worth pursuing* (e.g., Teece, 2020). These higher-order abilities, therefore, become a core antecedent of circular economy partnerships because they help organizations acquire the knowledge, resources, and routines required to shift from linear to circular value creation (Köhler et al., 2022). Other scholars (e.g., Sudusinghe & Seuring, 2022; Schepis et al., 2018) have also corroborated that circular economy initiatives almost always demand new network forms and cross-sector collaboration. In such multi-actor settings, firms must continually sense emerging opportunities, seize complementary assets, and reconfigure joint activities, precisely the sequence dynamic capabilities theory describes. Studies of manufacturing industries driven by advanced science (e.g., nanotechnology) confirm the link: Companies with stronger *collaboration-oriented* dynamic capabilities achieve superior innovation outcomes under circular economy pressures (Petricevic & Verbeke, 2019). Arguably, dynamic capabilities are not optional but a pre-condition for successful, circular economy collaboration. They enable firms in emerging economies to choose the right partners, align goals, and adapt the alliance as technologies, regulations, and market expectations evolve.

Collaboration platforms act as the infrastructure of capability building (Nham, 2022). They pool diverse resources so ecosystems, rather than isolated firms, can seize and transform. Achieving a circular economy transition is not the task of a single organization or policy agency; it demands concerted effort from government, business, and civil society actors alike (Ghisellini et al., 2016). Reflecting this need, scholars and practitioners have shown a growing interest in collaborative, multi-actor networks as vehicles for tackling complex sustainability challenges (Blomberg et al., 2023; Partidário et al., 2023). Schöggl et al. (2024) have argued that inter-organizational collaboration and digital technology incorporation are pivotal enablers in implementing circular economy-oriented practices. A similar pattern is present in the supply chain context, where technologies and green supply chain collaboration enhance circular economy practices (Farrukh Shahzad et al., 2025). Furthermore, Meier et al. (2023) have argued that blockchain technology can support supply chain collaboration and circular economy initiatives because it strengthens the traceability and dynamic capabilities critical for successful blockchain technology adoption.

Based on these assertions, it is evident that multi-actor involvement is essential for digital circular transformations. In these multi-stakeholder collaborations, platforms that formalize roles for NGOs, informal sector workers, and local governments lower the risk of capture by dominant incumbents and raise the legitimacy needed for subsequent policy support (e.g., Prasetyo, 2022). This can enhance the governance of natural resources, benefiting society (Ratner et al., 2022). Hence, multi-stakeholder collaborative platforms can engage diverse stakeholders to foster *transformative change* at a societal level.

5.4 Pillar 4—Policy Instruments Tailored to Low- and Middle-Income Contexts

Public policy is the lever that turns the early wins described in the first three pillars into economy-wide habits. In the language of dynamic capabilities theory, well-chosen policy mixes add an external transforming capability that locks in new routines while keeping them open to further learning (e.g., Schot & Edward Steinmueller, 2018). Designing such mixes can be harder in low- and middle-income economies than in developed economies because ministries face tight budgets, large informal sectors, and uneven data (e.g., Islam, 2025). There can be four aspects that make policy both realistic and robust, and each principle connects directly to the readiness, governance, and collaboration themes (first three pillars).

(a) First, policy instruments work best when they are coherent and mutually reinforcing. Ren and Albrecht (2023) have argued that command-and-control regulations and technology-push instruments are significant drivers for adopting circular economy innovations. They have asserted that potential interaction between policy instruments paves the way for desirable circular economy outcomes. Similar patterns emerge in Africa's plastic-packaging sector, where the circular plastic economy policy framework is coherently based on four elements such as extended producer responsibility for plastic products, social policies for informal section inclusion, common continental standards on design, and policy support for digital innovations (Schröder et al., 2023).
(b) Second, the mix has to be fiscally light but catalytic. KIPPRA's 2022 study demonstrates that Kenya can achieve plastic circularity through a policy mix that is fiscally light yet catalytic (Anomat & Ibrahim, 2022). It calls for re-introducing zero-tax import incentives on recycling machinery and other material recovery equipment, a one-off relief that lowers private entry costs without creating recurring fiscal burdens. Therefore, a fiscally light but catalytic policy mix matches the realities of a lower-income economy and maximizes environmental gains.
(c) Third, the policy must advance social inclusion because moving from a linear to a circular model is not just a technical exercise; it demands a profound shift in social norms and everyday behavior (Chizaryfard et al., 2021; Kovacic et al., 2020). Because circularity reshapes the way goods are produced, consumed, and discarded, it has significant implications for the social pillar of sustainable development, including inclusion, equity, and decent work (Padilla-Rivera et al., 2020). Most current studies define social outcomes only in broad terms and rarely tackle concrete issues such as who is included, who bears adjustment costs, or how benefits are shared (Eickhoff, 2024). Resultantly, social inclusion is a cornerstone of sustainable development and remains the least developed strand of the circular economy agenda in the context of developing countries (Ahmed et al., 2022; Belmonte-Ureña et al., 2021; Vazquez-Brust et al., 2020).
(d) Fourth, governments can multiply impact by investing in digital public goods (Nicholson et al., 2022). Digital technology now sits at the heart of many national development strategies, especially in low- and lower-middle-income countries that view it as a lever for faster social and economic progress (Heeks, 2020). Because digital tools are modular, reprogrammable, and built on shared standards, they can be recombined to serve multiple public purposes. This has led the UN and allied initiatives to promote digital public goods (DPGs) (open-source software, open data sets, open AI models, standards, and content that respect privacy and other international norms) as the common foundation on which new solutions can be built (United Nations, 2020).

6 Conclusion and Recommendation

In conclusion, this chapter has addressed that digital transformations are considered critical for emerging economies and organizations in these economies. The digitalization is considered a promising tool that can provide the robust route for reconciliation of economic growth with social sustainability and resource efficiency. To achieve this dual objective, digital transformation is not mere an end but a reflection of dynamic capabilities building process. Under the theoretical underpinning of dynamic capabilities, this chapter spotlights that governments and organization in emerging economies must develop the capability for sensing opportunities, resource mobilization, and system adaptation for symbolizing the social aspects and circular economy concerns simultaneously. This can be asserted as adaptive approach and without it, change driven by digital transformation for digital circular economy and social sustainability may be possible. In this regard, a four-pillar framework has been suggested consisting of four pillars named as (i) organizational readiness for digital circular transformation, (ii) ethical and participatory governance, (iii) multi-stakeholder collaboration platforms, and (iv) policy instruments tailored to low- and middle-income contexts. Organizational readiness is the reflection of sensing for digital circular transformation with a social impact. This readiness leads to develop the mindset for seizing the opportunity for ethical and participatory governance. This governance can lead to transform by involving multiple stakeholders in a collaborative manner. Lastly, this transformation is translated into a policy instrument for embedding the digital circular change with social sustainability balance.

There are managerial and policy implication for the proposed four-pillar framework. Policymakers in context of emerging or low- and middle-income economies can focus on reducing the digital divide by embedding the social evenness in circular economy policies. Managers and practitioners can align their digital circular initiates with community needs to avoid the digital disparity. Collectively, it can be argued that digital transformation can reinforce emerging economies to achieve circular targets and stronger social cohesion simultaneously.

References

Agarwal, A., & Assenova, V. A. (2024). Mobile money as a stepping stone to financial inclusion: How digital multisided platforms fill institutional voids. *Organization Science, 35*(3), 769–787. https://doi.org/10.1287/orsc.2022.16562

Agrawal, R., Wankhede, V. A., Kumar, A., Upadhyay, A., & Garza-Reyes, J. A. (2022). Nexus of circular economy and sustainable business performance in the era of digitalization. *International Journal of Productivity and Performance Management, 71*(3), 748–774.

Ahmad, N., & Zulkifli, A. M. (2022). Internet of Things (IoT) and the road to happiness. *Digital Transformation and Society, 1*(1), 66–94.

Ahmed, Z., Mahmud, S., & Acet, H. (2022). Circular economy model for developing countries: Evidence from Bangladesh. *Heliyon, 8*(5).

Al-Khatib, A. W. (2024). Enabling the circular economy in the digital transformation era: Evidence from an emerging country. *Kybernetes, 53*(2), 779–802.

Anomat, J., & Ibrahim, H. A. (2022). *Accelerating the circular economy in plastic waste management for Kenya*. Kenya Institute for Public Policy Research and Analysis, 2022. https://repository.kippra.or.ke/server/api/core/bitstreams/e6d6067d-93b7-46fe-8b46-3e8c6521e83d/content

Anshari, M., Hamdan, M., Ahmad, N., & Ali, E. (2025). Public service delivery, artificial intelligence and the sustainable development goals: Trends, evidence and complexities. *Journal of Science and Technology Policy Management, 16*(1), 163–181.

Arnstein, S. R. (1969). A ladder of citizen participation. *Journal of the American Institute of Planners, 35*(4), 216–224.

Atasu, A., Özdemir, Ö., & Van Wassenhove, L. N. (2013). Stakeholder perspectives on e-waste take-back legislation. *Production and Operations Management, 22*(2), 382–396.

Ateeq, A. (2024). Emerging economies and digital transformation: Opportunities and challenges. In *Business sustainability with artificial intelligence (AI): Challenges and opportunities: Volume 1* (pp. 129–136).

Behera, P., Haldar, A., & Sethi, N. (2023). Achieving carbon neutrality target in the emerging economies: Role of renewable energy and green technology. *Gondwana Research, 121*, 16–32.

Belmonte-Ureña, L. J., Plaza-Úbeda, J. A., Vazquez-Brust, D., & Yakovleva, N. (2021). Circular economy, degrowth and green growth as pathways for research on sustainable development goals: A global analysis and future agenda. *Ecological Economics, 185*, 107050. https://doi.org/10.1016/j.ecolecon.2021.107050

Blomberg, A., Kujala, J., & Heikkinen, A. (2023). Multi-stakeholder networks in a circular economy transition: A typology of stakeholder relationships. In *Stakeholder engagement in a sustainable circular economy: Theoretical and practical perspectives* (pp. 133–164). Springer International Publishing.

Breiter, K., Crome, C., Oberländer, A. M., & Schnaak, F. (2024). Dynamic capabilities for the Twin Transformation climb: A capability maturity model. *Information Systems Frontiers*, 1–22.

Cezarino, L. O., Liboni, L. B., Hunter, T., Pacheco, L. M., & Martins, F. P. (2022). Corporate social responsibility in emerging markets: Opportunities and challenges for sustainability integration. *Journal of Cleaner Production, 362*, 132224.

Chaudhuri, A., Subramanian, N., & Dora, M. (2022). Circular economy and digital capabilities of SMEs for providing value to customers: Combined resource-based view and ambidexterity perspective. *Journal of Business Research, 142*, 32–44.

Chen, F., Zhang, W., Chen, R., Jiang, F., Ma, J., & Zhu, X. (2024). Adapting carbon neutrality: Tailoring advanced emission strategies for developing countries. *Applied Energy, 361*, Article 122845.

Chizaryfard, A., Trucco, P., & Nuur, C. (2021). The transformation to a circular economy: Framing an evolutionary view. *Journal of Evolutionary Economics, 31*, 475–504.

Cohen, W. M., & Levinthal, D. A. (1990). Absorptive capacity: A new perspective on learning and innovation. *Administrative Science Quarterly, 35*(1), 128–152.

David, L. K., Wang, J., Brooks, W., & Angel, V. (2025). Digital transformation and socio-economic development in emerging economies: A multinational analysis. *Technology in Society, 81*, 102834.

Duc, K. N., & Dinh, N. C. (2024). Circular economy. *Navigating the Circular Age of a Sustainable Digital Revolution*, 1.

Eickhoff, H. (2024). The appeal of the circular economy revisited: On track for transformative change or enabler of moral licensing? *Humanities and Social Sciences Communications, 11*(1), 1–8.

Farrukh Shahzad, M., Liu, H., & Zahid, H. (2025). Industry 4.0 technologies and sustainable performance: Do green supply chain collaboration, circular economy practices, technological readiness and environmental dynamism matter? *Journal of Manufacturing Technology Management, 36*(1), 1–22.

Ferreira, N. C. M. Q. F., & Ferreira, J. J. M. (2025). The field of resource-based view research: Mapping past, present and future trends. *Management Decision, 63*(4), 1124–1153.

Florek-Paszkowska, A., & Ujwary-Gil, A. (2025). The digital-sustainability ecosystem: A conceptual framework for digital transformation and sustainable innovation. *Journal of Entrepreneurship, 21*(2), 116–137.

Ghisellini, P., Cialani, C., & Ulgiati, S. (2016). A review on circular economy: The expected transition to a balanced interplay of environmental and economic systems. *Journal of Cleaner Production, 114*, 11–32.

Gibson, E., Gazi, S., & Arner, D. W. (2024). Digital finance, financial inclusion and gender equality: Digital public goods, rearchitecting financial systems and economic empowerment of women.

Grego, M., Bartosiak, M., Palese, B., Piccoli, G., & Denicolai, S. (2025). Disentangling the 'digital': A critical review of information technology capabilities, information technology–enabled capabilities and digital capabilities in business research. *International Journal of Management Reviews, 27*(2), 238–260.

Guan, A. L. C., Manavalan, M., Ahmed, A. A. A., Azad, M. M., & Miah, S. (2022). Role of internet of things (IoT) in enabling productive work from home (wfh) for environmental volatiles. *Academy of Marketing Studies Journal, 26*(1), 1–11.

Guandalini, I. (2022). Sustainability through digital transformation: A systematic literature review for research guidance. *Journal of Business Research, 148*, 456–471.

Gupta, S., Kumar, V., & Karam, E. (2020). New-age technologies-driven social innovation: What, how, where, and why? *Industrial Marketing Management, 89*, 499–516.

Haesevoets, T., Roets, A., Steyvers, K., Verschuere, B., & Wauters, B. (2024). Towards a multi-faceted measure of perceived legitimacy of participatory governance. *Governance, 37*(3), 711–728.

Hariyani, D., Hariyani, P., Mishra, S., & Sharma, M. K. (2024). Leveraging digital technologies for advancing circular economy practices and enhancing life cycle analysis: A systematic literature review. *Waste Management Bulletin.*

Heeks, R. (2020). ICT4D 3.0? Part 1—The components of an emerging "digital-for-development" paradigm. *The Electronic Journal of Information Systems in Developing Countries, 86*(3), e12124.

Heeks, R. (2022). Digital inequality beyond the digital divide: Conceptualizing adverse digital incorporation in the global South. *Information Technology for Development, 28*(4), 688–704.

Islam, S. (2025). A systematic review of public budgeting strategies in developing economies: Tools for transparent fiscal governance. *American Journal of Advanced Technology and Engineering Solutions, 1*(01), 602–635.

Jamil, S. (2021). From digital divide to digital inclusion: Challenges for wide-ranging digitalization in Pakistan. *Telecommunications Policy, 45*(8), Article 102206.

Jawad, M., Naz, M., & Maroof, Z. (2021). Era of digital revolution: Digital entrepreneurship and digital transformation in emerging economies. *Business Strategy & Development, 4*(3), 220–228.

Kalkanci, B., Rahmani, M., & Beril Toktay, L. (2019). The role of inclusive innovation in promoting social sustainability. *Production and Operations Management, 28*(12), 2960–2982.

Kant, S., Adula, M., & Badi, S. (2025). Digital transition effect on green jobs by mediation of knowledge management in circular economy. In *Insights into digital business, human resource management, and competitiveness* (pp. 171–190). IGI Global Scientific Publishing.

Kanwal, P., Quttainah, M. A., & Kumar, A. (2025). Social impact of circular economy startups. In *Intersecting entrepreneurship, internationalization, and green innovation* (pp. 235–260). IGI Global.

Kaplinsky, R., & Kraemer-Mbula, E. (2022). Innovation and uneven development: The challenge for low-and middle-income economies. *Research Policy, 51*(2), Article 104394.

Kim, J., Latif, M., Song, J., Park, G., & Lee, J. (2025). Automated sorting of recyclable domestic waste. In *Harnessing automation and machine learning for resource recovery and value creation* (pp. 417–426). Elsevier.

Köhler, J., Sönnichsen, S. D., & Beske-Jansen, P. Towards a collaboration framework for circular economy: The role of dynamic capabilities and open innovation. *Business Strategy and the Environment, 31*(6), 2700–2713.

Koskinen, J., Knaapi-Junnila, S., Helin, A., Rantanen, M. M., & Hyrynsalmi, S. (2023). Ethical governance model for the data economy ecosystems. *Digital Policy, Regulation and Governance, 25*(3), 221–235.

Kovacic, Z., Strand, R., & Völker. T. (2020). *The circular economy in Europe: Critical perspectives on policies and imaginaries*. Taylor & Francis.

Leão, P., & da Silva, M. M. (2021). Impacts of digital transformation on firms' competitive advantages: A systematic literature review. *Strategic Change, 30*(5), 421–441.

Lee, P. S. (2024). *Data stewardship in action: A roadmap to data value realization and measurable business outcomes*. Packt Publishing Ltd.

Lember, V., Brandsen, T., & Tõnurist, P. (2019). The potential impacts of digital technologies on co-production and co-creation. *Public Management Review, 21*(11), 1665–1686.

Li, J., & Wei, J. (2025). Digital transformation and innovation in emerging economies: An introduction and future directions. *International Journal of Technology Management, 97*, 2.

Li, L., Zhou, H., Yang, S., & Teo, T. S. H. (2023). Leveraging digitalization for sustainability: An affordance perspective. *Sustainable Production and Consumption, 35*, 624–632.

Linders, D. (2012). From e-government to we-government: Defining a typology for citizen coproduction in the age of social media. *Government Information Quarterly, 29*(4), 446–454.

Maia, J. S., & Marin, H. d. F. (2021). Mobile apps for disadvantaged societies. *Acta Paulista de Enfermagem, 34*, eAPE002214.

Martínez-Caro, E., Cegarra-Navarro, J. G., & Alfonso-Ruiz, F. J. (2020). Digital technologies and firm performance: The role of digital organisational culture. *Technological Forecasting and Social Change, 154*, 119962.

Meier, O., Gruchmann, T., & Ivanov, D. (2023). Circular supply chain management with blockchain technology: A dynamic capabilities view. *Transportation Research Part e: Logistics and Transportation Review, 176*, Article 103177.

Merli, R., Preziosi, M., & Acampora, A. (2018). How do scholars approach the circular economy? A systematic literature review. *Journal of Cleaner Production, 178*, 703–722.

Molderez, I. (2020). Corporate social responsibility and the sustainable development goals (SDGs). In *Decent work and economic growth* (pp. 116–128). Springer International Publishing.

Möller, D. P. F., Wang, Q., Huang, L., Xiong, B., Guo, Y., Shi, T., & Zhang, R. (2025). Impact of digital transformation for sustainable circular economic environments. *Carbon Neutral Systems, 1*(1), 1–22.

Mustafa, G., Rafiq, W., Jhamat, N., Arshad, Z., & Rana, F. A. (2025). Blockchain-based governance models in e-government: A comprehensive framework for legal, technical, ethical and security considerations. *International Journal of Law and Management, 67*(1), 37–55.

Neligan, A., Baumgartner, R. J., Geissdoerfer, M., & Schöggl, J.-P. (2023). Circular disruption: Digitalisation as a driver of circular economy business models. *Business Strategy and the Environment, 32*(3), 1175–1188.

Neri, A., Negri, M., Cagno, E., Kumar, V., & Garza-Reyes, J. A. (2023). What digital-enabled dynamic capabilities support the circular economy? A multiple case study approach. *Business Strategy and the Environment, 32*(7), 5083–5101.

Nham, N. T. H. (2022). Making the circular economy digital or the digital economy circular? Empirical evidence from the European region. *Technology in Society, 70*, 102023.

Nicholson, B., Nielsen, P., Sahay, S., & Sæbø, J. I. (2022). Digital public goods platforms for development: The challenge of scaling. *The Information Society, 38*(5), 364–376.

Nyagadza, B., Pashapa, R., Chare, A., Mazuruse, G., & Hove, P. K. (2022). Digital technologies, Fourth Industrial Revolution (4IR) & Global Value Chains (GVCs) nexus with emerging economies' future industrial innovation dynamics. *Cogent Economics & Finance, 10*(1), 2014654.

Offenhuber, D. (2020). Towards ethical legibility: An inclusive view of waste technologies. In *The Routledge companion to smart cities*, pp. 210–224. Routledge.
Olaleye, I., Mokogwu, V., Olufemi-Phillips, A. Q., & Adewale. T. T. (2024). Unlocking competitive advantage in emerging markets through advanced business analytics frameworks. *GSC Advanced Research and Reviews, 21*(2), 419–426.
Olanrewaju, A. (2025). Social sustainability in construction. In *Sustainable construction management: Research and practice companion*, pp. 169–198. Springer, Singapore.
Padilla-Rivera, A., Russo-Garrido, S., & Merveille, N. (2020). Addressing the social aspects of a circular economy: A systematic literature review. *Sustainability, 12*(19), 7912.
Papa, A., Mital, M., Pisano, P., & Del Giudice, M. (2020). E-health and wellbeing monitoring using smart healthcare devices: An empirical investigation. *Technological Forecasting and Social Change, 153*, 119226.
Partidário, M., Monteiro, M. B., & Martins, R. (2023). Novel perspectives for multi-actor collaboration in strategic environmental assessment using ST4S. *Environmental Impact Assessment Review, 99*, Article 107023.
Petricevic, O., & Verbeke, A. (2019). Unbundling dynamic capabilities for inter-organizational collaboration: The case of nanotechnology. *Cross Cultural & Strategic Management, 26*(3), 422–448.
Pittri, H., Godawatte, G. A. G. R., Esangbedo, O. P., Antwi-Afari, P., & Bao, Z. (2025). Exploring barriers to the adoption of digital technologies for circular economy practices in the construction industry in developing countries: A case of Ghana. *Buildings, 15*(7), 1090.
Plekhanov, D., Franke, H., & Netland, T. H. (2023). Digital transformation: A review and research agenda. *European Management Journal, 41*(6), 821–844. https://doi.org/10.1016/j.emj.2022.09.007
Prasetyo, E. H. (2022). Legitimacy building of digital platforms in the informal economy: Evidence from Indonesia. *Journal of Entrepreneurship in Emerging Economies, 14*(6), 1168–1187.
Provenzano, V., & Seminara, M. R. (2024). The nexus between circular economy and sustainable development goals. In *International symposium: New metropolitan perspectives*, pp. 72–82. Springer.
Qadri, U. A., Ghani, M. B. A., Abbas, U., & Kashif, A. R. (2025). Digital technologies and social sustainability in the digital transformation age: a systematic analysis and research agenda. *International Journal of Ethics and Systems, 41*(1), 142–169. https://doi.org/10.1108/IJOES-08-2024-0239
Ranta, V., Aarikka-Stenroos, L., Ritala, P., & Mäkinen, S. J. (2018). Exploring institutional drivers and barriers of the circular economy: A cross-regional comparison of China, the US, and Europe. *Resources, Conservation and Recycling, 135*, 70–82.
Ratner, B. D., Larson, A. M., Barletti, J. P. S., ElDidi, H., Catacutan, D., Flintan, F., Suhardiman, D., Falk, T., & Meinzen-Dick, R. (2022). Multistakeholder platforms for natural resource governance: lessons from eight landscape-level cases. *Ecology and Society, 27*(2).
Redaputri, A. P., & Suhartanti, P. D. (2024). Circular economy in emerging countries: A systematic literature review of challenges and solution strategies for sustainability. In *Proceedings of the international conference of economics business and economics education science (ICE-BEES-24)*, p. 123. Springer Nature.
Ren, Q., & Albrecht, J. (2023). Toward circular economy: The impact of policy instruments on circular economy innovation for European small medium enterprises. *Ecological Economics, 207*, Article 107761.
Reypens, C., Lievens, A., & Blazevic, V. (2021). Hybrid orchestration in multi-stakeholder innovation networks: Practices of mobilizing multiple, diverse stakeholders across organizational boundaries. *Organization Studies, 42*(1), 61–83. https://doi.org/10.1177/0170840619868268
Roberts, N. (2015). Absorptive capacity, organizational antecedents, and environmental dynamism. *Journal of Business Research, 68*(11), 2426–2433.

Robertsone, G., & Lapiņa, I. (2023). Digital transformation as a catalyst for sustainability and open innovation. *Journal of Open Innovation: Technology, Market, and Complexity, 9*(1), Article 100017.

Rodríguez-González, R. M., Madrid-Guijarro, A., & Maldonado-Guzmán, G. (2023). Digital organizational culture and absorptive capacity as precursors to supply chain resilience and sustainable performance. *Journal of Cleaner Production, 420*, 138411.

Rushambwa, T. (2024). Rethinking social policy in the era of digital transformation: Social policy and reinvigorated social contracts, a qualitative analysis. Available at SSRN 4974999.

Rydzewski, P. (2025). Digital inequality and sustainable development. *Problemy Ekorozwoju, 20*(1), 96–108.

Sang, T. M., & Anh, N. Q. (2025). Adopting digital technology for sustainable performance: Circular economy practice in the transportation sector. *Discover Sustainability, 6*(1), 166.

Schepis, D., Ellis, N., & Purchase, S. (2018). Exploring strategies and dynamic capabilities for net formation and management. *Industrial Marketing Management, 74*, 115–125.

Schöggl, J.-P., Stumpf, L., & Baumgartner, R. J. (2024). The role of interorganizational collaboration and digital technologies in the implementation of circular economy practices—Empirical evidence from manufacturing firms. *Business Strategy and the Environment, 33*(3), 2225–2249.

Schot, J., & Edward Steinmueller, W. (2018). Three frames for innovation policy: R&D, systems of innovation and transformative change. *Research Policy, 47*(9), 1554–1567.

Schröder, P., Oyinlola, M., Barrie, J., Fwangkwal, B., & Abolfathi, S. (2023). Making policy work for Africa's circular plastics economy. *Resources, Conservation and Recycling, 190*, Article 106868.

Sharma, S. (2020). *Digital development: An interpretive policy analysis of privacy and social inclusion in world bank technology projects* [PhD dissertation].

Sudusinghe, J. I., & Seuring, S. (2022). Supply chain collaboration and sustainability performance in circular economy: A systematic literature review. *International Journal of Production Economics, 245*(2022), 108402.

Taherdoost, H. (2024). *Digital transformation roadmap: From vision to execution*. CRC Press. https://doi.org/10.1201/9781003471226

Teece, D. J. (2007). Explicating dynamic capabilities: the nature and microfoundations of (sustainable) enterprise performance. *Strategic Management Journal, 28*(13), 1319–1350. https://doi.org/10.1002/smj.640

Teece, D. J. (2020). Hand in glove: Open innovation and the dynamic capabilities framework. *Strategic Management Review, 1*(2), 233–253. https://doi.org/10.1561/111.00000010

Teece, D. J., Pisano, G., & Shuen, A. (1997). Dynamic capabilities and strategic management. *Strategic Management Journal, 18*(7), 509–533.

Toşa, C., Paneru, C. P., Joudavi, A., & Tarigan, A. K. M. (2024). Digital transformation, incentives, and pro-environmental behaviour: Assessing the uptake of sustainability in companies' transition towards circular economy. *Sustainable Production and Consumption, 47*, 632–643.

Trad, A. T. (2024). Societal or organizational transformation projects: The sustainable digital r(E)volution for navigating the circular age (SDE4NCA). In *Navigating the circular age of a sustainable digital revolution* (pp. 121–174). IGI Global.

United Nations. *Report of the secretary-general: Roadmap for digital cooperation* [Internet]. United Nations. Available from: https://www.un.org/en/content/digital-cooperation-roadmap/

Uwuigbe, U., Issah, O., Ranti, U. O., Zubeiru, M., Anaba, S., & Seidu, A. J. S. (2024). Circular economy: A bibliometric review of research in emerging economies (2010–2024). *International Journal of Energy Economics and Policy, 15*(1), 77–89.

van Eechoud, T., & Ganzaroli, A. (2023). Exploring the role of dynamic capabilities in digital circular business model innovation: Results from a grounded systematic inductive analysis of 7 case studies. *Journal of Cleaner Production, 401*(2023), 1–12.

Varriale, V., Camilleri, M. A., Cammarano, A., Michelino, F., Müller, J., & Strazzullo, S. (2025). Unleashing digital transformation to achieve the sustainable development goals across multiple sectors. *Sustainable Development, 33*(1), 565–579.

Vazquez-Brust, D., Piao, R. S., de Melo, M. F. D. S., Yaryd, R. T., & Carvalho, M. M. (2020). The governance of collaboration for sustainable development: Exploring the "black box". *Journal of Cleaner Production, 256*, 120260. https://doi.org/10.1016/j.jclepro.2020.120260

Xu, N., Shahzad, F., & Rui, H. (2025). Unveiling environmental excellence: The confluence of green Industrial Internet of Things, circular economy practices and dynamic capabilities in manufacturing firms. *Journal of Manufacturing Technology Management, 36*(2), 479–503.

Zahra, S. A., & George, G. (2002). Absorptive capacity: A review, reconceptualization, and extension. *Academy of Management Review, 27*(2), 185–203.

Zhang, M., Liu, R., & Sun, H. (2025). Achieving carbon-neutral economies through circular economy, digitalization, and energy transition. *Scientific Reports, 15*(1), 13779.

Chapter 8
Reimagining Growth: ICT-Driven Green Growth Transition in Emerging BRICS Economies

Shikha Daga, Remy Jonkam Oben, and Kyla L. Tennin

Abstract Protecting the environment and the attainment of sustainable development goals are top global priorities in the present scenario. Rising emissions across the globe have led to a rise in interest toward studies that focus on growth that is not only economic but is also sustainable. An important factor that is documented to make the transition is the adoption of clean sources of energy. Renewable energy (RE) adoption is thus an important highlight of contemporary research focusing on sustainable growth. The concept is even more important for emerging economies that are scaling high on the economic wealth generation and hence need to decouple the same from environmental deterioration. In addition to various factors that can promote RE adoption, ICT usage is also one of them. To identify the interlinkage between the two, the study has used annual data from the year 2000 to 2021 for the emerging economies namely "Brazil, Russia, India, China, and South Africa." Advanced panel techniques are used to identify short- and long-run estimations and integration. Using cointegration and CCEMG, the study identified the presence of a long-run equilibrium relationship between ICT imports and renewable energy adoption practices. Accordingly, the findings of this study offer valuable insights for policymakers, providing an evidence-based foundation for formulating policies that leverage digital technologies to foster sustainable energy transitions. Emphasizing the integration of ICT in national energy and innovation strategies can facilitate not only environmental sustainability but also inclusive economic development. As such, the study contributes to the broader discourse on how digital transformation can serve as a critical enabler of green economy practices and long-term ecological resilience.

S. Daga (✉)
Department of Commerce, PGDAV College, University of Delhi, New Delhi, India
e-mail: Shikha.menani@pgdav.du.ac.in

R. J. Oben
Department of Business Administration, Near East University, Nicosia, Türkiye
e-mail: remyjonkam.oben@neu.edu.tr

K. L. Tennin
College of Doctoral Studies, University of Phoenix, Arizona, USA
e-mail: KLTennin@email.phoenix.edu

K. Kumar et al. (eds.), *AI-Driven Circular Economy for Carbon Neutrality in Emerging Economies*, Environmental Footprints and Eco-design of Products and Processes,
https://doi.org/10.1007/978-3-032-30298-4_8

Keywords Circular economy · Green growth · BRICS · ICT · Sustainable Development Goals · Renewable energy adoption

1 Introduction

Exponential rise in economic wealth without paying due attention to the associated environmental impact has led to a significant change in the overall climatic condition. The condition has become so severe that it has led to prompt measures by global institutions, be it the Sustainable Development Goals (SDGs) at the global level or the enhanced environmental regulations at the microlevel. It has also led to increased interest among academicians in the domain of the study, where there are multiple studies focusing on the factors that impact the environment, along with highlighting the ones that help in mitigating emissions. A study by Oben et al. (2025), for instance, highlighted how economic growth, financial development, and fossil fuel energy consumption have a significant impact on emission rise, with human development leading to its mitigation. A study by Agarwal and Ojha (2024) further highlighted the importance of digitalization and big data analytics to promote sustainability in manufacturing. The traditional linear model of "take-make and dispose" has cast its irrevocable impact on the global economy, promoting academicians, policymakers, and corporations to make a paradigm shift toward the circular economy (CE). It is a regenerative system aimed at minimizing waste, extending the supply chain, and reusing materials to reduce pressure on the overall limited resources. CE is based on the principle of 3Rs: reduce, reuse, and recycle, which helps in promoting sustainable development. While the concept of CE is important for promoting green growth, its implementation is not the same in all economies. This is because emerging economies that are in a state of transition face dual goals of economic growth and protecting the environment. It not only puts pressure on its limited resources but is also expected to delay the path toward CE because of infrastructural hurdles, financial constraints, policy incohesiveness, and lack of awareness. There is thus a divide between the developed and emerging economies in terms of the adoption of CE principles which needs to be explored to fast-track its implementation on a global level. A study by Oben et al. (2024), for instance, highlighted how encouraging sustainable fintech solutions across African countries can help promote sustainable growth. In the enhanced globalization and increased trade across nations, ICT import of goods in the BRICS helps in bridging the infrastructural gaps by assisting with smart grids and digital energy systems that help in promoting renewable energy adoption. The present chapter explores how the import of goods helps emerging economies like the BRICS to attain the goal of CE by making a transition toward clean energy sources.

The last few years have seen tremendous work in the domain of environmental protection and sustainable development. It has given rise to the identification of various factors that have an impact on the adoption of sustainable practices. Whether it is the financial constraints, regulatory rigidity, lack of necessary infrastructure, and the absence of technical support, all have a magnifying impact on long-term growth. The factors become all the more important for economies that are hard-pressed for managing their socioeconomic needs along with the adoption of sustainability practices. A study by Daga et al. (2026) analyzed the impact of various economic, technological, political, and financial variables on renewable energy adoption across the top 20 carbon emitters and documented heterogeneous impact using GMM and quantile regression. It is therefore necessary to identify the long-run equilibrium and integration across the various determinants of green energy transition across the set of economies that are at the same level of development. The study fills the literature gap by identifying the role of ICT import of goods in promoting RE adoption across the emerging BRICS nations. By enabling predictive analytics, intelligent automation, and optimized resource flows, AI technologies offer the tools necessary to transition toward more circular economic models even in the fashion industry (Gazzola et al., 2025). BRICS is one of the major contributors to global GDP and emissions, and hence, efforts to promote RE adoption in these economies can produce significant overall effects. Analyzing the integration of ICT import of goods and RE adoption among emerging economies is imperative as they are growing at a rate that demands an increased energy use and which, if backed by clean sources of energy, can bring the necessary changes in the overall environmental quality. An important factor that can contribute in adoption of clean sources of energy is the contemporary ICT-enabled infrastructure that helps in the adoption and use of green sources of energy.

Despite the theoretical background that proves a significant impact of ICT practices on RE adoption, the literature related to its effectiveness on the BRICS is still at a nascent stage. Also, emerging economies are different from their developed counterparts in terms of technological efficiency, awareness levels, and regulatory environment, and hence, a separate study is mandated. Recognizing the gap, the present chapter empirically investigates the relationship between ICT imports of goods and RE adoption in BRICS over a period of time.

To empirically test the same, the study has made use of ICT goods imports as it represents the technological infrastructure necessary for deploying and operating AI systems. ICT imports of goods measure BRICS economies' access to AI-enabling infrastructure, computers, servers, and software systems critical for machine learning applications and green growth acceleration. For measuring RE adoption, the study has used renewable energy consumption as a percentage of total energy consumption. Renewable energy consumption is also one of the pillars for circular economy (CE) practices because it reflects a shift away from the linear "take-make-dispose" model toward more sustainable, resource-efficient systems which is the core principle of the circular economy. To account for other omitted variables and cross-sectional heterogeneity, the study has used control variables to account for economic wealth proxied by GDP, level of urbanization, trade openness, and industrialization.

For empirical analysis, the study has used a dynamic panel data framework with tests for measuring cross-sectional dependency and cointegration modeling to identify how ICT goods imports impact renewable energy adoption in the BRICS economies. Through this approach, the study aims to capture both the dynamic equilibrium relationships between ICT and RE performance. The next section provides a detailed literature review of the existing studies that highlight the diverse factors that impact RE adoption.

2 Literature Review

The motivation for this research is derived from three interrelated global challenges, namely ecological overshoot, technological disruption, and economic inequality. This is because the continuous race to escalate wealth generation has put immense pressure on natural resources, especially in economies like BRICS, where rapid economic growth intersects with environmental challenges (Ali et al., 2025). Also, the pressure on emerging economies to grow economically while adhering to international climate commitments is also burdening and well-documented. This section highlights the challenges and its documented literature and how use of ICT and digitalization can assist in promoting green economy principles across the BRICS.

Literature has documented multiple roles of ICT usage in RE adoption and transition toward a sustainable economy; for instance, technological innovation can lead to enhanced use of resources, thereby minimizing waste and promoting the 3Rs. Information and technological use is also documented to enhance energy efficiency, thereby optimizing resource use. AI and the Internet of Things are documented to improve the circular economy (Ali et al., 2025). These enhanced ICT infrastructure backed by increased ICT import of goods facilitates contemporary technology that leads to efficient use of materials, minimizes waste, and enhances the lifecycle of products, all of which are core tenets of the circular economy (CE). AI and deep learning technologies have the potential to increase energy systems' sustainability and efficiency, especially those that rely on renewable energy sources like solar and wind (Talaat et al., 2025).

The role of ICT in promoting RE adoption is well documented across the literature as highlighted by Frikha and Mrad (2025), wherein AI is highlighted to facilitate renewable energy integration, digital twin applications, and compliance with international sustainability frameworks. It is infact considered a key driver of the energy transition and sustainable economic development wherein AI investments are documented to stimulate renewable energy production, while AI readiness and institutional ecosystems are found to drive renewable energy consumption (Vasilescu et al., 2025). Results are supported by Georgescu et al. (2025) wherein artificial intelligence investments are documented to have a significant and positive impact on renewable energy consumption in both the short and long term. Even advanced methodology like machine learning documents a positive effect of renewable energy consumption on environmental degradation, with GDP per capita and urbanization negatively affecting the same (Tezcan, 2025). The role of financial inclusion and globalization in promoting renewable energy consumption is also documented in the literature (Salakpi et al., 2025).

There are studies that have focused on ICT imports and exports and their impact on carbon emissions. A study, for instance, used different definitions of ICT, for example, ICT import of goods and ICT export of goods separately and also the net effect, wherein it highlighted that when the net trading balance is taken, it brings a reduction in the carbon emissions across the European Union countries. But when the same is measured using ICT import of goods and its export relative to the total import and export of the country, the results are not the same and in fact it can be the opposite where it can lead to an increased level of emissions (Emous & Krušinskas, 2026). The factor is thus important to be considered in the studies related to sustainable development, especially in the contemporary world where there is a rise in digitalization, adoption of AI, and enhanced import and export of ICT.

This ICT and green economy integration is more popular in technologically advanced and rich economies because of their robust digital infrastructure and regulatory framework, all of which promote technological innovation. Cetin et al. (2025), for instance, highlighted how technological innovation helps in reducing the ecological footprints. However, while BRICS nations are racing high on the technological front, they are documented to lack the cohesive framework needed to align technological innovation with sustainability goals. The disconnect underscores the need for clear identification of the short- and long-run interconnectedness and interrelationship between ICT and RE adoption, and this is the main objective of this study. Various studies have explored the relationship between ICT and RE adoption. Study by Chishti et al. (2026), for instance, documented that AI has positive effects on green energy security across the short, medium, and long run. Studies further provide that decision-makers need to restructure energy and environmental policies by prioritizing AI technologies (Hamid et al., 2026). Results are supported by Qiu et al. (2026), which highlights a positive impact of AI development on carbon neutrality, the relationship promoting AI development across the developing economies. Studies also document that renewable energy and different measures of CE significantly improve environmental quality (Majeed & Luni, 2020).

The literature supports how economies with higher ICT penetration tend to perform better on sustainability parameters. The studies further provide how AI-based decision-making helps in supporting efficiency in supply chain alertness, thereby promoting CE practices and promoting sustainable development. This is necessary as an important aspect of CE adoption is the use of renewable energy. The results are also highlighted across the literature as AI is found to consistently stimulate RE adoption (He et al., 2025). The role of enhanced ICT infrastructure is not limited to green energy transition at the macrolevel but is also important for the greening of the organizations. AI-enabled tools are found to influence green employee behavior, thereby empowering employees to engage in eco-friendly practices that contribute to sustainability (Najam Shaikh et al., 2025).

It thus signifies how a shift from non-renewable and polluting energy sources to sustainable alternatives can assist in reducing the overall environmental impact. The transition to renewable energy in BRICS countries is particularly important given their reliance on fossil fuels and the corresponding environmental degradation. In addition to the ICT use (backed by enhanced import of ICT goods) and RE adoption, various other factors impact the relationship that have to be controlled for in the model to get the true relationship between the variables of interest. For instance, the

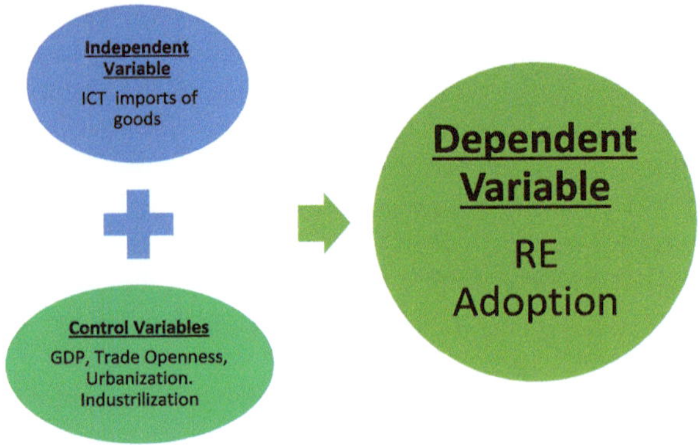

Fig. 1 Variables used in the study

relationship between economic variables like economic wealth measured by GDP, trade openness depicting the level of globalization of an economy, urbanization, a proxy for energy demand and usage, and industrialization, a major user of energy sources, all are critical factors that need to be taken care of while addressing the ICT–RE relationship (Fig. 1).

As documented by the literature, developed economies have stringent environmental regulations that encourage businesses to invest in environmental protection practices. In contrast, regulations across emerging economies like BRICS are presumed to be in their nascent stage and are still evolving. Therefore, policy reforms that incentivize sustainable innovation and ensure regulatory compliance are essential. Therefore, it is important to identify the dynamic relationship between ICT import of goods and RE adoption, which is the main objective of this study. To achieve this, the analysis tests the following hypothesis:

H1: ICT import of goods has a positive long-run impact on RE adoption across BRICS countries.

Tested using CCEMG to capture cross-sectional dependence and heterogeneous impacts.

H2: There exists a long-run cointegrating relationship between the ICT import of goods and RE adoption.

Tested using the Westerlund cointegration test.

H3: Common correlated shocks influence both the ICT import of goods and RE adoption.

CCEMG estimators are used to account for unobserved common factors.

The next section provides a detailed research methodology used for data analysis.

3 Research Methodology

To analyze the relationship between ICT import of goods and RE adoption across the BRICS economies, the study has made use of annual data from 2000 to 2021. All the data are extracted from world development indicators, and a brief description of the same is provided in Table 1. The data have been taken till 2021, as this is the last stream of data that are available. For identifying RE adoption, use of renewable sources of energy is used. The measure also reflects the transition from linear fossil fuel dependency to closed-loop renewable systems, and hence, the study has used it because it is closely linked to both CE adoption and sustainable development and also finds support in the existing literature to assist in the transition toward a green economy.

Based on the above variables, the present study has used panel econometric techniques like the common correlated effects mean group (CCEMG) model and the cointegration test. The advanced econometric techniques take care of cross-sectional dependence, heterogeneity, and endogeneity in the data. A detailed flowchart depicting the progression of the various methods used in the study is provided in Fig. 2. As a first step in the panel data spread across a diverse set of economies, the study tested for cross-sectional dependence at the first stage. This is necessary to identify inter-country correlations and the presence of the same makes the first generational unit root invalid. To account for this, the study first conducted Pesaran cross-sectional dependence (CD) test to determine the presence of such dependence across units. Identification of CD among the variables is then followed with testing of cointegration to identify the long-term interdependence. The cointegration test

Table 1 Description of variables

Variables	Description	Literature support
Dependent variable		
Renewable Energy Adoption	Renewable energy use as a % of total final energy consumption	Tezcan (2025)
Independent variable		
ICT import of goods	ICT goods imports as a % of total goods imports	Authors' Proxy
Control variables		
Economic wealth	GDP (constant 2015 US$)	Daga et al. (2025)
Trade openness	Trade (as a % of GDP)	Aldegheishem (2024)
Urbanization	Urban population as a % of the total population	Abd El-Aal et al. (2024)
Industrialization	Industry (including construction) measured by value-added as a % of GDP	Jangid et al. (2025)

Source Data collected from World Development Indicators, https://databank.worldbank.org/source/world-development-indicators

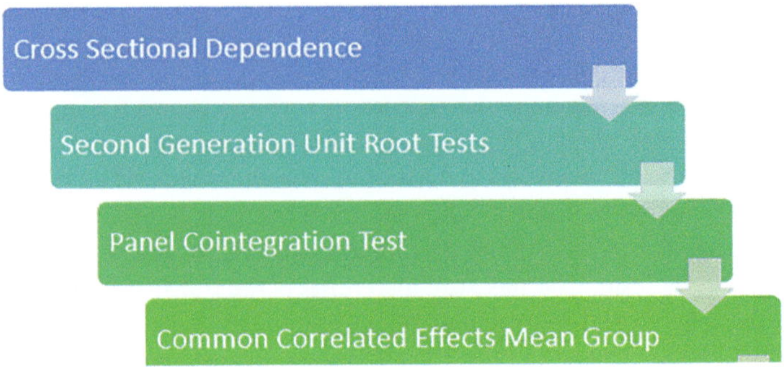

Fig. 2 Econometric progression for data analysis

that is used in the study is Westerlund cointegration as it directly examines panel error-correction behavior, handling cross-sectional dependence and mixed $I(0)/I(1)$ integration across BRICS, thereby validating the estimation of long-run relationships. This step is particularly important in our study because it confirms that ICT import of goods and RE indicators move together in the long run, and any short-run deviations will adjust over time. After checking for cointegration, the next step in the empirical testing is to use the common correlated effects mean group (CCEMG) to identify the long-run coefficients. The methodology that is used in the chapter is not just technically rigorous but is also contextually appropriate for the research question. This is because both ICT import of goods and RE are not only global but are also interconnected, and their relationship is influenced by various unobserved global trends, international policies, technological diffusion, and economic linkages. Methods like CCEMG are thus uniquely suited to this context because they control for common shocks and heterogeneous responses. This approach thus moves beyond mere correlation to uncover causal inferences and adjustment pathways.

4 Results and Discussion

The first step in the analysis of the panel data is to provide descriptive statistics that highlight the basic parameters of the study, namely the average, median, deviation, and skewness of the data (Table 2).

Descriptive statistics reveal substantial heterogeneity across BRICS economies typical of emerging market panels. GDP demonstrates extreme dispersion reflecting China's economic dominance within the group, while renewable energy adoption shows wide variation across countries and time. ICT goods imports exhibit moderate penetration as a share of total trade, consistent with technology import dependence in developing economies. Positive skewness across key variables underscores the structural disparities that characterize BRICS composition, validating the need for panel methods that accommodate cross-sectional heterogeneity.

Table 2 Descriptive statistics

	ICT	RE	GDP	*T*	UP	Ind
Mean	11.293	22.68364	2,626,377,618,384.60	44.29922	60.30432	29.89234
Median	8.955	14.9	1,403,335,647,500.68	46.40263	62.629	28.21052
Maximum	26.07	50	15,851,276,922,873.10	68.09391	87.317	47.5574
Minimum	3.68	3.2	221,691,970,228.10	22.10598	27.667	18.1885
Std. Dev	5.878908	16.87277	3,455,697,146,267.57	11.81116	19.14061	7.858065
Skewness	1.335762	0.307678	2.36	− 0.321434	− 0.339223	0.952766
Kurtosis	0.371401	− 1.598779	4.84	− 0.90517	− 1.189654	− 0.037058

Source Authors' computation

The descriptive statistics presented in Table 2 provide critical insights into the distributional characteristics of the variables used in the study. There is heterogeneity across the variables, and hence, it requires the use of robust panel estimation techniques like CCEMG and cointegration. These techniques not only efficiently capture the cross-sectional dependence but also provide for dynamic interactions. The second step in the panel estimation where there are multiple explanatory variables is to capture the possibility of correlations among them. For that, a correlation matrix is computed (Table 3).

The correlation values show a high association between a few variables like GDP, ICT, and industry. The next step in the panel estimation after checking for the correlation is to cross-check the presence of cross-sectional dependence using CD tests. Table 4 highlights the various CD test results.

The results from the Breusch–Pagan LM, Pesaran scaled LM, and Pesaran CD tests all provide significant results for the presence of cross-sectional dependence. The results provide for strong interconnectedness between the BRICS economies. It therefore refutes the use of first-generation unit root test. The next step is thus to proceed for second-generation unit root testing techniques (Table 5). The CIPS and CADF are employed further to test for unit root testing and also because they account for cross-sectional dependence. The test results ensure that the results are robust in the presence of inter-country linkages.

Table 3 Correlation matrix

	ICT	RE	GDP	*T*	UP	Ind
ICT	1.0000					
RE	− 0.1408	1.0000				
GDP	0.7907	− 0.1234	1.0000			
T	0.0001	− 0.7521	− 0.1092	1.0000		
UP	− 0.1875	− 0.1142	− 0.1432	− 0.2631	1.0000	
Ind	0.7912	− 0.3713	0.6422	0.3852	− 0.4435	1.0000

Source Authors' computation

Table 4 Cross-sectional dependence test

Test	Statistic	*p*-value	Interpretation
Breusch–Pagan LM	30.6956	0.0007	Reject null → There is cross-sectional dependence
Pesaran scaled LM	4.62767	0.0000	Reject null → There is cross-sectional dependence
Pesaran CD	4.46401	0.0000	Reject null → There is cross-sectional dependence

Source Authors' computation

Table 5 Unit root test results

Variable	CIPS test statistic	CADF test statistic
RE	– 1.84	– 1.26*
GDP	– 1.52	0.74
ICT	– 2.14**	– 1.69**
Trade	– 1.38	0.89
Urb	– 1.61	0.52
Ind	– 1.29	1.13
ΔRE	– 2.47**	– 1.53*
ΔGDP	– 4.76***	– 10.82***
ΔICT	– 3.28***	– 5.41***
ΔTrade	– 2.39**	– 1.47*
ΔUrb	– 5.13***	– 12.67***
ΔInd	– 3.94***	– 7.28***

Authors' construction

Note *, **, and *** show the rejection of the null hypothesis at the 10%, 5%, and 1% levels of statistical significance, respectively. CADF and CIPS, respectively, stand for the cross-sectionally augmented Dickey–Fuller and cross-sectionally augmented IPS (Im, Pesaran, and Shin) tests (Pesaran, 2007)

Source Authors' Computation

Results of the second-generation unit root test provide for mixed integration levels with certain variables being stationary at the level and others at the first difference. While the dependent variable that is renewable energy adoption is found to be integrated of order one, the main independent variable that is the ICT goods imports is found to be integrated of order zero, and all other control variables are integrated of order 1.

Due to the mixed integration levels, traditional panel regression models fail, and hence, we have made use of cointegration techniques to identify long-run equilibrium relationships between the variables.

The next step here is to proceed with the Westerlund cointegration test (Westerlund, 2007) to identify the presence of long-run equilibrium between the variables. The test also helps in identifying whether the variables move together over time in spite of facing short-run fluctuations. Results of the cointegration are presented in Table 6. As the dependent variable is $I(1)$ and the independent and control variables show a mixed integration involving both $I(0)$ and $I(1)$, the Westerlund (2007) error correction-based cointegration test is more appropriate compared to the Johansen cointegration as the latter requires uniform integration across all variables and also is not apt in the case of cross-sectional dependence. Westerlund not only helps in accommodating the heterogeneous integration but also takes care of the cross-sectional dependence that is confirmed in the study. The use of Westerlund cointegration thus helps in identifying the long-run equilibrium (cointegration) detection across the BRICS economies.

Table 6 Cointegration test results

	Test statistic
Westerlund (2007)	4.37***
Modified Dickey–Fuller	– 0.82
Dickey–Fuller	– 3.91***
Augmented Dickey–Fuller	– 5.46***
Unadjusted modified Dickey–Fuller	– 3.87***
Unadjusted Dickey–Fuller	– 5.73***

Authors' construction
Note *, **, and *** indicate statistical significance at the 10%, 5%, and 1% levels, respectively
Source Authors' computation

The Westerlund cointegration, which is a second-generation cointegration test, shows the presence of cointegration. The results provide for a stable long-run relationship between renewable energy adoption and ICT goods imports across the BRICS economies. The results are justified by the significant values of Dickey–Fuller, augmented Dickey–Fuller, unadjusted modified Dickey–Fuller, and unadjusted Dickey–Fuller. The results validate the presence of panel cointegration despite different integration levels and the presence of cross-sectional dependence.

On confirming the cointegration, the next step is to go for the common correlated effects mean group (CCEMG) model. These estimators are particularly suitable for panel data with cross-sectional dependence and heterogeneity, enabling robust estimation of long-run relationships. The use of CCEMG ensures that unobserved common factors and individual-specific characteristics across countries are properly accounted for. Also, following the progression from Westerlund cointegration that confirmed the presence of long-run relationship, it is imperative to identify the estimates which are provided by the CCEMG. Results of the same are highlighted in Table 7.

Table 7 provides the results of the CCEMG model that efficiently deals with cross-sectional dependence in the panel data estimation. The dependent variable here is the renewable energy adoption, and the main explanatory variable is the ICT imports of goods, controlling for GDP, trade openness, urbanization, and industrialization.

Table 7 Common correlated effects mean group model

Variable	Estimate	Std. Error	z-value	Pr(> |z|)	Significance
ICT	– 0.0345	0.1935	– 0.1780	0.8587	
T	– 0.1207	0.1313	– 0.9192	0.3580	
Urb	5.2812	9.3807	0.5630	0.5734	
Ind	– 0.2067	0.1944	– 1.0635	0.2876	
GDP	– 33.5163	9.1277	– 3.6719	0.0002	***

Source Authors' computation

Results of the CCEMG highlight a negative and significant relationship between GDP and RE adoption while the relationship between all other variables is found to be insignificant in promoting RE adoption. The inverse relationship between GDP and RE adoption documents that economic expansion is not accompanied by a green energy transition. This can be because in emerging economies the transition toward green energy is still slow and the race for wealth generation is greater which is generally met by traditional sources of energy. The results also point toward the emergent need to decouple growth from emissions and promote development that is both economic and green.

5 Implications

The role of ICT import of goods in promoting RE adoption across the emerging economies is a topic that requires due attention. This is because it not only assists in promoting cleaner sources of energy but also helps in making the transition toward sustainable development smooth. The implications are mainly meant for the policymakers, but it has equal importance for other domains too. The present section highlights a few of the implications as derived from the chapter. The long-run equilibrium identified in the chapter between ICT and RE adoption highlights the strategic role that ICT adoption can play in promoting green growth via the enhanced RE usage. The role of technological innovation, represented by the import of ICT goods, provides for a positive impact on the RE that provides for its critical role in integrating technology across the national level energy and development strategies. This can be done via the incentives provided on enhanced ICT usage. It can be done through favorable tariff structures and/or streamlined procedures for the import of ICT related equipment. Due to the rising demand of ICT tools across emerging economies, it becomes imperative to have a conducive regulatory environment for their import. It will also assist in fostering innovation at the domestic level in the long run. Also, to ensure its widespread use, it is necessary to enhance technical awareness and literacy among the masses as the ICT adoption is mainly hindered because of a lack of qualified people.

Policymakers need to identify ICT tools not only as a productivity tool but also as a sustainability enabler that is capable of optimizing resource use along with monitoring energy flows, and supporting smart grid infrastructure. Lastly, to identify and enhance the impact of ICT on overall sustainable development, it is necessary to have an efficient monitoring and evaluation system. This requires the development of new indicators and datasets that can assist in tracking technological inflows, usage patterns, and their environmental impact.

The chapter has several economic implications in addition to the policymaking highlighted above. The inverse relationship between GDP and RE adoption documented in the chapter provides for a careful assessment of economic growth and its impact on the adoption of green sources of energy. This necessitates a reevaluation of the development models especially across the emerging economies. This can be done by incentivizing green business models and environmentally responsible entrepreneurship. It is also necessary to ensure that investments in ICT are not only directed toward economic productivity but also assist in promoting eco-efficiency in manufacturing, energy, and urban development.

6 Conclusion

Climate change has sparked a debate across academics and policymaking regarding the steps that can be taken to enhance measures toward sustainability in addition to maintaining the wealth generation that is required to fulfill the socioeconomic objectives of the economies. One important measure that is being talked about at the global level is the transition toward RE adoption at every level. It will not only reduce pressure on scarce resources but will also enhance environmental quality. It will also ensure in enhancing the adoption of CE principles that can foster green growth. The RE principles are backed by energy efficiency and focus on recycling, reuse, and reduction. To ensure that CE principles are adopted at the global level, it is necessary to make a paradigm shift toward contemporary technology that can assist in its quick and efficient adoption. It becomes all the more important for emerging economies that have to balance their growth with sustainability measures.

7 Limitations and Scope for Future Research

The study though being the gap breaker in the ICT–RE nexus, still no study is free from certain inherent limitations. The current chapter, for instance, is based on certain proxies for the ICT usage and RE adoption; the proxies, though suitable, can be enhanced further and made more exhaustive. As discussed in the literature, there are other comprehensive measures of green growth as renewable energy use is just a part of the same. Future studies can create a composite index of various measures that promote green growth to identify the relationship more exhaustively. Another important limitation is using ICT imports of goods; the measure, though theoretically sound, can be elaborated further, wherein an index that combines various measures of ICT adoption can be used to find the robustness of the results. Secondly, the availability of data on an annual basis and till 2021 also requires further improvement in terms of enhanced frequency and real-time availability of data. Based on the limitations, future studies can focus on a diverse set of economies, making a comparison between emerging and developed ones and stating if the relationship holds in general or if it is specific to the level of development. Secondly, while the present study made use of cointegration for establishing short- and long-run equilibrium, future studies can focus on advanced econometric methodology that can lead to enhanced and more robust results.

References

Abd El-Aal, M. F., Hegazy, M. S., & Abdelsamiea, A. T. (2024). Role of economic expansion, energy utilization and urbanization on climate change in Egypt based on artificial intelligence. *International Journal of Energy Economics and Policy, 14*(3), 127–137. Econjournals. https://doi.org/10.32479/ijeep.15607

Agarwal, A., & Ojha, R. (2024). Prioritizing implications of Industry-4.0 on the sustainable development goals: A perspective from the analytic hierarchy process in manufacturing operations. *Journal of Cleaner Production, 444*. Elsevier Ltd. https://doi.org/10.1016/j.jclepro.2024.141189

Aldegheishem, A. (2024). The impact of air transportation, trade openness, and economic growth on CO_2 emissions in Saudi Arabia. *Frontiers in Environmental Science, 12*. Frontiers Media SA. https://doi.org/10.3389/fenvs.2024.1366054

Ali, M. U., Khan, I., & Khan, H. (2025). Leveraging industry 4.0 technologies and industrial symbiosis: Advancing circular economy practices in BRICS economies. *Journal of Environmental Management, 383*. Academic Press. https://doi.org/10.1016/j.jenvman.2025.125471

Cetin, M., Ozturk, I., Sumerli Sarigul, S., Murshed, M., & Kilavuz, E. (2025). Nexus between technological innovation and environmental pollution in selected OECD countries. *Natural Resources Forum, 49*(1), 841–862. John Wiley and Sons Inc. https://doi.org/10.1111/1477-8947.12458

Chishti, M. Z., Ali, A., Javaid, M. Q., & Özkan, O. (2026). Achieving sustainable development with the application of artificial intelligence: Effects on global green energy security amid geopolitical risk. *Technology in Society, 85*. Scopus. https://doi.org/10.1016/j.techsoc.2025.103201

Daga, S., Tao, Z., & Ahmad, N. (2026). Renewable energy adoption in high-emission economies: Drivers and implications for green growth. In V. Kandpal, W. Leal Filho, K. Frizzo, N. Aguilar-Rivera, & T. Choudhury (Eds.), *Green horizons: Advancing climate solutions through green finance, circular economy, and ESG* (pp. 445–461). Springer Nature Switzerland. https://doi.org/10.1007/978-3-032-03484-7_23

Daga, S., Yadav, K., Yadav, B., Bregoli, D., Liscio, M. C., & Sospiro, P. (2025). Tourism-driven carbon emissions and the innovation response: A cross-country analysis of extended QUAD economies. *Sustainable Futures, 10*, Article 101189. https://doi.org/10.1016/j.sftr.2025.101189

Emous, R. V., & Krušinskas, R. (2026). Is digitalization leading to CO_2 emission cutting? *Journal of Economic Studies, 53*(9), 1–18. https://doi.org/10.1108/JES-03-2025-0147

Frikha, M. A., & Mrad, M. (2025). AI-driven supply chain decarbonization: Strategies for sustainable carbon reduction. *Sustainability (Switzerland), 17*(21). Scopus. https://doi.org/10.3390/su17219642

Gazzola, P., Pavione, E., Hillebrand, R., Vota, V., & Rosa, R. (2025). The circular economy and the role of technology in the fashion industry: A comparison of empirical evidence. *Sustainability (Switzerland), 17*(7). Multidisciplinary Digital Publishing Institute (MDPI). https://doi.org/10.3390/su17073104

Georgescu, I., Yazıcı, A. M., Bayram, V., Öztırak, M., Toy, A., & Dogan, M. (2025). Governing the green transition: The role of artificial intelligence, green finance, and institutional governance in achieving the SDGs through renewable energy. *Sustainability (Switzerland), 17*(12). Scopus. https://doi.org/10.3390/su17125538

Hamid, S., Wang, K., & Zhang, X. (2026). Unraveling the role of artificial intelligence, eco-innovation efficiency, and stringent environmental policies in environmental sustainability: Is the load capacity curve hypothesis true in G7 economies? *Technology in Society, 85*. Scopus. https://doi.org/10.1016/j.techsoc.2025.103187

He, C., Tu, Y., Li, X., & Dai, W. (2025). Do artificial intelligence investments, financial development, and energy security risks promote renewable energy transition? Evidence from the United States. *Sustainability (Switzerland), 17*(24). Scopus. https://doi.org/10.3390/su172411067

https://databank.worldbank.org/source/world-development-indicators

Jangid, H., Rao, N. V. M., & Bal, D. P. (2025). Revisiting the load capacity curve hypothesis: evidence from fintech, renewable energy and environmental sustainability in G-20 economies. *Technological Sustainability*, 1–18.

Majeed, M. T., & Luni, T. (2020). Renewable energy, circular economy indicators and environmental quality: A global evidence of 131 countries with heterogeneous income groups. *Pakistan Journal of Commerce and Social Sciences (PJCSS), 14*(4), 866–912.

Najam Shaikh, S., Sohu, J. M., Shaikh, S., Zahra Kherazi, F., & Soomro, S. (2025). Artificial intelligence adoption, green HRM, and employee behavior: Driving environmental sustainability in SMEs with organizational culture as a moderator. *SAGE Open, 15*(4). Scopus. https://doi.org/10.1177/21582440251392673

Oben, R. J., Seraj, M., & Zihni Eyüpoğlu, Ş. (2025). The role of human development, financial technology, economic growth, and fossil fuel energy consumption in environmental sustainability: Novel insights from the world's leading carbon emitters using MMQR. *Environment, Development and Sustainability*. https://doi.org/10.1007/s10668-025-06572-y

Oben, R., Seraj, M., & Eyüpoğlu, ŞZ. (2024). Toward environmental sustainability in Africa: The impact of financial technology, human development, and renewable energy consumption across income groups. *Environmental Quality Management, 34*(2), Article e22347. https://doi.org/10.1002/tqem.22347

Qiu, H., Kang, J., Li, X., Zhang, L., & Lei, B. (2026). The multiple impacts of artificial intelligence on carbon neutrality: Evidence From China. *SAGE Open, 16*(1). Scopus. https://doi.org/10.1177/21582440251408407

Pesaran, M. H. (2007). A simple panel unit root test in the presence of cross-section dependence. *Journal of Applied Econometrics, 22*(2), 265–312. https://doi.org/10.1002/jae.951

Salakpi, A., Mireku, K., & Domeher, D. (2025). Financial inclusion and globalization: Catalysts for renewable energy adoption in Sub-Saharan Africa. *Journal of Economics and Development, 27*(4), 322–340.

Talaat, F. M., Kabeel, A. E., & Shaban, W. M. (2025). Powering a sustainable future: AI-driven integration of renewables for optimized grid management. *Sustainable Futures, 10*. Scopus. https://doi.org/10.1016/j.sftr.2025.100821

Tezcan, N. (2025). Machine learning methods for investigating the nexus between FDI and environmental degradation: Evidence from Sub-Saharan African countries. *Journal of Capital Markets Studies*, 1–15.

Vasilescu, L., Sichigea, M., Sitnikov, C., & Mihai, L.-S. (2025). Nexus between artificial intelligence, renewable energy, and economic development: A multi-method approach. *Economies, 13*(9). Scopus. https://doi.org/10.3390/economies13090271

Westerlund, J. (2007). Testing for error correction in panel data. *Oxford Bulletin of Economics and Statistics, 69*(6), 709–748.

Chapter 9
From Algorithms to Action: AI's Contribution to Achieving the SDGs

Yasser Mahfooz, Obaidur Rahman, and Manik Batra

Abstract In 2015, the United Nations introduced 17 interconnected global sustainable goals (SDGs), which aim to make a world which is inclusive for all. This study explores the collaboration of AI and circular economy principles to advance the UN SDGs. On meticulously examining the recent research in the domain, this research indicates the AI's contribution toward the progress of multiple SDGs while managing localized risk factors in emerging economies. The analysis highlights the correlation of AI, SDGs, and circular economy in order to speed up the growth of emerging economies while achieving the carbon neutrality goals. The barriers to its large-scale implication include data accessibility, infrastructure limitations, and limited funding. The analysis led to identification of key determinants of success to test the deployment of AI and the need for holistic framework for cross-border integration of AI to SDGs. This research advances the domain by exploring the integration of AI and circular economy principles to achieve the SDG targets in an emerging economy.

Keywords Artificial · Intelligence · SDGs · Circular · Economy · Emerging

1 Introduction

In 2015, United Nations introduced "The Sustainable Development/Evolution Goals (SDGs)," which works on five-pillar framework of People, Planet, Peace, Prosperity, and Partnerships. The 17 interconnected global objectives with 169 main targets aim

Y. Mahfooz
Leicester Castle Business School, De Montfort University, Leicester, UK
e-mail: yasser.mahfooz@dmu.ac.uk

O. Rahman (✉)
School of Business, Galgotias University, Greater Noida, India
e-mail: obaidkhan.mba@gmail.com

M. Batra
School of Business Studies, Sharda University, Greater Noida, India
e-mail: manik.batra@sharda.ac.in

K. Kumar et al. (eds.), *AI-Driven Circular Economy for Carbon Neutrality in Emerging Economies*, Environmental Footprints and Eco-design of Products and Processes,
https://doi.org/10.1007/978-3-032-30298-4_9

to end poverty, protect planet, ensure prosperity, and reduce inequalities by 2030 (Bexell & Jönsson, 2017). Kroll et al. (2019) described SDGs as universal approach which requires collaboration between government, businesses, civil societies, and economies with a focus of leaving no one behind. Alongside these growth opportunities, the potential bottlenecks are also identified by researchers. Ecker et al. (2020) highlighted issues like limited data availability for all, ensuring the quality of monitoring of the progress, and Zhou et al. (2021) highlighted financial constraints, especially of emerging economies. It limits the resource allocation power of developing nations toward the sustainable goals. Researchers believe that rapid advancement of artificial intelligence (AI) technologies give rise to novel opportunities to accelerate the efforts toward SDGs (Singh, 2024). SDGs, on one hand, provide an effective roadmap where AI provides the approach and tools; a synergetic confluence of the two can help in achieving these sustainable goals (Kulkov et al., 2024). However, a thorough assessment is necessary at each step to find potential risks and identify key opportunities. This study investigates the effectiveness of implementing AI approaches and tools to ensure sustainable growth across all 17 SDGs.

2 Literature Review

The symbiosis among AI, SDGs, and the circular economy principles has been a key area of attention for researchers. Vinuesa et al. (2020) reviewed the mapping of AI's potential to all 17 SDGs. It is found that AI can assist in achieving specific SDG targets, especially environmental and energy related. It also highlighted the structural risks like algorithmic bias, high energy requirement of AI systems, and inequality amplification. The research was primarily conceptual and treat SDGs unevenly. Sakubu (2025) further enhances it by suggesting global deployment of AI for actual long-term impact across various geographies.

3 AI and Sustainable Development

Initial research on AI was mainly focused on its social impact, ethical concerns, and more on its economic impact. With time, researchers started exploring the positive impact of AI on social, environmental, and economic areas. Several studies have explored AI's capabilities to achieve the 17 SDGs, identifying its direct and indirect contributions (Fan et al., 2023; Meitei et al., 2025; Sætra, 2021; Visvizi, 2022; Yadav et al., 2024). Sætra (2021) identified that traditional ESG frameworks do not capture the sustainable implications of AI effectively. Hence, proposed a structured approach to evaluate effects of AI on ESG using SDG criteria to improve transparency of corporate reporting. However, this study lacks field testing of the framework and how it influences actual SDG outcomes. Visvizi (2022) explored the influence of AI on politics and society. The study concluded that AI can serve as a powerful tool for addressing complex societal challenges and that its ethical, inclusive, and transparent adoption will help in achieving reduced inequalities (SDG 10).

Fan et al. (2023) focused on achieving sustainable goals using AI and deep learning (DL) models. It is found that AI and DL models together can contribute to 134 of the 169 SDG targets. The key successes of this study include renewable energy optimization and exposure modeling in environmental health. This further highlighted the importance of regulatory oversight and ethical deployment of AI and DL models. However, the explainability and reliability of AI and DL models are questionable. Yadav et al. (2024) discussed how AI enhances data collection, monitoring, and analytics for SDG progress measurement. It underlined that AI can help in identifying actionable insights, enabling better evidence-based decision and policymaking. Meitei et al. (2025) explored varied research and delivered a bibliometric evaluation of those studies. It showed the growth in number of publications on AI and SDGs; SDG 3 (health), SDG 7 (clean energy), and SDG 11 (sustainable cities) remained attractive areas for research. But mainly the short-term impact of AI on SDGs is explored to an extent, paving pathway for future research. Tautiva et al. (2025) and Meneses Cerón et al. (2024) highlighted the persistent bottlenecks, including regulatory, ethical, and methodological issues related to AI's implementation. The study underlined the need for transparency, trust, and universal standards in AI applications tied to SDGs. AI, on one hand, offers immense future scope and opportunities while on the other hand, it also intensifies the growing concern of increased energy consumption, biased algorithms deepening the persistent inequalities, and privacy concerns (Crawford, 2021). AI is "twin-edged" in nature, and these studies highlighted that AI offers transformative scope toward SDGs, but the scope turns to reality when AI is deployed with transparency, maintains ethical oversight, and cooperates with cross-disciplines. These studies served as the base for our investigation. We have identified the need for holistic evaluation frameworks to capture both positive and negative effects of AI. Hitherto, a central perspective is needed for the responsible expansion of AI with risk mitigation strategies.

4 AI and the Circular Economy

The idea of the circular economy reached to mainstream agenda due to the growing concerns for resource scarcity and environmental degradation. The initial studies in this domain were focused on dealing with waste and pollution management, products and material usage, and regeneration of natural systems (Masanpally et al., 2025; Morseletto, 2020). The Ellen MacArthur Foundation & Mckinsey report (2019) highlighted how AI enhances the circular economy (Tutore et al., 2024). It stated that AI helps in designing products, operating circular business models, and developing circular infrastructure.

Lacy and Rutqvist (2015) explained that AI can optimize product design for increased durability and recyclability, predict equipment failure to extend product lifespans, augment waste sorting and recycling processes through computer vision, and facilitate reverse logistics and material recovery. Jabbour et al. (2020) and Raj et al. (2020) explored the AI's ability to improve efficiency, transparency, and decision-making across value chains by facilitating various circular economy strategies, including reuse, repair, remanufacturing, and recycling. These literary works also highlighted the ethical issues of AI such as—algorithmic bias, data privacy concerns, and lack of transparency. AI may enhance the circular economy but only if ethical and governance risks are addressed. The literature in this domain is mostly conceptual with limited real-world influence of AI on circular economy initiatives, especially in diverse geographical contexts.

5 Emerging Economies: Context and Challenges

Adoption of sustainable growth and the circular economy principles by emerging economies arises varied opportunities but with this arises a unique set of hurdles (Mishra & Kumar, 2024). Singh et al. (2024) identified these hurdles as rapid urbanization, large informal sectors, limited infrastructure, varying levels of technological adoption and considerable socioeconomic inequalities. Sharma et al. (2024), Kumar et al. (2024), Mishra et al. (2023), Kumar (2022), Kumar et al. (2021a, 2021b) highlighted the need for context-specific solutions to address local resource availability and cultural practices.

World Economic Forum (2018) emphasized that where developed nations established digital infrastructures from scratch, the emerging economies can adopt advanced technologies without going through intermediate stages of growth, giving them the edge due to "leapfrogging" opportunities. Despite this, the constraints related to digital divide, data governance, and the need for capacity building still persist in the system. UNCTAD (2019) highlighted the importance of inclusive AI growth that considers the specific needs and vulnerabilities of emerging economies.

6 Carbon Neutrality and AI

Carbon neutrality refers to "net-zero emissions" and aims at reducing the greenhouse gases and their effects on the environment. Many nations have pledged to reach carbon neutrality by 2050 to align with the Paris Agreement. It is also the central pillar of climate action (SDG 13), and an impactful circular economy should aim at achieving carbon neutrality.

The literature in this domain, namely Kumar et al. (2024), Agrawal and Kumar (2024), Tarkar and Kumar (2024), Singh et al. (2025a), and Agrawal et al. (2026), highlighted the strategies for decarbonization across energy, industry, transport, and agriculture sectors. IEA (2020) elaborated AI's contribution to carbon neutrality. It helps in optimizing renewable energy grids by preventing overloads and balancing supply and demand, predicts energy demand, and improves carbon capture technologies and energy efficiency. AI also helps in monitoring and verifying carbon emissions, supporting carbon accounting and facilitating carbon trading mechanisms (Singh et al., 2025b). However, the high energy requirements of AI itself, which generates carbon emissions (Strubell et al., 2019), highlights a crucial paradox and the need for research into "green AI" approaches to enhance energy efficiency and implement sustainable computing (Sharma et al., 2025).

7 Research Gaps and Objectives

The research gaps identified based on the review of aforesaid literature are as follows:

1. **Lack of holistic framework to understand AI's contribution to SDGs**: Though the individual connection between AI, SDGs, circular economy, and emerging economies exists, there is a lack of an integrated framework that interconnects all these elements and delivers a holistic understanding of AI's contribution.
2. **Regulatory approaches for emerging economies are unexplored**: In emerging economies, the specific policy and governance models, which ensures responsible AI deployment (in the context of circular economy and SDGs), are underexplored.
3. **Lack of scalability of AI-driven sustainable solutions**: Can successful AI-driven sustainable solutions be expanded to other emerging economies and if they are adaptable to each country's local needs and context?

This sets the stage for a deeper investigation into how AI can drive progress toward SDGs within a circular economy, specifically for dynamic environment of emerging economies. This study will focus on optimizing resource flow, enhancing energy efficiency, promoting sustainability in food systems and agriculture, and developing sustainable cities while addressing socioeconomic and ethical concerns.

8 Research Methodology

This research adopts the systematic literature review (SLR) approach, which is widely recognized as a transparent and rigorous pathway to identify, evaluate, and synthesize existing research relevant to the topic (Kitchenham & Charters, 2007).

An exhaustive exploration was conducted across key scholarly databases such as Google Scholar, Scopus, Web of Science, and specific repositories of organizations like the United Nations, World Bank, and leading AI research institutes; peer-reviewed articles, conference proceedings, book chapters, and reputable reports from international organizations published primarily in English, focusing on the intersection of AI with at least two of the core themes (SDGs, circular economy, emerging economies, carbon neutrality). To capture the recent advancements, the publications from the last decade (2015–2025) were prioritized alongside some works from earlier periods.

Keywords and phrases used in the search included: *"Artificial Intelligence," "AI," "Machine Learning," "Deep Learning," "Sustainable Evolution Goals," "SDGs," "Circular Economy," "Carbon Neutrality," "Climate Change Mitigation," "Emerging Economies," "Developing Countries," "Resource Efficiency," "Waste Management," "Renewable Energy" and combinations thereof. Suitable Boolean operators (AND, OR) were employed for refining search queries* (Alharbi & Stevenson, 2020).

Data Collection: Data were derived from relevant research works which included:

1. The specific AI technologies
2. The SDGs
3. The circular economy principles
4. The geographical and localized context
5. Benefits, challenges, and ethical considerations of AI.

These integrated research works on AI & SDG helped us to explore patterns and sequences and identify the pertinent research gaps, which sets the stage for further investigation and analysis.

9 Analysis and Synthesis

9.1 Optimizing Resource Flows and Waste Management

Efficient management of resources and waste aligns with SDG 12, i.e., Responsible Consumption and Production. The circular economic principles aim to manage resources efficiently as its one of the core outcomes. Solaja (2024) stated AI's capabilities can act as a catalyst in emerging economies to optimize resources and waste management. Computer vision and ML algorithms can identify different types of materials (plastics, metals, organic waste) from the waste stream with high accuracy (Erkinay Ozdemir et al., 2021). This gives rise to automated sorting systems, improving the quality of recycled materials and reducing contamination (Daga et al., 2025). In emerging economies, this can formalize the informal waste collection, enhance efficiency of the system, and create safer working conditions for waste pickers (Kathpal et al., 2021). For example, in AI-powered smart waste management facility, machines working on sorting mechanism can be used to separate the recyclable material from mixed solid waste. This can increase efficiency by identifying and segregating valuable recyclable materials from the waste, which could have end up in the landfills. It results in lower methane emissions as organic waste will reach the landfills and finally decomposes, advancing the carbon neutrality targets of the facility.

Collaboration of AI algorithms with IoT devices like sensors can assess data from machineries and forecast the failures, thus facilitating proactive maintenance reducing delay and inefficiency (Ayeni, 2025). This elongates the lifecycle of products and equipment, reducing the need for new production and hence conserving resources (Haque et al., 2024). Dsilva et al. (2025) explained the AI's contribution in minimizing downtime and optimizing production cycles. It explored how AI models help in predicting machine breakdown in a manufacturing facility, which minimizes downtime and thereby reduces waste material and energy consumption. This boosts the resource efficiency in emerging economies as they have limited resources. Spirito (2024) mentioned AI can facilitate the collection, repair, remanufacturing, and recycling of products. Though supply chains for reverse logistics are too complex, ML models can predict the best routes, reducing the emissions by reducing usage of fuel. ML models can also help in identifying opportunities for material recovery and what to recover to gain maximum value (Xia et al., 2023). Reintroduction of products in the value chain can be too complex for emerging economies with fragmented logistics network, but AI-driven platforms can help it. Alongside, AI can also predict the demand of recycled material, ensuring the balance between demand and supply and make recycling profitable for long-term support to circular economy.

9.2 Enhancing Energy Efficiency and Renewable Energy Integration

SDG 7 focuses on transitioning to clean energy and increase its affordability. AI can play an influencing role to accelerate this change. Ukoba et al. (2024) emphasized that AI algorithms can analyze real-time energy consumption patterns, weather forecasts, and renewable energy generation to optimize grid operations. AI can also predict energy demand, balance supply and demand, and integrate intermittent renewable sources more effectively (Rao et al., 2024; Tripathi et al., 2022), thereby improving grid resilience, minimizing blackouts, and minimizing energy waste, especially in emerging economies where grids might be less robust. Ohalete et al. (2023), Said and Alanazi (2023), and Lipu et al. (2021) studied this domain and concluded that AI can use meteorological data to predict the output of solar panels and wind turbines, enabling better forecasting and integration into the grid. AI can also determine the best placement and orientation of the energy assets to attain maximum efficiency. Agupugo et al. (2024) explained how AI-powered microgrids can contribute to carbon neutrality. In emerging economies, the remote communities can be benefitted by AI-powered microgrids as it offers reliable and affordable clean energy, reducing their dependency on fossil fuels contributing to carbon neutrality goals. Bajwa et al. (2024) highlighted the competency of AI-driven structural management systems. These systems can augment heating, ventilation, and air conditioning based on occupancy, external weather, and historical data, aiming to reduce the overall energy consumption. AI can also optimize production processes to minimize energy waste.

In emerging economies, all this together can help to cater rising energy demand due to rapid industrialization.

10 Sustainable Agriculture and Food Systems

In emerging economies, AI can revolutionize agriculture by improving food security and encouraging sustainable land adoption (SDG 2 & SDG 15). AI-driven drones can monitor crop health, food, soil conditions, and water needs with precision (Aarif et al., 2025; Srivastava et al., 2018). ML algorithms further recommend optimal irrigation, fertilization, and pest control strategies, minimize resource utilization, and maximize yields (Oyenuga et al., 2025). AI can boost Life on Land (SDG 15) and Clean Water & Sanitation (SDG 6) by reducing the chemical runoff and water depletion, respectively. In emerging economies, the agricultural productivity may be low and climate change can adversely impact it. AI has the power to predict food demand and optimize inventory management in supply chains, reducing post-harvest losses and retail food waste (Olawale et al., 2025; Srivastava et al., 2021, 2024). Computer vision also be used to sort and grade the produce. Food loss at different phases of supply chain may also occur in emerging economies. AI is expected to play a crucial part in bettering food security (SDG 2).

11 AI and Sustainable Cities

Creating more sustainable and liveable urban environments is challenging for rapidly growing emerging economies. Deep and Verma (2023) in their research examined how AI can help reducing this hurdle. AI analyzes vast datasets related to traffic patterns, public transport usage, energy consumption, and waste generation to make informed urban planning decisions. This includes optimizing public transportation routes, managing traffic congestion, and designing energy efficient buildings and districts (Qiao et al., 2023). Dikshit et al. (2023) contributed the understanding of reduction of carbon emissions in emerging economies by using AI to reduce commute time and increase effectiveness of resource deployment, improving overall air quality. Upadhyaya et al. (2025) added AI can also be employed to improve early warning systems for natural disasters by analyzing meteorological data and satellite imagery. This can protect lives and livelihood in vulnerable emerging economies by allowing time for early evacuation. Assessing damage and coordinating aid using AI can help in post-disaster recovery (Hanwacker, 2025).

12 Addressing Socioeconomic Dimensions and Ethical Considerations

Successful implementation of AI in emerging economies requires addressing the socioeconomic and ethical dimensions of SDGs which correlates with SDG 10 (Reduced Inequalities) and SDG 16 (Peace, Justice and Strong Institutions). One of the biggest limitations in the modern world is "digital divide." AI can be considered transformative if people across social strata have access to digital infrastructure, affordable internet, and digital literacy (Božić, 2023; Kaith & Sachdeva, 2024; Lutz, 2019). However, this raises concerns about data governance, data security, and ownership (Liang et al., 2022). In this regard, robust data governance frameworks are essential to protect individual rights and prevent misuse.

The next set of limitations includes algorithmic bias. AI algorithms can spread or intensify prevailing social prejudice unless they are designed and trained on diverse and representative datasets. Ensuring equality, accountability, and transparency in AI systems is essential to avoid aggravating inequalities in emerging economies. By removing inequalities and enhancing digital inclusivity, the AI will help in reskilling of humans and creation of new jobs. AI can only automate certain tasks, and humans have immense opportunities in areas like AI evolution data science and maintenance of AI systems. Investing in education and reskilling programs is vital to prepare the workforce in emerging economies for the AI-driven long-term horizon (Agrawal et al., 2026; George, 2024). For better results and optimum growth, the AI also needs to follow local contextualization. AI solutions must be tailored to the specific cultural, social, and economic contexts of emerging economies. For successful implementation and adoption of AI, optimal utilization of local expertise and community engagement is crucial.

The evidence available so far suggests that AI possesses varied powerful tools which can contribute in building more efficient, circular, and carbon–neutral economy. In context of emerging economies, these tools can accelerate the progress toward SDGs by optimizing resource deployment, enhancing clean energy systems, improving agriculture, and building sustainable cities. However, this potential is still underexplored and requires examinations of its limitations to ensure responsible, inclusive, and ethical deployment of AI.

13 Key Findings

Based on the analysis of AI's contribution to SDG, the following important observations were identified:

Strong connection between AI, SDGs and Circular economy: The analysis highlights the AI's connection to SDG and circular economy. The research explains how AI helps in waste management, improving reverse logistics, developing green energy, ensuring food security, and building sustainable cities.

Through these applications, AI aligns circular economy principles with the five-pillar framework of SDGs.

Limitations of emerging economies: Emerging economies face varied hurdles like fragmented communities, renewable energy limitations, digital divide, complex supply chains for reverse logistics, financial constraints, and many more. The effective implementation of AI can help emerging economies to move faster and catch the global pace of growth, fulfilling the idea of "leaving no one behind." By using advance AI technologies, emerging economies can centralize its decentralized ecosystem.

Carbon neutrality and AI: Implementing AI can facilitate the carbon neutrality goals, linking SDG 7 and SDG 13. The AI's role in this area is mainly to reduce the dependency on fossil fuels by generating, controlling, and optimizing clean energy to deliver it among all at affordable price.

Key determinants of success: Though AI offers differentiated benefits, its reliability can be tested through its ethical deployment, inclusive approach, and governance transparency. Various key risks like bias algorithms, deepening inequalities, high energy consumption by AI models and tools, digital divide, and privacy concerns need to be tackled to accomplish SDG targets using AI.

Need for Holistic framework: Existing ESG reporting standards are not efficient enough to reflect AI's ripple effect on SDG. For long-term impact and cross-border growth, a centralized and holistic framework is needed to test the AI's effect (positive or negative) toward the sustainable growth targets. It must be mixed with common risk mitigation strategies to deal with the common issues effectively.

14 Discussion and Implications

The data collected from various literary sources provide compelling evidence of AI's capability for achieving the SDG targets. This study underlines that adoption and continued improvement of AI can considerably push emerging economies. It yields them better sustainable benefits and helps them move toward circular economy. Implementation of circular economic principles will help emerging economies to achieve carbon neutrality goals. This study also highlights the need for key policies to reduce the potential barriers that may arise during its implementation.

AI's capabilities enhance with the quality of data it is trained on. Hence, AI can provide us with necessary intelligence to close material loops effectively. This paper explored how AI can sort valuable recyclable materials from waste using computer vision. It exponentially increases recycling purity and prevents recyclable waste from ending up in landfills, where it would generate methane (greenhouse gas). This highlights that AI's implementation can potentially reduce carbon emissions. This investigation mentions AI can improve reverse logistics supply chain by creating efficient systems for collecting and remanufacturing products. AI can also extend the lifespan of machineries by using predictive maintenance, which reduces the demand for new production. This simply links to carbon neutrality as better machinery health and reduced production demand result in lower carbon emissions. In emerging economies, this reduced energy consumption enhances grid stability and helps industries in lowering their carbon footprints. The research further explored how AI is beneficial in agriculture. This study suggests that we can use AI technologies to reduce the utilization of carbon-intensive fertilizers. It can reduce the harmful effects of excessive fertilizers, improving food quality and crop yield which contributes directly to SDG 2 (Food security). Minimizing the carbon-intensive fertilizers and food waste helps in moving ahead toward the carbon neutrality goals. AI offers immense capabilities, but there arises various barriers in between. The success of AI initiatives relies on its strength to be inclusive, transparent, and climate-resilient. Many nations lack robust digital infrastructure to facilitate advanced application of AI and tools. The digital divide and quality of data are cited as its biggest barriers, which further intensifies by financial and investment gaps. Lack of local expertise in emerging economies hinders the successful deployment of AI-driven circular solutions. It is crucial to bridge these gaps to build resilient and equitable, carbon–neutral, long-term AI solutions. To achieve it, investment in reskilling programs and tailoring AI solutions to local contexts are essential.

15 Conclusion

This study explores the supporting evidences for the AI's collaboration with Sustainable Evolution Goals. It is studied that AI is capable to enable progress across 134 SDG targets out of 169 total targets. It uncovers that AI, when aligned with circular economic principles, promotes attainment of SDGs. AI helps emerging economies to catch up the global pace, by moving faster than before toward the SDGs. Lu and Liao (2025) used carbon reduction as key measure to understand the benefits of AI. It is considered beneficial if for every 1% rise in AI application causes a decline of 0.0395% in carbon emissions. UNGASC (2024) talks about the current trajectory of SDG progress, where just 17% of the goals are on track. It necessitates urgent action to accelerate implementation of AI-driven circular economy solutions to realize carbon neutrality while promoting sustainable advancement.

The study further explores the relation of AI in optimizing resource flow, enhancing energy efficiency, promoting sustainability in food systems and agriculture, and developing sustainable cities while addressing socioeconomic and ethical concerns. Based on the assessment of a variety of literary works, the study found a strong symbiotic relationship between AI, SDGs, and circular economy; upliftment of emerging economies using AI in order to fulfill "leaving no one behind" condition; achieving carbon neutrality using AI; the need to evaluate success using key determinants, i.e., governance, transparency, and inclusivity of the AI models; and lastly, the need for holistic framework to explore the AI's capabilities and bottlenecks on global standards.

16 Directions for a Carbon–Neutral Future

AI is advancing at fast pace and so should the governing policies. AI's implications on SDGs can be maximized by uniting both the technological innovation and adaptive governance. Without holistic framework and governing policies, AI may broaden the inclusivity, ethical, and privacy concerns. Success can be captured when AI moves fast and the ethical arrangements and advancements of AI moves faster. Along with this, building human and institutional capacity is crucial to leverage the AI's capabilities toward sustainable growth. Emerging economies need to upskill human resource while improving their existing system. Due to rapidly advancing technologies, there arises a need for a flexible governance framework to safeguard social and environmental priorities, ensuring inclusive and sustainable growth. Researchers are expected to examine the long-term impacts of AI across varied cultures, economies, and institutions. They can formalize optimal strategies for integrating AI into existing structures. Importantly, the future research is expected to explore the implementation of AI so as to advance inclusivity and sustainable growth without intensifying the existing inequalities or creating new limitations. The publications explored during this investigation provides evidence that AI can act as a powerful accelerator for achieving SDG targets if paired with appropriate governance and implementation strategies.

References

Aarif, K. O., Mohammed, Alam, A., & Hotak, Y. (2025). Smart sensor technologies shaping the future of precision agriculture: Recent advances and future outlooks. *Journal of Sensors, 2025*(1), Article 2460098.

Agrawal, D., & Kumar, K. (2024). Achieving SDGs through MSMEs: An empirical assessment of environmental consideration initiatives in India. In *Green transition impacts on the economy, society, and environment* (pp. 358–376). IGI Global.

Agrawal, D., Kumar, K., & Rahman, O. (2026). ESG investment strategy. In V. Ratten (Ed.) *International encyclopedia of business management* (pp. 302–308). Academic Press. https://doi.org/10.1016/B978-0-443-13701-3.00300-5

Agupugo, C. P., Barrie, I., Makai, C. C., & Alaka, E. (2024). AI learning-driven optimization of microgrid systems for rural electrification and economic empowerment. *Engineering Science & Technology Journal, 5*(9), 2835–2851.

Alharbi, A., & Stevenson, M. (2020). Refining Boolean queries to identify relevant studies for systematic review updates. *Journal of the American Medical Informatics Association, 27*(11), 1658–1666.

Ayeni, O. (2025). Integration of Artificial Intelligence in predictive maintenance for mechanical and industrial engineering. *International Research Journal of Modernization in Engineering Technology and Science, 7*(3), 1–23.

Bajwa, H. S., Kumar, R., & Kaur, S. (2024). AI-driven building management systems for energy optimization. *Smart Cities Technology, 9*(2), 89–106.

Bexell, M., & Jönsson, K. (2017). Responsibility and the United Nations' sustainable development goals. *Forum for Development Studies, 44*(1), 13–29.

Božić, V. (2023). Artificial intelligence as the reason and the solution of digital divide. *Language Education and Technology, 3*(2).

Crawford, K. (2021). *Atlas of AI: Power, politics and the planetary costs of artificial intelligence.* Yale University Press.

Daga, S., Kumar, K., Yadav, K., & Singh, P. (2025). Financing and investment in a circular economy: Paving the way. *Blending Approaches Towards Sustainability Focus*, 17.

Deep, G., & Verma, J. (2023). Embracing the future: AI and ML transforming urban environments in smart cities. *Journal of Artificial Intelligence, 5*, 57.

Dikshit, S., Atiq, A., Shahid, M., Dwivedi, V., & Thusu, A. (2023). The application of artificial intelligence to optimize the routing of vehicles and curtail traffic congestion in urban areas. *EAI Endorsed Transactions on Energy Web, 10*, 1–13.

Dsilva, J., Francis, G., & Pillai, R. (2025). Sustainability: A global perspective. *Zero Carbon Industry, Eco-Innovation and Environmental Sustainability*, 123.

Ecker, O., Breisinger, C., & Al-Riffai, P. (2020). Data quality challenges in SDG monitoring. *Development Policy Review, 38*(2), 234–251.

Erkinay Ozdemir, M., Ali, Z., Subeshan, B., & Asmatulu, E. (2021). Applying machine learning approach in recycling. *Journal of Material Cycles and Waste Management, 23*(3), 855–871.

Fan, Z., Yan, Z., & Wen, S. (2023). Deep learning and artificial intelligence in sustainability: A review of SDGs, renewable energy and environmental health. *Sustainability, 15*(18), Article 13493.

George, A. S. (2024). Reimagining India's engineering education for an AI-driven future. *Partners Universal International Research Journal, 3*(3), 1–20.

Hanwacker, L. S. (2025). The role of artificial intelligence in disaster recovery. *Journal of Business Continuity & Emergency Planning, 18*(2), 167–179.

Haque, R., Bajwa, A., Siddiqui, N. A., & Ahmed, I. (2024). Predictive maintenance in industrial automation: A systematic review of IOT sensor technologies and AI algorithms. *American Journal of Interdisciplinary Studies, 5*(1), 1–30.

International Energy Agency. (2020). *AI and energy: Implications for energy security and climate* (IEA Report).

Jabbour, C. J. C., Jabbour, A. B. L. S., & Foropon, C. (2020). AI and circular economy: A systematic review. *Journal of Cleaner Production, 268*, Article 122186.

Kaith, A., & Sachdeva, G. (2024). Bibliometric insights into green human resource management: A connection with business ethics and values. *PURUSHARTHA-A journal of Management, Ethics and Spirituality, 17*(1), 51–63. https://doi.org/10.21844/16202117104

Kathpal, S., Kumar, K., Kumar, P., & Mishra, S. (2021). Building a low carbon Indian economy. *IOP Conference Series: Earth and Environmental Science, 795*(1), Article 012023.

Kitchenham, B., & Charters, S. (2007). *Guidelines for performing systematic literature reviews in software engineering* (Technical Report EBSE 2007-001). Keele University and Durham University.

Kroll, C., Warcoin, L., & Pradhan, P. (2019). Sustainable Development Goals (SDGs): Are we successful in turning trade-offs into synergies? *Palgrave Communications, 5*(1), 1–11.

Kulkov, I., Kulkova, J., Rohrbeck, R., Menvielle, L., Kaartemo, V., & Makkonen, H. (2024). Artificial intelligence-driven sustainable development: Examining organizational, technical and processing approaches to achieving worldwide goals. *Sustainable Advancement, 32*(3), 2253–2267.

Kumar, K. (2022). The uptake of sustainability disclosure practices by environmentally polluting industries in India. In *The equal pillars of sustainability* (pp. 199–224). Emerald Publishing Limited.

Kumar, K., Kumari, R., & Kumar, R. (2021a). The state of corporate sustainability reporting in India: Evidence from environmentally sensitive industries. *Business and Society Review, 126*(4), 513–538.

Kumar, K., Singh, R., Sharma, A., & Lehri, B. P. S. (2024). Climate change adaptation strategies for achieving net-zero economy. In *Transition towards a sustainable future: Net zero policies and environmental sustainability* (pp. 163–183). Springer Nature Singapore.

Kumar, P., Kumar, K., Mathur, S., & Singh, M. (2021). Aligning developmental programs to implement SDGs in India: Limitations and the way forward. *Journal of Contemporary Issues in Business & Government, 27*(6).

Lacy, P., & Rutqvist, J. (2015). *Waste to wealth: The circular economy advantage* (Vol. 91). Palgrave Macmillan.

Liang, W., Tadesse, G. A., Ho, D., Fei-Fei, L., Zaharia, M., Zhang, C., & Zou, J. (2022). Advances, hurdles and opportunities in creating data for trustworthy AI. *Nature Machine Intelligence, 4*(8), 669–677.

Lipu, M. S. H., Miah, M. S., Hannan, M. A., Hussain, A., Sarker, M. R., Ayob, A., Saad, M. H. M., & Mahmud, M. S. (2021). Artificial intelligence-based hybrid forecasting approaches for wind power generation: Progress, hurdles and prospects. *IEEE Access, 9*, 102460–102489.

Lu, Y., & Liao, Z. (2025). The influence of AI application on carbon emission intensity of industrial enterprises in China. *Scientific Reports, 15*(1), Article 12585.

Lutz, C. (2019). Digital inequalities in the age of artificial intelligence and big data. *Human Behavior and Emerging Technologies, 1*(2), 141–148.

Macarthur, E. L. L. E. N., & Heading, H. E. A. D. I. N. G. (2019). How the circular economy tackles climate change. *Ellen MacArthur Found, 1*(3), 1–71.

Masanpally, M., Bhattacharjee, T., & Butt, T. E. (2025). Sustainable development goals and the implications of business responsibility. *Environmental Progress & Sustainable Energy, 44*(1), Article e14517.

Meitei, A. J., Rai, P., & Rajkishan, S. S. (2025). Application of AI/ML techniques in achieving SDGs: A bibliometric study. *Environment, Development and Sustainability, 27*(1), 281–317.

Meneses Cerón, L. Á., van Klyton, A., Rojas, A., & Muñoz, J. (2024). Climate risk and its effect on the cost of capital—A systematic literature review. *Sustainability, 16*(23), Article 10727.

Mishra, P., & Kumar, K. (2024). Uncovering the sustainability reporting: Bibliometric analysis and future research directions. *International Journal of Disclosure and Governance, 21*(4), 520–542.

Mishra, P., Sant, T. G., & Kumar, K. (2023). Carbon disclosure and organization performance: A literature review. *Sustainability and Climate Change, 16*(4), 302–317.

Morseletto, P. (2020). Restorative and regenerative: Exploring the concepts in the circular economy. *Journal of Industrial Ecology, 24*(4), 763–773.

Ohalete, N. C., Aderibigbe, A. O., Ani, E. C., Ohenhen, P. E., Daraojimba, D. O., & Odulaja, B. A. (2023). AI-driven solutions in renewable energy: A review of data science applications in solar and wind energy optimization. *World Journal of Advanced Research and Reviews, 20*(3), 401–417.

Olawale, R. A., Olawumi, M. A., & Oladapo, B. I. (2025). Sustainable farming with machine learning solutions for minimizing food waste. *Journal of Stored Products Research, 112*, Article 102611.

Oyenuga, M. O., Singh, R., Apata, S. B., Khan, S., & Kumar, V. (2025). Net zero technology and smart agriculture. In *Smart technologies for climate change and net zero policies: Practical approaches towards sustainability* (pp. 339–350). Springer Nature Switzerland.

Qiao, P., Zhou, Y., & Wang, Q. (2023). Simulation and application of low carbon heating buildings based on genetic algorithm optimization PID. In *2023 International Conference on Integrated Intelligence and Communication Systems (ICIICS)* (pp. 1–5). IEEE.

Raj, A., Dwivedi, G., & Sharma, A. (2020). Artificial intelligence for circular economy implementation. *Resources, Conservation and Recycling, 162*, Article 105046.

Rao, B. N., Praveen, M., & Babu, D. R. (2024). A review on the contribution of AI in optimizing renewable energy grid management. *International Journal of Scientific Research in Engineering and Management, 8*(11), 1–13.

Sætra, H. S. (2021). A framework for evaluating and disclosing the ESG related impacts of AI with the SDGs. *Sustainability, 13*(15), Article 8503.

Said, Y., & Alanazi, A. (2023). AI-based solar energy forecasting for smart grid integration. *Neural Computing and Applications, 35*(11), 8625–8634.
Sakubu, D. (2025). Challenges of artificial intelligence today and future implications for society and the world. *World Journal of Advanced Research and Reviews.*
Sharma, A., Chaudhary, M., Kumar, K., & Daga, S. (2024). Promoting net-zero economy for sustainable development: Practice-based view. In *Net zero economy, corporate social responsibility and sustainable value creation* (pp. 101–116). Springer Nature Switzerland.
Sharma, A., Kumar, K., & Dwesar, R. (2025). "Keenly aware of environmental issues, yet wavering in conviction": Understanding the factors influencing the consumers' willingness to pay for green energy. *Sustainable Futures, 9*, Article 100427.
Singh, R. (2024). Artificial intelligence technologies for sustainable development. *Technology for Development Quarterly, 12*(3), 45–62.
Singh, R., Kumar, K., & Khan, S. (2024). A comprehensive view of artificial intelligence (AI)–based technologies for sustainable evolution goals (SDGs). *Artificial Intelligence Enabled Management: An Emerging Economy Perspective, 183*, 9783111172408–012.
Singh, R., Mushtaq, S., Kumar, K., & Rahman, O. (2025). Creating sustainable workplace: Enhancing HRM strategies in the digital circular economy. In *Sustainable innovations and digital circular economy* (pp. 129–145). Springer Nature Singapore.
Singh, R., Sharma, A., & Kumar, K. (2025). Navigating carbon neutrality: Strategies for sustainable forthcoming period and environmental challenges. In *Climate neutrality through smart eco-innovation and environmental sustainability* (pp. 69–80). Springer Nature Switzerland.
Solaja, O. M. (2024). Unleashing the power of AI: Revolutionizing plastic waste management for sustainable development in developing nations. *Waste Technology, 12*(1), 28–38.
Spirito, C. (2024). *Artificial intelligence applications in reverse logistics, how technology could improve return and waste management creating value* [Doctoral dissertation, Politecnico di Torino].
Srivastava, S. K., Anshul, A., Pathak, P., & Bansal, J. (2018). Responsible consumption for curbing food wastage: An exploratory enquiry. *PURUSHARTHA-A Journal of Management, Ethics and Spirituality, 11*(2), 104–116.
Srivastava, S. K., Singh, S., & Srivastava, M. (2021). Food waste at household and social gatherings: Drivers and possible remedies. *PURUSHARTHA-A Journal of Management, Ethics and Spirituality, 14*(2), 117–133.
Srivastava, S. K., Sinha, A. K., & Kumar, S. (2024). Workplace spirituality empowered by spiritual leadership to promote organizational effectiveness: A bibliometric analysis. *PURUSHARTHA-A journal of Management, Ethics and Spirituality, 17*(1), 1–22. https://doi.org/10.21844/16202117101
Strubell, E., Ganesh, A., & McCallum, A. (2019). Energy and policy considerations for deep learning in NLP. In *Proceedings of the 57th annual meeting of the association for computational linguistics*, pp. 3645–3650.
Tarkar, P., & Kumar, K. (2024). Sustainable human resource management and sustainable performance: Mediating role of organization citizenship behavior. *The Bottom Line.*
Tautiva, J. A. D., Barros-Celume, S., Lecuna, A., & Barragan-Rodriguez, C. (2025). Navigating climate pressures: A bibliometric and systematic literature review of climate change strategies in small and medium-sized enterprises. *Heliyon, 11*(13).
Tripathi, V., Singh, K., Parihar, S., & Srivastava, S. K. (2022). A bibliometric analysis on supply chain collaboration. *PURUSHARTHA-A journal of Management, Ethics and Spirituality*, 15(2), 66–103. https://doi.org/10.21844/16202115205
Tutore, I., Parmentola, A., Di Fiore, M. C., & Calza, F. (2024). A conceptual model of artificial intelligence effects on circular economy actions. *Corporate Social Responsibility and Environmental Management, 31*(5), 4772–4782.
Ukoba, K., Olatunji, K. O., Adeoye, E., Jen, T. C., & Madyira, D. M. (2024). Optimizing renewable energy systems through artificial intelligence: Review and long-term horizon prospects. *Energy & Environment, 35*(7), 3833–3879.

UNCTAD, I. (2019). Guidance on Core indicators for entity reporting on the contribution towards the attainment of the Sustainable Development Goals. https://unctad.org/en/PublicationsLibrary/diae2019d1_en.pdf

United Nations General Assembly Second Committee. (2024, October 10). *World on track to attain only 17 per cent of 2030 agenda targets, speakers warn, as Second Committee takes up sustainable advancement reports* (UN Press Release GAEF/3604). https://press.un.org/en/2024/gaef3604.doc.htm

Upadhyaya, A. N., Debnath, P., Rani, P., Selvakumar, P., Manjunath, T. C., & Bhattacharya, S. (2025). AI-based disaster prediction and early warning systems. In *AI and emerging technologies for emergency response and smart cities* (pp. 55–74). IGI Global.

Vinuesa, R., Nerini, H. F., Fathallah, H., Diepeveen, K., Battiston, S., Crespi, F., Caliste, D., et al. (2020). The contribution of artificial intelligence in achieving the Sustainable Progress Goals. *Nature Communications, 11*(1), Article 233.

Visvizi, A. (2022). Artificial intelligence (AI) and sustainable development goals (SDGs): Exploring the impact of AI on politics and society. *Sustainability, 14*(3), Article 1730.

World Economic Forum. (2018). *The future of jobs report 2018.*

Xia, H., Han, J., & Milisavljevic-Syed, J. (2023). Predictive modeling for the quantity of recycled end-of-life products using optimized ensemble learners. *Resources, Conservation and Recycling, 197*, Article 107073.

Yadav, P., Tudela, L. A. M., & Marco-Lajara, B. (2024). The contribution of AI in assessing and achieving the sustainable growth goals (SDGs). In *Issues of sustainability in AI and new-age thematic investing* (pp. 1–17). IGI Global.

Zhou, N., Fridley, D., McNeil, M., Zheng, N., Ke, J., & Levine, M. (2021). *China's energy and carbon emissions outlook to 2050* (Technical Report). Lawrence Berkeley National Laboratory.

Chapter 10
The Role of Technology and Innovation in Carbon Accounting and Climate Policy: A Bibliometric Mapping of Knowledge Evolution and Structural Fragmentation

M. I. M. Riyath and Sushma Maurya

Abstract This bibliometric analysis maps technology and innovation research in carbon accounting and climate policy using 387 Scopus-indexed documents (2000–2025). Co-citation network analysis, keyword co-occurrence mapping, and thematic positioning characterize the field's intellectual structure, connectivity patterns, and topical evolution. Temporal patterns separate into three descriptively distinct periods: formative (2000–2014, 33 documents), consolidation (2015–2020, 133 documents), and expansion (2021–2025, 221 documents). Lower citation density in recent cohorts is consistent with topical diffusion and citation-window truncation effects rather than quality variation. Co-citation analysis under the specified network settings identifies carbon market economics as a densely connected hub (link strengths 104–125), whereas life cycle assessment, agricultural systems, and policy frameworks appear as weakly connected clusters; this indicates limited co-citation linkage in the corpus and does not imply relative substantive importance. Source concentration shows the Journal of Cleaner Production accounting for 49.6% of publications, positioning environmental and technology outlets as primary venues. Thematic mapping places "climate change" as a motor theme, with "carbon emissions" and "energy efficiency" in niche positions; placement reflects co-word structure under the keyword specification used. Overlay visualization highlights terms with later average publication years, indicating recency of usage rather than intrinsic importance. Overall, the findings provide descriptive evidence of knowledge structure within the analytic corpus rather than causal claims about governance effectiveness, subject to limitations typical of single-database bibliometric designs.

M. I. M. Riyath
Department of Accountancy and Finance, Faculty of Management and Commerce, South Eastern University of Sri Lanka, Oluvil, Sri Lanka
e-mail: riyath@seu.ac.lk

S. Maurya (✉)
Department of Commerce, Janki Devi Memorial College, University of Delhi, Delhi, India
e-mail: sushma@jdm.du.ac.in

K. Kumar et al. (eds.), *AI-Driven Circular Economy for Carbon Neutrality in Emerging Economies*, Environmental Footprints and Eco-design of Products and Processes,
https://doi.org/10.1007/978-3-032-30298-4_10

Keywords Bibliometric analysis · Carbon accounting · Climate technology · Climate governance · Knowledge structure · Research evolution

1 Introduction

Carbon accounting remains difficult to operationalize at scale because reported emissions are often incomplete and weakly comparable across firms and reporting regimes. The problem is most visible for Scope 3 value-chain emissions, where coverage is uneven and methodological choices materially shift reported totals (Hettler & Graf-Vlachy, 2024; Klaaßen & Stoll, 2021). For preparers, data availability and method selection remain binding constraints rather than marginal refinements (World Resources Institute, 2022). In parallel, measurement capacity has expanded through digital platforms and computational methods, including blockchain-enabled traceability, satellite monitoring, and AI/NLP applications that reduce collection costs and increase temporal granularity (Alonso-Robisco et al., 2025; Oladeji & Mousavi, 2023; Perdana & Tan, 2024; Zhang, 2025). Policy architectures have also broadened the scope of carbon governance, including economy-wide commitments under the Paris Agreement (Seo, 2017). These developments expand the feasible set of monitoring and reporting practices, but they do not by themselves resolve comparability, verification, and integration problems that determine whether carbon information is decision-usable.

The literature linking technology and innovation to carbon accounting and climate governance has expanded rapidly, but growth does not imply cumulative integration. Contributions on measurement methods, carbon market design, corporate governance and disclosure, and sectoral mitigation often develop within partially separate communities, with limited cross-referencing and limited conceptual overlap. This separation matters for both scholarship and practice. It constrains theory accumulation when adjacent streams adopt different vocabularies, units of analysis, and validation standards. It also complicates translation from measurement advances to governance outcomes when accounting methodologies, verification routines, and policy instruments are discussed in parallel rather than in coupled frameworks. Claims of "fragmentation" are common in this domain, yet they are frequently asserted rather than measured. What is missing is an evidence-based description of how the literature is structured, where connectivity is strong or weak, and how that structure has shifted as the corpus has expanded. Measuring this structural separation requires systematic mapping of citation and co-word networks across the corpus.

Narrative reviews are poorly matched to this task because the corpus is interdisciplinary, large, and heterogeneous in methods and outlets. A field-spanning bibliometric approach provides a tractable alternative. Bibliometric mapping does not adjudicate the substantive correctness of individual measurement approaches or policies; it describes knowledge structure using observable linkages in the scholarly record. In this setting, fragmentation can be operationalized as weak connectivity across clusters in co-citation networks, separation of topical areas in co-word and thematic maps, and limited overlap between governance-oriented and operational-methodology vocabularies. These diagnostics are particularly useful when methodological advances (e.g., remote sensing, AI-based estimation) diffuse quickly but governance uptake depends on standards, verification practices, and institutional design that may be anchored in different research communities.

This study addresses four related questions. First, how has the intellectual structure of technology and innovation research in carbon accounting and climate policy evolved over 2000–2025? Second, where are the principal discontinuities between technological innovation streams and carbon governance streams, and how stable are these discontinuities over time? Third, what do diffusion and collaboration patterns (authors and institutions) imply about how knowledge travels within the domain? Fourth, which themes appear central, which are specialized, and which are declining, and what does this thematic organization imply for integration priorities? We analyze 387 Scopus-indexed documents from 131 sources using co-citation network analysis, keyword co-occurrence mapping, thematic mapping, temporal trend diagnostics, and collaboration network analysis. Procedures are reported to support transparency and replication, including parameter choices used to construct networks and thematic structures.

The results provide a structured account of both continuity and separation in the knowledge base. The temporal profile distinguishes periods in which influence is concentrated in a smaller set of contributions and later periods in which output expands and citations per document are lower, a pattern consistent with topical diffusion and entry of applied work that accumulates citations more slowly (Shackleton et al., 2023; Thanasi-Boçe & Hoxha, 2025; Yang, 2025). Network and thematic maps locate carbon market economics as a central organizing area, while several methodological and governance-related strands occupy more weakly connected positions in the co-citation space. The thematic map places Climate Change as a dominant theme while indicating specialization in Carbon Emissions and Energy Efficiency and a declining Carbon Footprint theme. Keyword networks further indicate emerging fronts including renewable energy optimization, AI-enabled emissions modeling, and verification of carbon sequestration. These patterns are descriptive: they identify where cross-stream integration is limited within the scholarly record rather than establishing that any stream is substantively superior.

The paper contributes in two ways. Substantively, it replaces broad claims about fragmentation with measurable diagnostics of connectivity, thematic organization, and diffusion in the literature linking technology and innovation to carbon accounting and climate governance. Methodologically, it provides a field map that can be used as a baseline for targeted synthesis, including research that explicitly couples policy architectures and standards with implementable accounting, verification, and interoperability routines. As with bibliometric studies generally, the design inherits limitations related to database coverage and citation-age effects; we address these constraints through transparent reporting of data construction and mapping parameters and interpret the evidence as knowledge-structure diagnostics rather than causal claims about governance outcomes.

2 Methodology

We constructed a Scopus-based bibliometric dataset using a structured search and screening protocol consistent with PRISMA-style reporting (explicit identification, screening, and eligibility decisions). Scopus was selected because it indexes a wide range of peer-reviewed outlets spanning accounting, management, and adjacent sustainability domains. The search strategy was designed to jointly capture five elements: (i) carbon accounting activities, (ii) climate policy and governance terms, (iii) technology and innovation terms, (iv) sustainability and emissions constructs, and (v) analytics and monitoring language. The exact search string was as follows:

> TITLE-ABS-KEY (carbon AND (account* OR measur* OR track* OR report* OR manag*)) AND TITLE-ABS-KEY (climat* AND (polic* OR strateg* OR govern* OR regulat*)) AND TITLE-ABS-KEY (technolog* OR innovat* OR digital* OR platform* OR system* OR "artificial intelligence" OR "machine learning") AND TITLE-ABS-KEY (sustainab* OR environment* OR "greenhouse gas*" OR emission* OR "carbon footprint*" OR ESG OR "net zero") AND TITLE-ABS-KEY ("data analy*" OR analytic* OR monitor* OR assess* OR metric* OR dashboard*).

The query was executed on July 8, 2025, for the period 2000–2025, and returned 6460 records. To concentrate on disciplinary relevance, we restricted subject areas to Business, Management and Accounting, yielding 442 documents (174 sources; 1534 authors). We excluded non-English items ($n = 6$). Eligibility screening then removed 49 records that did not meet inclusion criteria (conference proceedings and non-article formats such as editorials, commentaries, and notes). The resulting analytic corpus comprised 387 documents. Identification, screening, and exclusion counts are reported in the PRISMA-style flow summary and are consistent with the arithmetic $442 - 6 - 49 = 387$.

Analyses were conducted in R using bibliometrix (Aria & Cuccurullo, 2017) and in VOSviewer (Van Eck & Waltman, 2010). We apply four complementary procedures. First, co-citation analysis characterizes the intellectual base by identifying groups of references that are cited together. Second, keyword co-occurrence mapping describes topical structure by estimating which terms co-appear across documents. Third, temporal trend diagnostics summarize changes in output and citation accumulation over time. Fourth, co-authorship network analysis describes collaboration patterns across authors and countries. Where citation comparisons span publication years, we report year-adjusted indicators (e.g., citations-per-year variants alongside total citations) to reduce mechanical advantages for older publications and avoid interpreting raw citation totals as directly comparable across cohorts. The following sections present results in two stages. Section 3 establishes temporal, source, and author patterns using standard bibliometric indicators (output volume, citation counts, collaboration metrics, and source concentration). Section 4 maps intellectual structure through co-citation networks, keyword co-occurrence, and thematic positioning, revealing where the corpus is internally connected and where integration remains limited.

Table 1 reports descriptive characteristics of the final dataset. The corpus contains 387 documents published in 131 sources (2000–2025), with an annual growth rate of

Table 1 Descriptive summary of bibliometric dataset (2000–2025)

Category	Description	Results
Timespan	Publication years	2000–2025
Sources	Journals, books, etc	131
Documents	Total publications	387
Annual growth rate (%)	Average annual increase	15.9%
Average document age	Mean age of documents (years)	4.79
Average citations per doc	Mean citations per document	31.55
Total references	References across all documents	23,925
Keywords plus (ID)	Indexed keywords	2348
Author keywords (DE)	Author-assigned keywords	1464
Authors	Total contributing authors	1387
Single-authored Docs	Documents with one author	45
Authors of single-authored docs	Unique authors of solo publications	44
Co-authors per doc	Mean number of authors per doc	3.9
International co-authorship (%)	Cross-country collaborations	38.5%
Document types	Articles	309
	Book chapters	53
	Reviews	25

Note DE = Descriptors provided by authors; ID = Keywords Plus, algorithmically generated by indexing services

15.9%. Average document age is 4.79 years, and the mean citation count is 31.55 per document. Document types comprise 309 articles (79.8%), 53 book chapters (13.7%), and 25 reviews (6.5%), supported by 23,925 cited references, 1464 author keywords, and 2348 keywords plus identifiers. Authorship is distributed across 1387 authors, including 45 single-authored documents (11.6%). The mean number of co-authors per document is 3.9, and the international co-authorship rate is 38.5%.

3 Performance Analysis

3.1 Temporal Publication and Citation Dynamics

Figure 1 reports annual publication counts and citation accumulation for the study corpus over 2000–2025 (citations recorded as of the extraction date). The pattern is descriptively separable into three periods. The formative period (2000–2014) is small in volume (33 documents) but includes several early contributions that remain highly cited in the dataset (e.g., Becken & Patterson, 2006; Berg & Lindholm, 2005). Papers such as Tseng et al. (2013) (406 citations) and Jones et al. (2013) (321 citations) function as citation anchors. Because these studies entered the literature earlier, their total citations reflect both influence and longer exposure time; accordingly, their prominence is interpreted as early structural salience within the corpus rather than evidence of superior substantive validity. The subsequent period (2015–2020) combines higher output with concentrated citation accumulation. In the late-2010s, a relatively small number of publications accounts for a large share of total citations attributed to those cohorts. For example, Fernández Fernández et al. (2018) (477 citations) are a major contributor to the citation totals of its publication year. From 2021 to 2025, publication volume rises sharply (221 documents; 57.1% of the corpus). Observed citations per document are lower for these recent cohorts and are spread across a wider set of papers (e.g., Chen et al., 2022; Huovila et al., 2022). This pattern is reported as a descriptive shift in citation density rather than a tested "inverse correlation." Two non-exclusive mechanisms are plausible: topical diffusion into more specialized and application-oriented subareas, and citation-window truncation for recent publications—particularly in 2024–2025—which mechanically depresses observed citation totals. Subsequent network and thematic analyses therefore focus on where influence concentrates and where topical separation persists as the corpus expands.

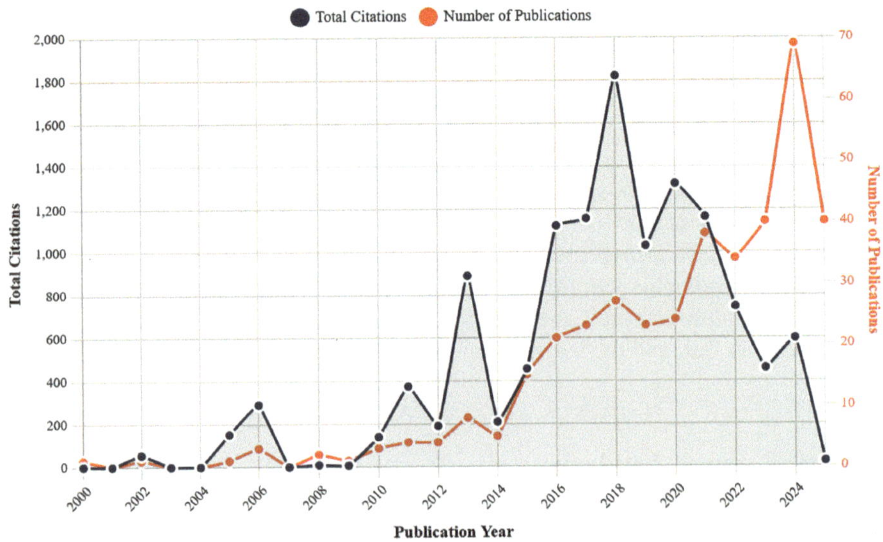

Fig. 1 Temporal citation impact in carbon accounting and climate governance research (2000–2025). Annual publications (right axis) and total citations to papers published in each year (left axis), as of 8 July 2025

3.2 Source Impact and Publication Concentration

Temporal expansion occurred through concentrated outlet structures. Table 2 reports source-level concentration and impact indicators. The outlet structure is highly skewed. Journal of Cleaner Production (JCP) accounts for 192 of 387 documents (49.6%) and 8757 of 23,543 citations (37.2%), implying 45.6 citations per article within the dataset. JCP also has the highest within-source indices ($h = 55$; $g = 86$; $m = 2.29$), consistent with its role as the primary publication venue for this literature. This concentration is substantively relevant because the field's core publication infrastructure is located mainly in environmental and technology-facing journals rather than traditional accounting outlets, which may influence framing and terminology. To summarize cross-outlet heterogeneity, Table 2 groups sources descriptively by volume, tenure, and citation indices. JCP is the sole "dominant" outlet and shows sustained activity from 2002 to 2025; its g/h ratio (1.56) indicates citations are spread across a relatively broad set of JCP papers rather than concentrated in a small number of outliers. In the recent period (2020–2025), JCP adds 98 publications relative to its pre-2020 baseline of 94 (+ 104%), reinforcing its centrality. A second set of outlets shows growth in the later period, including Technological Forecasting and Social Change (TFSC) and Business Strategy and the Environment (BSE). TFSC adds nine publications in 2020–2025 relative to its pre-2020 baseline (+ 180%), and BSE adds six (+ 200%). These patterns indicate selective diffusion into strategy and innovation venues while overall concentration remains anchored in JCP. Other sources display mixed trajectories. Sustainability Accounting, Management and Policy Journal shows no recent growth in this dataset ($m = 0.36$). Socio-Economic Planning Sciences enters recently and records $h = 3$ ($m = 0.75$) over a short window, reflecting uptake within a limited time span. Overall, the seven most productive outlets contribute 62.3% of publications while representing 5.3% of sources (7 of 131), indicating that field visibility is channeled through a small set of journals with partial spillover into adjacent communities.

3.3 Author-Level Bibliometric Insights

Within these outlets, citation intensity varies sharply across authors. Table 3 summarizes author-level patterns in citation intensity, tenure, and co-authorship structure in the indexed corpus. Citation intensity varies across the listed authors. WANG Y records the highest citations-per-publication (TC/NP = 78.5) and an h-index of 5 within the dataset, with an increasing citation trend over the observation window. GÖSSLING S shows a similar citations-per-publication profile (TC/NP = 75.7) with a stable trend, despite fewer publications in the corpus. Higher-output profiles exhibit lower TC/NP values, including WANG X (17.2) and ZHANG Y (11.7). This pattern is consistent with two non-exclusive factors affecting citation accumulation within a fixed window: (i) recency effects, because more recent publications have had less time

Table 2 Source-level bibliometric indicators and publication growth in carbon accounting research (2000–2025)

Source	*h*-index	*g*-index	*m*-index	TC	*N*	Start year	Timespan (yrs)	Recent growth (2020–2025)	Tier
Journal of Cleaner Production	55	86	2.29	8757	192	2002	24	98 (+ 104%)	Dominant
Technological Forecasting and Social Change	12	14	1.09	490	14	2015	11	9 (+ 180%)	Emerging
Business Strategy and the Environment	6	9	0.67	479	9	2017	9	6 (+ 200%)	Emerging
Clean Technologies and Environmental Policy	5	9	0.31	151	9	2010	16	7 (+ 350%)	Specialized
Journal of Sustainable Tourism	7	8	0.35	604	8	2006	20	4 (+ 100%)	Specialized
Sustainability Accounting, Management and Policy Journal	5	5	0.36	114	5	2012	14	0 (0%)	Niche
Socio-Economic Planning Sciences	3	4	0.75	52	4	2022	4	4 (New)	Emerging

Note TC = Total Citations; n = Number of Publications; Timespan = Years since first publication in dataset. Recent growth reflects number of publications from 2020 to 2025 with percentage increase from the 2020 baseline. Tier classification: Dominant (high overall impact), Emerging (recent and rapid growth), Specialized (consistent moderate impact), Niche (limited growth or theoretical stagnation). Source selection required $\geq$ 4 publications and $\geq$ 3 years of contribution to accrue citations; and (ii) topical dispersion into applied and policy-adjacent areas where citations may accrue more slowly. For authors with missing index values (e.g., h- or m-index), these omissions reflect information constraints within the dataset and the observation window rather than an inference about influence.

Co-authorship is a dominant structural feature in the listed profiles. Fractionalized authorship credit (FA)—the sum of fractional contributions across an author's NP papers—is materially below NP for most authors, indicating that output is primarily produced in multi-author teams under fractional counting. Co-authorship reliance (CR = 1 − FA/NP) is high for WANG Y (0.838), WANG X (0.835), ZHANG Y (0.808), ZHANG X (0.800), and BASTIANONI S (0.830). GÖSSLING S shows lower reliance by the same measure (CR = 0.639), implying a higher average fractional share per paper in this corpus. These indicators describe publication structure in the indexed corpus and do not support claims about attribution beyond what fractional counting captures. Tenure and recency fields differentiate longer-standing participation from newer entry within the observation window. ZHANG X shows the longest active span in the dataset, whereas WANG X and ZHANG Y concentrate activity in the post-2020 period. Taken together, Table 3 indicates (i) concentrated citation intensity among a small subset of contributors and (ii) a strongly team-produced publication structure, alongside newer higher-output profiles whose citation trajectories remain developing within the available time window.

3.4 *Global Versus Local Citation Impact in Carbon Accounting*

These collaboration patterns do not fully explain within-corpus versus global citation uptake. Table 4 contrasts global citation impact (total citations and citations per year) with local citation uptake, defined as citations received within the study corpus (i.e., citations from the 387 documents). The comparison separates broad visibility in the wider literature from incorporation into the corpus's internal citation architecture. The distinction is descriptive: A low or zero local count does not imply low quality; it indicates limited within-corpus uptake given the corpus boundary and the citation window. Several highly cited publications show strong global diffusion but limited within-corpus uptake. Fernández Fernández et al. (2018) record 477 total citations (59.6 citations/year; normalized TC = 7.07), and Tseng et al. (2013) record 406 total citations (31.2 citations/year; normalized TC = 3.65); both receive 0 local citations in the corpus (local share = 0.00%). Under Table 4 definitions, this pattern indicates broad diffusion across adjacent sustainability and environmental literatures without routine citation within the focal technology–carbon accounting–climate governance corpus captured by the search strategy.

By contrast, some publications show non-zero local uptake relative to their global footprint. Alsaifi et al. (2020) receive 155 total citations (25.8 citations/year; normalized TC = 2.82) and three local citations (local share = 1.94%). Lewandowski (2017)

Table 3 Author impact and co-authorship structure

Author	NP (Articles)	FA (Articles fractionalized)	CR = 1 − FA/NP	CI = FA − NP	*h*-index	TC/NP	*m*-index	*g*/*h*	Start year	Recent activity	Career span (yrs)	Citation trend
WANG Y	6	0.970	0.838	−5.030	5	78.5	0.50	1.20	2016	2024	9	Increasing
ZHANG X	5	1.000	0.800	−4.000	3	64.0	0.27	1.67	2015	2025	11	Stable
WANG X	6	0.992	0.835	−5.008	3	17.2	0.60	2.00	2021	2025	5	Increasing
ZHANG Y	6	1.152	0.808	−4.848	–	11.7	–	–	2021	2025	5	Increasing
GÖSSLING S	3	1.083	0.639	−1.917	3	75.7	0.30	1.00	2016	2024	9	Stable
BASTIANONI S	3	0.510	0.830	−2.490	3	52.0	0.33	1.00	2017	2019	3	Mixed

Note NP = number of publications (full counting). FA = *Articles Fractionalized*, the sum of fractional authorship credits across NP papers (FA can exceed 1). CR = co-authorship reliance (1 − FA/NP) NP (higher = stronger reliance on multi-author production). CI = collaboration index FA − NPFA − NPFA − NP (more negative = stronger co-authorship intensity under fractional counting). Career span = = = Recent Activity − Start Year + 1. Missing indices (—) reflect insufficient data within the observation window.

Table 4 Citation impact stratification and global–local uptake (within-corpus citations)

Document	Year	Total citations (TC)	TC per year	Norm TC	Local cites	Local share (%)	Norm local	Local efficiency	Diffusion pattern	Impact type
Fernández Fernández Y (J Clean Prod)	2018	477	59.6	7.07	0	0.00	0.0	—	Broad external	Global foundation
Tseng M-L (J Clean Prod)	2013	406	31.2	3.65	0	0.00	0.0	—	Broad external	Global foundation
Zhang X (J Clean Prod)	2015	181	16.5	5.95	1	0.55	7.5	7.50	Broad external	Dual impact
Alsaifi K (Bus Strategy Environ)	2020	155	25.8	2.82	3	1.94	24.0	8.00	Emerging dual	Local leader
Lewandowski S (Bus Strategy Environ)	2017	159	17.7	3.16	2	1.26	11.5	5.75	Emerging dual	Dual impact
Becken S (J Sustainable Tour)	2006	278	13.9	2.87	1	0.36	3.0	3.00	Broad external	Global foundation

Note Local Cites are citations received from within the 387-document corpus. Local Share (%) = Local Cites ÷ Total Citations × 100. Norm values are normalized citation counts (as defined in the study). Local Efficiency = Norm Local ÷ Local Cites (undefined when Local Cites = 0). Diffusion Pattern is descriptive (e.g., Broad External = low/zero local uptake; Emerging Dual = non-zero local uptake). Impact Type labels use the table's thresholds (e.g., Global Foundation: TC $\geq$ 250; Local Leader: Local Cites $\geq$ 3) shows a comparable mixed profile (159 total citations; two local citations; local share = 1.26%). Zhang et al. (2015) illustrate sensitivity to small local counts: Despite strong global impact (181 citations; normalized TC = 5.95), one local citation yields a local share of 0.55%, which is informative about within-corpus uptake under the study boundary but not a basis for strong claims about consolidation. The time dimension further clarifies the global–local contrast. Older contributions such as Becken and Patterson (2006) have longer exposure time (278 total citations over 19 years; 13.9 citations/year; normalized TC = 2.87) but limited within-corpus uptake (one local citation; local share = 0.36%). Because local citations are counted within a finite corpus, local uptake is mechanically constrained by corpus size and by the extent to which the citing documents overlap with the focal themes and keywords. Table 4 indicates asymmetry between global visibility and within-corpus consolidation. Some publications function as globally prominent reference points without being frequently mobilized inside the focal corpus, whereas others exhibit higher relative local uptake and appear more directly incorporated into within-domain citation accumulation under the study boundary. This motivates the subsequent network analyses, which locate where influential work sits in co-citation and thematic structures and where cross-cluster connectivity is limited.

The performance indicators in Sect. 3 establish publication volume, source concentration, and citation patterns. They do not, however, reveal how the corpus is intellectually structured—which references form shared foundations, how topics cluster, and where integration is limited. Section 4 addresses these questions through network and thematic mapping. Co-citation analysis (4.2) identifies groups of references cited together, revealing the intellectual base. Keyword co-occurrence (4.3) and thematic mapping (4.1) describe topical organization and boundaries between sub-communities. These approaches operationalize “fragmentation” as measurable connectivity patterns rather than as assertion.

4 Science Mapping

4.1 Thematic Evolution in the Co-word Network

Figure 2 positions themes on a two-dimensional thematic map using Callon centrality (x-axis; relevance) and Callon density (y-axis; development). Centrality reflects the extent to which a theme connects to other themes in the co-word network, whereas density reflects internal cohesion among the keywords grouped within a theme. The four quadrants—motor, basic, niche, and emerging/declining—are used here as descriptive categories based on thematic map position. The motor quadrant (high centrality, high density) is anchored by climate change, with sustainable development and renewable energy located in the same region. Their placement indicates themes that are both internally cohesive and strongly connected to the broader network, serving as integrative frames through which multiple subtopics co-occur. The basic

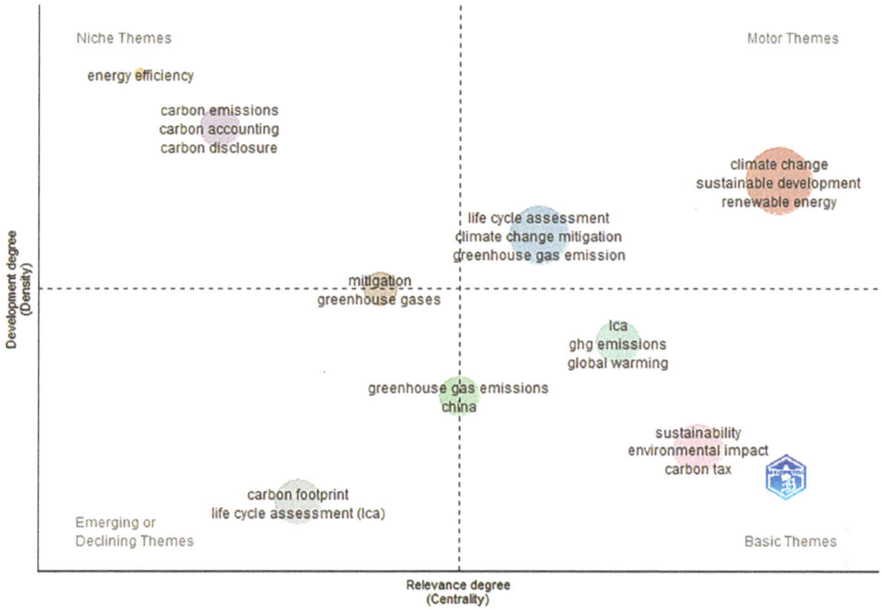

Fig. 2 Thematic map

quadrant (high centrality, low density) contains sustainability, environmental impact, and carbon tax. These terms appear as broad connectors that link multiple strands of work but do not form tightly bounded, internally dense sub-communities in the co-word structure.

The niche quadrant (low centrality, high density) includes energy efficiency and a grouped set of reporting/measurement terms (carbon emissions, carbon accounting, carbon disclosure). Their location indicates specialized conversations with strong within-theme development but weaker linkage to the field's integrative vocabulary. The emerging/declining quadrant (low centrality, low density) contains carbon footprint and life cycle assessment (LCA). In this co-word map, these themes appear peripheral under the study's keyword specification and observation window. This interpretation is strictly structural: It reflects thematic map position and does not imply low substantive importance. Several themes appear near quadrant boundaries, including life cycle assessment, climate change mitigation, and emissions-related terms (e.g., greenhouse gas emission, GHG emissions, global warming). Their intermediate placement suggests partial overlap between methodological vocabulary and policy framing in the co-word network, motivating the subsequent mapping analyses. These thematic positions reflect keyword co-occurrence patterns. The subsequent co-citation analysis reveals whether this thematic structure corresponds to shared intellectual foundations or whether citation architecture is organized differently.

4.2 Co-citation Analysis

The co-citation network characterizes the intellectual base of the analytic corpus by grouping references that are cited together. Table 5 reports seven clusters and their within-network link-strength ranges. The resulting structure is uneven: One cluster forms a dense hub, several clusters are weakly connected, and a small set of policy, standards, and strategy references appear as single-item clusters under the clustering specification. Cluster 1 (carbon market economics) forms the network hub. It contains seven references and exhibits link-strength values far above those of other clusters (104–125; Table 5), indicating a repeatedly co-cited backbone within the analytic corpus. The cited items span 1995–2012 and combine a general forecasting-validation method with EU, ETS-focused empirical studies and carbon-price modeling contributions. Diebold and Mariano (1995) provide a widely used approach for comparing predictive accuracy in forecasting contexts. Kanen (2006) frames emissions trading systems as market instruments with pricing dynamics. EU ETS-focused contributions examine price drivers and structural change (Alberola et al., 2008), operating mechanisms and economic effects (Zhang & Wei, 2010), stochastic allowance-price models (Benz & Trück, 2009), and volatility behavior relevant to risk assessment (Feng et al., 2011). Zhu (2012) introduces an additional modeling direction through machine-learning-oriented prediction approaches. High link strength in Cluster 1 indicates dominance within the mapped citation architecture, not dominance in policy-efficacy terms.

Cluster 2 (corporate disclosure and governance) groups four references spanning 2008–2020 (Table 5). The strand is anchored by early empirical work on climate-related disclosure as a corporate reporting phenomenon (Stanny & Ely, 2008). It links corporate carbon exposure to capital-market responses in the context of emissions trading policy (Chapple et al., 2013), outlines managerial integration of carbon metrics into management systems (Schaltegger & Csutora, 2012), and consolidates evidence on governance determinants and the financial implications of carbon performance and disclosure (Velte et al., 2020). In Table 5, Cluster 2's link-strength range (1–4) is substantially lower than Cluster 1's, indicating a more specialized co-citation base under this specification. Cluster 3 (agricultural and food systems) contains three references (2016–2018) focused on soil-land-based mitigation and food-system impacts (Table 5). Paustian et al. (2016) articulate soil- and land-based mitigation pathways. Huang et al. (2018) contribute evidence relevant to management practices and their greenhouse gas and yield implications. Poore and Nemecek (2018) provide a comparative framework for food-system environmental impacts used in supply-chain and dietary-impact discussions. The cluster's link-strength range (1–2) indicates limited co-citation connectivity beyond its specialized base under the present specification.

Cluster 4 (lifecycle assessment methodology) contains two references representing LCA infrastructure and methodological framing (Table 5). Wernet et al. (2016) provide the ecoinvent database foundation (version 3), supporting lifecycle inventory-based applications. Earles and Halog (2011) address consequential LCA,

Table 5 Co-citation clustering of foundational references in the analytic corpus

Cluster	Representative works	Works (n)	Citation range	Link-strength range	Thematic focus
1	Alberola et al. (2008), Benz and Trück (2009), Diebold and Mariano (1995), Feng et al. (2011), Kanen (2006), Zhang and Wei (2010), Zhu (2012)	7	4–5	104–125	Carbon market economics
2	Chapple et al. (2013), Schaltegger and Csutora (2012), Stanny and Ely (2008), Velte et al. (2020)	4	4	1–4	Corporate disclosure and governance
3	Huang et al. (2018), Paustian et al. (2016), Poore and Nemecek (2018)	3	4	1–2	Agricultural and food systems
4	Earles and Halog (2011), Wernet et al. (2016)	2	4–5	2	Life cycle assessment methodology
5	Adoption of the Paris Agreement (2015)	1	5	0	International climate policy
6	IPCC Guidelines (2006)	1	5	0	Measurement standards
7	Lee (2012)	1	4	0	Corporate strategy

Note Clusters are derived from the co-citation network of cited references in the analytic corpus ($N = 387$). Citation range reports how often each reference is cited within the analytic corpus. Link strength refers to co-citation connectivity in the VOSviewer network under the selected counting and normalization options. Single-item clusters with link strength 0 indicate cited references that do not form measurable co-citation linkages sufficient to join a multi-item cluster under this specification

emphasizing system-wide effects rather than attributional footprints alone. The cluster's link strength is reported as 2 (Table 5), indicating a small, specialized co-citation base in this clustering solution. Clusters 5–7 are single-reference clusters with link strength reported as 0 in Table 5: Adoption of the Paris Agreement (2015), the IPCC Guidelines for national greenhouse gas inventories (2006), and Lee (2012) on corporate strategy. Under the co-citation specification used, these items are cited but do not form measurable co-citation linkages sufficient to integrate them into multi-item clusters. This is a structural result of the co-citation specification and thresholds and does not imply low substantive importance. Table 5 indicates that carbon market economics supplies a highly shared intellectual base, while several operational-method and governance-related foundations remain weakly connected in the co-citation space. "Fragmentation" is used here strictly as a network descriptor referring to limited co-citation connectivity across clusters, not as an outcome claim about governance effectiveness.

4.3 Keyword Co-occurrence Synthesis

Keyword co-occurrence analysis (Fig. 3) was conducted on 47 terms meeting a $\geq$ 15 occurrences threshold, selected from 3555 unique keyword strings in the dataset. The network partitions into four clusters that differentiate (i) decision and policy evaluation language, (ii) life cycle and environmental assessment methods, (iii) climate-governance framing, and (iv) sectoral/application-oriented accounting. As a co-word structure, the map indicates a field organized around a central climate-emissions vocabulary with specialized methodological and applied sub-communities. In the overlay visualization, colors reflect average publication year. One cluster captures decision and appraisal vocabulary that links carbon metrics to policy and management choices. High-frequency/high-connectivity items include energy efficiency (38 occurrences; total link strength 226), decision-making (30; 173), and carbon capture (21; 100). These terms co-occur with evaluation-oriented constructs visible in the network (e.g., cost and investment language), consistent with carbon accounting outputs being mobilized as inputs to appraisal rather than as stand-alone measurements. Overlay timing places several items in this decision-support domain earlier in average publication year than newer transition-oriented topics.

A second cluster concentrates the methodological assessment spine and is dominated by life cycle and impact-assessment terms. The backbone consists of life cycle (94; 679), life cycle assessment (50; 327), and life cycle assessment (LCA) (51; 403). Broader impact categories such as environmental impact (50; 388) and eutrophication (15; 117) indicate that the methodological stream frequently situates carbon accounting within multi-impact assessment frameworks. Terms such as

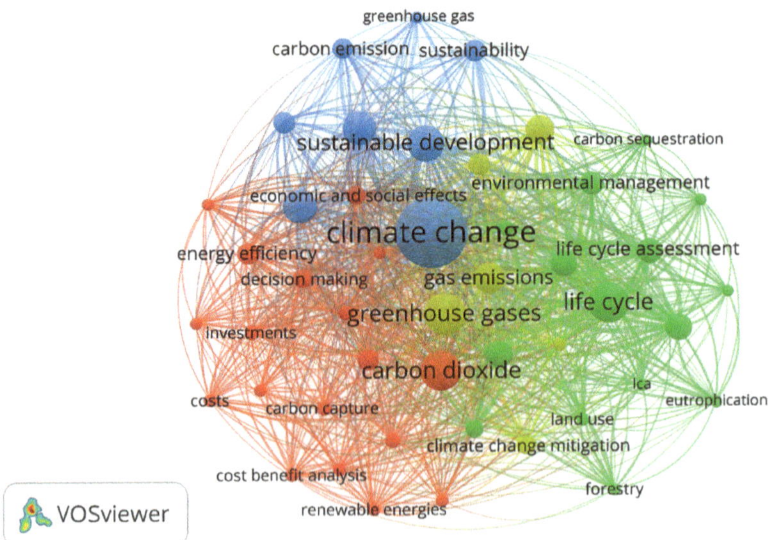

Fig. 3 Keyword co-occurrence

fertilizers (15; 124) and carbon sequestration (15; 71) also appear in this space, consistent with applications covering both emissions sources and removals within LCA-oriented accounting. The duplication of "life cycle assessment" strings reflect keyword heterogeneity rather than distinct concepts and motivates harmonization in preprocessing where feasible. The network's integrative hub is anchored by climate change (217; 996), the highest-occurrence and highest-link-strength term, linking governance framing to measurement and application vocabularies. Closely connected terms include sustainable development (83; 430), emission control (72; 418), and the emissions terms carbon emission (35; 133) and carbon emissions (37; 167), with sustainability (39; 132) functioning as a bridging label. An application-oriented cluster connects this hub to specific contexts and mitigation activity, anchored by greenhouse gases (99; 682) and gas emissions (63; 470), with sectoral terms such as agriculture (17; 140) and applied-accounting language such as carbon footprint (56; 305) alongside climate change mitigation (30; 196). Overlay timing places several applied emissions strings later in average publication year, consistent with recent growth in application-oriented work within the corpus. Across clusters, connectivity is highest around integrative climate/emissions terms, while methodological and sectoral vocabularies remain identifiable as sub-communities, providing a descriptive basis for locating where cross-stream bridging is limited in the mapped corpus.

4.4 Directions for Future Studies

The maps reveal three structural gaps. First, weak cross-linking exists between LCA/impact-assessment vocabularies and governance/decision vocabularies (Sects. 4.2, 4.3). Second, policy and standards documents appear as isolated citations rather than co-cited foundations (Sect. 4.2). Third, recent application terms—renewable energy, carbon sequestration, risk assessment—show unclear integration status in the keyword network (Sect. 4.3). The following directions target these interfaces, framing research problems where limited connectivity indicates integration opportunities rather than established research streams. The co-occurrence overlay (Fig. 4) highlights keywords with comparatively later average publication years—renewable energy, carbon emission(s), sustainability, carbon sequestration, greenhouse gas emission(s), decision-making, global warming, and risk assessment—indicating where term usage has expanded most recently within the mapped corpus. Because overlay colors capture recency of usage rather than intrinsic importance, the directions below are framed as research problems at interfaces where the maps indicate limited cross-stream connectivity.

A first set of problems concerns time-resolved measurement and attribution. (1) For renewable energy and electricity emissions, research can develop accounting approaches that capture temporal and locational variation in grid intensity and procurement timing, replacing reliance on average factors that may obscure marginal effects; this includes formalizing time-resolved attribution rules (e.g., hourly

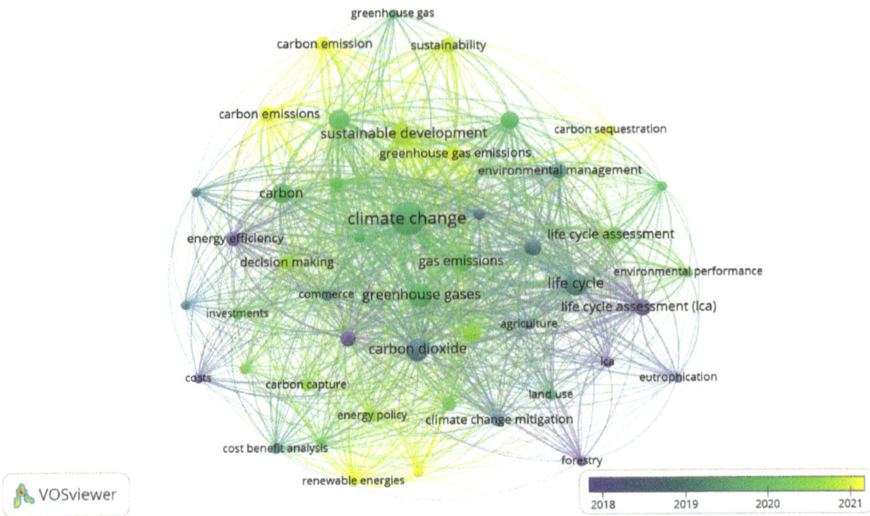

Fig. 4 Co-occurrence analysis: overlay

matching) and evaluating how alternative rules change measured abatement, comparability, and incentives across regions and market designs. (2) For sequestration and removals, research can strengthen MRV for permanence, additionality, leakage, and uncertainty, with particular emphasis on uncertainty quantification that separates measurement error, model uncertainty, and baseline error; high-value contributions would integrate remote sensing with ground-truth validation, specify transparent baselines, and report uncertainty in decision-usable formats rather than as residual caveats.

A second set of problems concerns estimation under constraints and risk-relevant integration. (3) For Scope 3, a frontier is scalable predictive estimation under persistent data gaps, using machine learning or hybrid engineering–statistical models that rely on proxy variables (sector, geography, technology, firm size) while retaining interpretability and auditability; key tests include validation against supplier disclosures, transferability across industries, and sensitivity of comparability and target-setting to model choice. (4) The overlay prominence of risk assessment motivates frameworks that connect carbon accounting information to decision-relevant physical and transition risk metrics used in finance and enterprise risk management, including joint representation of hazard exposure, transition pathways (e.g., carbon pricing scenarios), and market responses, and aggregation from asset to firm to portfolio with explicit uncertainty propagation and sensitivity analysis.

A third set of problems concerns organizational response, robust decisions, and multi-dimensional governance. (5) Carbon management systems remain weakly evidenced on behavioral response; studies can test whether measurement and feedback systems (digital feedback, choice architecture, incentive design) change travel, commuting, procurement, or operational behaviors using designs capable of causal inference and reporting persistence and spillovers. (6) Relatedly, decision-making and risk language supports adaptation of robust decision approaches under deep uncertainty—dynamic adaptive policy pathways, real options, stochastic optimization—demonstrating when they change decisions relative to static optimization and under what governance conditions, they are implementable. (7) Because sustainability appears as a recent and connecting term, research can extend beyond single-metric carbon optimization to multi-criteria decision analysis incorporating biodiversity, water, and social dimensions, focusing on transparent weighting, normalization, treatment of non-compensatory thresholds, and compatibility with reporting requirements. (8) Finally, AI-enabled monitoring and governance tools require work that couples technical design (forecasting, anomaly detection, automated reconciliation) with interoperability and accountability constraints, including data ontologies for platform interoperability, evaluation of model drift and monitoring error, and assurance routines specifying when automation is appropriate versus when human review is required. These directions are most relevant where the maps indicate weak cross-linking—particularly at interfaces between LCA/impact-assessment vocabularies, governance/decision vocabularies, and sectoral application domains.

5 Conclusion

This study maps technology and innovation research in carbon accounting and climate policy using a bibliometric analysis of 387 Scopus-indexed documents (2000–2025). It traces the field's intellectual base and development, the linkage between technology-oriented and governance-oriented strands, and changes in keyword usage. The corpus shows rapid growth (annual growth rate 15.9%) with integrative themes coexisting alongside specialized sub-communities that remain weakly connected in co-citation and co-word spaces. The findings address four guiding questions. (1) The intellectual structure shifts from concentrated early influence (2000–2014; including Tseng et al., 2013) to a more topically dispersed recent literature (2021–2025) with lower citations-per-document, consistent with topical diffusion and citation-age effects. (2) Technology–governance connectivity remains limited: co-citation analysis reveals a dominant carbon market economics hub (link strength 104–125) alongside weakly connected clusters, with policy frameworks (e.g., Paris Agreement; IPCC Guidelines) appearing as isolated nodes—cited but not routinely co-cited with empirical foundations. (3) Diffusion proxies indicate rapid expansion under a concentrated publication infrastructure (Journal of Cleaner Production: 49.6%), suggesting that growth has occurred through a limited set of dominant outlets rather than broad dispersion across sources. (4) Thematic mapping positions climate change as an integrative motor theme, while assessment vocabulary remains distinct from decision/policy framing; carbon footprint is peripheral under the keyword specification used.

An overlay of emerging themes identifies eight near-term integration opportunities: (1) time-resolved electricity accounting, (2) sequestration MRV with uncertainty quantification, (3) scalable Scope 3 estimation, (4) integrated risk modeling, (5) behavioral mechanisms in emissions management, (6) robust decision frameworks, (7) multi-criteria sustainability approaches, and (8) AI-enabled monitoring with auditable design. Limitations are typical of bibliometric designs. The corpus is restricted to a single database and the keyword strategy used, and citation indicators are sensitive to citation age, field differences, and database coverage. Accordingly, the findings are interpreted as descriptive evidence of knowledge structure rather than causal claims about policy effectiveness or corporate outcomes. Robustness can be strengthened by triangulating databases, testing alternative keyword harmonization strategies, and combining bibliometric mapping with targeted qualitative synthesis of standards, protocols, and implementation evidence across regulatory settings.

References

Alberola, E., Chevallier, J., & Chèze, B. t. (2008). Price drivers and structural breaks in European carbon prices 2005–2007. *Energy Policy, 36*(2), 787–797. https://doi.org/10.1016/j.enpol.2007.10.029

Alonso-Robisco, A., Carbo, J. M., Kormanyos, E., & Triebskorn, E. (2025). Houston, we have a problem: Can satellite information bridge the climate-related data gap? *Latin American Journal of Central Banking*, 100173. https://doi.org/10.1016/j.latcb.2025.100173

Alsaifi, K., Elnahass, M., & Salama, A. (2020). Carbon disclosure and financial performance: UK environmental policy. *Business Strategy and the Environment, 29*(2), 711–726. https://doi.org/10.1002/bse.2426

Aria, M., & Cuccurullo, C. (2017). Bibliometrix: An R-tool for comprehensive science mapping analysis. *Journal of Informetrics, 11*(4), 959–975.

Becken, S., & Patterson, M. (2006). Measuring national carbon dioxide emissions from tourism as a key step towards achieving sustainable tourism. *Journal of Sustainable Tourism, 14*(4), 323–338. https://doi.org/10.2167/jost547.0

Benz, E., & Trück, S. (2009). Modeling the price dynamics of CO_2 emission allowances. *Energy Economics, 31*(1), 4–15. https://doi.org/10.1016/j.eneco.2008.07.003

Berg, S., & Lindholm, E.-L. (2005). Energy use and environmental impacts of forest operations in Sweden. *Journal of Cleaner Production, 13*(1), 33–42. https://doi.org/10.1016/j.jclepro.2003.09.015

Chapple, L., Clarkson, P. M., & Gold, D. L. (2013). The cost of carbon: Capital market effects of the proposed emission trading scheme (ETS). *Abacus, 49*(1), 1–33. https://doi.org/10.1111/abac.12006

Chen, H.-M., Kuo, T.-C., & Chen, J.-L. (2022). Impacts on the ESG and financial performances of companies in the manufacturing industry based on the climate change related risks. *Journal of Cleaner Production, 380*, Article 134951. https://doi.org/10.1016/j.jclepro.2022.134951

Conference of the Parties. (2015). Conference of the Parties. Twenty-first session, Paris, 30 November to 11 December 2015. https://unfccc.int/resource/docs/2015/cop21/eng/l09r01.pdf

Diebold, F. X., & Mariano, R. S. (1995). Comparing predictive accuracy. *Journal of Business & Economic Statistics, 13*(3), 253–263. https://doi.org/10.1080/07350015.1995.10524599

Earles, J. M., & Halog, A. (2011). Consequential life cycle assessment: A review. *The International Journal of Life Cycle Assessment, 16*(5), 445–453. https://doi.org/10.1007/s11367-011-0275-9

Feng, Z.-H., Zou, L.-L., & Wei, Y.-M. (2011). Carbon price volatility: Evidence from EU ETS. *Applied Energy, 88*(3), 590–598. https://doi.org/10.1016/j.apenergy.2010.06.017

Fernández Fernández, Y., Fernández López, M. A., & Olmedillas Blanco, B. (2018). Innovation for sustainability: The impact of R&D spending on CO_2 emissions. *Journal of Cleaner Production, 172*, 3459–3467. https://doi.org/10.1016/j.jclepro.2017.11.001

Hettler, M., & Graf-Vlachy, L. (2024). Corporate scope 3 carbon emission reporting as an enabler of supply chain decarbonization: A systematic review and comprehensive research agenda. *Business Strategy and the Environment, 33*(2), 263–282.

Huang, Y., Ren, W., Wang, L., Hui, D., Grove, J. H., Yang, X., Tao, B., & Goff, B. (2018). Greenhouse gas emissions and crop yield in no-tillage systems: A meta-analysis. *Agriculture, Ecosystems & Environment, 268*, 144–153. https://doi.org/10.1016/j.agee.2018.09.002

Huovila, A., Siikavirta, H., Antuña Rozado, C., Rökman, J., Tuominen, P., Paiho, S., Hedman, Å., & Ylén, P. (2022). Carbon-neutral cities: Critical review of theory and practice. *Journal of Cleaner Production, 341*, Article 130912. https://doi.org/10.1016/j.jclepro.2022.130912

Intergovernmental Panel on Climate Change. (2006). *2006 IPCC guidelines for national greenhouse gas inventories*. https://www.ipcc-nggip.iges.or.jp/public/2006gl/

Jones, P. T., Geysen, D., Tielemans, Y., Van Passel, S., Pontikes, Y., Blanpain, B., Quaghebeur, M., & Hoekstra, N. (2013). Enhanced Landfill Mining in view of multiple resource recovery: A critical review. *Journal of Cleaner Production, 55*, 45–55. https://doi.org/10.1016/j.jclepro.2012.05.021

Kanen, J. L. M. (2006). *Carbon trading and pricing*. Environmental Finance Publications.

Klaaßen, L., & Stoll, C. (2021). Harmonizing corporate carbon footprints. *Nature Communications, 12*(1), 6149.

Lee, S. Y. (2012). Corporate carbon strategies in responding to climate change. *Business Strategy and the Environment, 21*(1), 33–48. https://doi.org/10.1002/bse.711

Lewandowski, S. (2017). Corporate carbon and financial performance: The role of emission reductions. *Business Strategy and the Environment, 26*(8), 1196–1211. https://doi.org/10.1002/bse.1978

Oladeji, O., & Mousavi, S. S. (2023). Towards AI-driven integrative emissions monitoring & management for nature-based climate solutions. *arXiv preprint* arXiv:2312.11566

Paustian, K., Lehmann, J., Ogle, S., Reay, D., Robertson, G. P., & Smith, P. (2016). Climate-smart soils. *Nature, 532*(7597), 49–57. https://doi.org/10.1038/nature17174

Perdana, A., & Tan, S. (2024). Harnessing technologies and data to accelerate and operationalize environmental, social, and governance (ESG) initiatives. In *Digital transformation in accounting and auditing: Navigating technological advances for the future* (pp. 347–375). Springer.

Poore, J., & Nemecek, T. (2018). Reducing food's environmental impacts through producers and consumers. *Science, 360*(6392), 987–992. https://doi.org/10.1126/science.aaq0216

Schaltegger, S., & Csutora, M. (2012). Carbon accounting for sustainability and management. Status quo and challenges. *Journal of Cleaner Production, 36*, 1–16. https://doi.org/10.1016/j.jclepro.2012.06.024

Seo, S. N. (2017). Beyond the Paris Agreement: Climate change policy negotiations and future directions. *Regional Science Policy & Practice, 9*(2), 121–141. https://doi.org/10.1111/rsp3.12090

Shackleton, S., Taylor, A., Gammage, L., Gillson, L., Sitas, N., Methner, N., Barmand, S., Thorn, J., McClure, A., Cobban, L., Jarre, A., & Odume, O. N. (2023). Fostering transdisciplinary research for equitable and sustainable development pathways across Africa: What changes are needed? *Ecosystems and People, 19*(1). https://doi.org/10.1080/26395916.2022.2164798

Stanny, E., & Ely, K. (2008). Corporate environmental disclosures about the effects of climate change. *Corporate Social Responsibility and Environmental Management, 15*(6), 338–348. https://doi.org/10.1002/csr.175

Thanasi-Boçe, M., & Hoxha, J. (2025). Blockchain for sustainable development: A systematic review. *Sustainability, 17*(11), 4848. Retrieved 2025/05/25, from https://doi.org/10.3390/su17114848

Tseng, M.-L., Chiu, S. F., Tan, R. R., & Siriban-Manalang, A. B. (2013). Sustainable consumption and production for Asia: Sustainability through green design and practice. *Journal of Cleaner Production, 40*, 1–5. https://doi.org/10.1016/j.jclepro.2012.07.015

Van Eck, N., & Waltman, L. (2010). Software survey: VOSviewer, a computer program for bibliometric mapping. *Scientometrics, 84*(2), 523–538.

Velte, P., Stawinoga, M., & Lueg, R. (2020). Carbon performance and disclosure: A systematic review of governance-related determinants and financial consequences. *Journal of Cleaner Production, 254*, Article 120063. https://doi.org/10.1016/j.jclepro.2020.120063

Wernet, G., Bauer, C., Steubing, B., Reinhard, J., Moreno-Ruiz, E., & Weidema, B. (2016). The ecoinvent database version 3 (part I): Overview and methodology. *The International Journal of Life Cycle Assessment, 21*(9), 1218–1230. https://doi.org/10.1007/s11367-016-1087-8

World Resources Institute. (2022). *Trends show companies are ready for scope 3 reporting with US climate disclosure rule*. Retrieved 28/07/2025 from https://www.wri.org/update/trends-show-companies-are-ready-scope-3-reporting-us-climate-disclosure-rule

Yang, B. (2025). Global carbon neutrality solution. In B. Yang (Ed.), *The carbon emission liability mechanism: A new solution for global carbon neutrality* (pp. 293–361). Springer Nature Singapore. https://doi.org/10.1007/978-981-96-3612-9_10

Zhang, X. (2025, 2025/06/13). Design and implementation of energy efficiency audit, carbon footprint tracking and sustainable development path planning system for green economy enterprises based on blockchain technology. In: *2025 3rd Cognitive models and artificial intelligence conference (AICCONF)*. https://doi.org/10.1109/AICCONF64766.2025.11064234

Zhang, X., Luo, L., & Skitmore, M. (2015). Household carbon emission research: An analytical review of measurement, influencing factors and mitigation prospects. *Journal of Cleaner Production, 103*, 873–883. https://doi.org/10.1016/j.jclepro.2015.04.024

Zhang, Y.-J., & Wei, Y.-M. (2010). An overview of current research on EU ETS: Evidence from its operating mechanism and economic effect. *Applied Energy, 87*(6), 1804–1814. https://doi.org/10.1016/j.apenergy.2009.12.019

Zhu, B. (2012). A novel multiscale ensemble carbon price prediction model integrating empirical mode decomposition, genetic algorithm and artificial neural network. *Energies, 5*(2), 355–370.

Chapter 11
Conceptualizing Climate Change Vulnerability and Adaptation: A CIMO-Based Strategic and Policy Perspective

Priyanka Verma, Arif Hasan, Mini Agrawal, and Shakeel Ahmad Sofi

Abstract The socio-ecological systems are being increasingly strained by climate change all over the world. This has seen vulnerability become a critical subject of interest among both the policy makers and strategists. Several initiatives of adaptation have been initiated in the recent years. However, vulnerability remains widespread in many areas, and to a great extent is associated with a disjointed policy formulation, inadequate contextualization, and lack of alignment between plans and actions. One of the main issues is that the literature available is inclined to view vulnerability, adaptation, and policy as independent issues. This division constrains knowledge of what works, who works, and in what circumstances. The chapter fills that gap by offering a conceptual framework of examining climate change vulnerability and adaptation. It employs the organizing structure of the Context-Intervention-Mechanism-Outcome (CIMO) logic model. The framework is based on the literature of existing climate governance and strategies. It theorizes vulnerability as a context-specific state of affairs that is determined by socio-economic systems, institutional power, and environmental vulnerability. It establishes the major strategic and policy interventions which are regulatory tools, institutional reforms and community-based interventions. It also evaluates the processes by which these interventions operate, i.e., capacity development, risk awareness, coordination, and social learning. The CIMO framework transcends linear models of adaptation policy by revealing analytically the

P. Verma
Amity University Punjab, Mohali, India

A. Hasan (✉) · M. Agrawal
Amity University Madhya Pradesh, Gwalior, India
e-mail: ahasan@gwa.amity.edu

M. Agrawal
e-mail: magrawal@gwa.amity.edu

S. A. Sofi
INTI University, Malaysia, Malaysia

K. Kumar et al. (eds.), *AI-Driven Circular Economy for Carbon Neutrality in Emerging Economies*, Environmental Footprints and Eco-design of Products and Processes,
https://doi.org/10.1007/978-3-032-30298-4_11

causal links between interventions, mechanisms, and outcomes. It emphasizes that the context-sensitive design is crucial and demonstrates how an adversarial response is produced by a lack of match between interventions and the vulnerability conditions. The chapter provides a theoretically subjected and structured instrument to the policy makers, planners and researchers involved in the development, assessment and scaling of adaptation strategies in various national and governance settings.

Keywords Climate change vulnerability · Climate adaptation · Sustainability · Policy and strategy · CIMO framework · Adaptive capacity · Climate governance · Resilience · Context-sensitive policy · Sustainable development

1 Introduction

One of the major issues of twenty-first-century global concern is climate change. Its impact can be observed already in all regions of the world. They feature increased frequency of extreme weather, faster degradation of the ecology, as well as the increase of socio-economic disturbance (UNDRR, 2019). These impacts are, however, not equal. There is a wide variety of communities with different exposure to hazards, resources at their disposal, and the efficacy of the institutions. This is why vulnerability has become the key concept in climate science and policy (Adger, 2006; Cutter et al., 2003; Sarkodie et al., 2022). Vulnerability does not just mean exposure to climatic hazards. It is also influenced by social, economic, and institutional environments that define the ability of communities to prepare, respond to, and recover due to climate-related stresses (Heltberg et al., 2009; Lampis, 2013; Smit & Wandel, 2006). To be able to come up with effective strategies of adapting to this broader sense of vulnerability, one must understand it. It is also essential to sustainable development within a period of increasing climate doubt. Regardless of this realization, most of the policy responses are reactive and fragmented. They are concerned with momentary technical solutions. They are not paying enough attention to the contexts of governance, the actual socio-economic conditions, and the process of institutions that motivate the real-life outcomes (Magnan et al., 2016; OECD, 2020). A more logical structure is required. It must combine vulnerability situations, the interventions of adaptation, causality, and the observed consequences under one analytical framework.

The chapter addresses this need by using the Context-Intervention-Mechanism-Outcome (CIMO) logic model as the tool of analysis. CIMO was initially created in management and policy sciences. It offers a systematic process of examining the way particular interventions have an outcome in particular settings, via recognizable mechanisms (Denyer et al., 2008; Pundir et al., 2023). Using it in the context of climate adaptation, it changes the focus of the question of what works to one, which is more effective: what works, whom, and under what circumstances? The chapter makes contributions to climate policy and sustainability literature through providing a theoretically based framework of more coherent, effective, and equitable strategies of adaptation.

2 Theoretical Foundations of Climate Change Vulnerability

Vulnerability is a concept that has evolved over the last few decades. The initial strategies were much concerned with physical exposure to climatic stressors. This has been expanded in more recent scholarships. Vulnerability has now been identified to be a socially constructed state of being that develops as a result of the interaction between environmental processes and social systems (Adger, 2006; Cutter et al., 2003; McLaughlin, 2011). This change has a serious implication on adaptation policy design and evaluation.

There are three interdependent dimensions of vulnerability that are determined by contemporary frameworks. The former is the exposure: the nature and the strength of the climate stimuli to which a system is exposed. The second one is sensitivity: how much a system is influenced by climate variability. The third one is adaptive capacity: how individuals, communities, and institutions can adapt to the effects of climate (Boswell et al., 2012; Smit & Wandel, 2006). Vulnerability is a dynamic construct of these three dimensions, in a larger social, economic, and institutional context (Downing et al., 2005; Soares et al., 2012).

Vulnerability can be well differentiated to two other similar concepts which are resilience and adaptation. The concept of resilience is the level of social-ecological systems to accommodate disruptions without disrupting the fundamental functions (Folke, 2006; Zorita et al., 2025). Adaptation is defined as willful changes in practices, institutions, or even policies related to real or expected impacts of climate (Fazey et al., 2010). In other words, vulnerability is the state which is targeted by the adaptation. The buffer that restricts damage in the process of the reduction is called resilience. It is crucial to consider vulnerability as a governance-mediated process of learning, socio-political transformation, to understand the effectiveness of long-term policies (Conway et al., 2019; Ostrom, 2010).

Socio-economic studies prove that vulnerability is unevenly distributed. Vulnerability is increased by poverty, inequality based on gender in resource access, geographic marginalization, and lack of social safety nets. This holds irrespective of a situation where physical exposure is comparable across communities (Adger, 2006; Heltberg et al., 2009; Waiba & Dorji, 2024). In low- and middle-income nations, these obstacles are exacerbated by the weak institutions, informal economies, and ineffective infrastructure (Bank, 2021). Particular socio-economic and institutional contexts of the place should be of central focus in any meaningful adaptation framework.

The research in governance gives an additional insight. The impact of institutional alignment, intergovernmental, and policy coherence on vulnerability outcomes are all measurable (OECD, 2020; UNDRR, 2019). In disjointed governance, a lack of clarity, or weak cross-sectoral communication, policy responses would be symbolic or counterproductive (Adodoadji-Dogbe & Urban, 2023). Vulnerability cannot thus be termed as an environmental issue alone. It is also a government issue. To effectively tackle it, there is a need to take coordinated policy measures that relate climate targets to larger development ones (Bank, 2021). All these theoretical backgrounds provide a justification to the use of the CIMO framework on climate adaptation research (Denyer et al., 2008).

3 Rationale for Using the CIMO Framework

The vulnerability to climate change is a product of a complicated interaction between climatic stress factors, societal frameworks, institutions, and policy decisions. Such complexity poses a basic dilemma to the study of adaptation. The conventional methods have been based on linear or output-based models. These models are capable of recording the existing interventions. They do not, however, often describe the reasons why the same intervention may be successful in one environment and not in the other (Berrang-Ford et al., 2021; Nalau et al., 2024). This is a crucial gap in light of the imminent requirement of translating adaptation knowledge into context-specific policy and practice.

The CIMO system was created in order to deal with this sort of issue. CIMO, which started in the field of management and policy research, describes the way certain interventions produce the results of specific processes that are inherent to the context (Denyer et al., 2008; Pundir et al., 2023). It is not merely the description of what happens. It asks how and why. This renders CIMO highly applicable to climate adaptation studies, where context-dependence is one of the characteristics of a successful approach (Adger, 2006; Haase, 2022; McLaughlin, 2011).

CIMO structures analysis based on four aspects, namely context, intervention, mechanism, and outcome. All the elements are different but related to one another. This framework is useful in enabling researchers and practitioners to go beyond their fundamental issue of what works. It allows a more policy-relevant and richer question: what works, who and in what circumstances? All these factors advantageously impact the health of the target population (Berrang-Ford et al., 2021; Conway et al., 2019). One of the core lessons of CIMO is that results are endowed by mechanisms rather than interventions. Community-based measures, policy instruments, and institutional reforms alone do not yield results. They are acted on by mobilizing the latent processes like capacity building, social learning, coordination and behavior change. It is these processes that lead to the realization of whether the outcomes of adaptation are actually improved (Folke, 2006; Smit & Wandel, 2006). Table 1 gives the four CIMO components when used in climate change vulnerability and adaptation.

Table 1 CIMO framework for conceptualizing climate change vulnerability and adaptation

CIMO element	Conceptual focus	Key dimensions	Illustrative examples
Context (C)	Climate change vulnerability settings	Exposure, sensitivity, adaptive capacity, state of governance settings	Climate-sensitive livelihoods, poor institutions, urban informality
Intervention (I)	Strategic and policy responses	Regulatory tools, institutional reforms, planning tools, community-based policies	Climate adaptation policies, early warning systems, local adaptation strategies
Mechanism (M)	Causal processes triggered by interventions	Capacity building, learning, coordination, behavioral change	Improved risk awareness, institutional learning, collective action
Outcome (O)	Adaptation and sustainability outcomes	Reduced vulnerability, improved resilience, alignment with sustainable development	Improvement in adaptive capacity, long-term climate resilience

Source Author's conceptualization based on Denyer et al. (2008) and climate adaptation literature

4 Context (C): Climate Change Vulnerability Settings

Context in CIMO framework is considered to be the underlying circumstances under which adaptation measures are formulated and executed. There are considerable differences in the contexts of vulnerability between regions, sectors, and social groups. Exposure, sensitivity, and adaptive capacity are the major factors of vulnerability that are always defined in the context of larger social, economic, and institutional frameworks (Boswell et al., 2012). The perception of context is necessary to determine adaptation pathways that are meaningfully local and have fairness.

Among the forces that promote climate vulnerability, there are the socio-economic conditions. All of these factors, including income levels, reliance on livelihoods that are susceptible to climate, education, gender disparity, and social protection quality, influence the response of communities to climate risk. Vulnerability is increased by poverty and economic informality on relatively low physical exposure. This is especially the case in the developing world, where the financial safety nets and social support systems do not suffice (Sarkodie et al., 2022; Waiba & Dorji, 2024). These are not just simple contextual backgrounds of socio-economic factors. They are constitutive in themselves of vulnerability. The other important contextual dimension is environmental exposure. It also entails the extent to which a community or system is exposed to risks like floods, drought, excessive heat, and rising sea levels. Fragile ecosystems or natural-resource-based areas are particularly vulnerable (Halofsky & Peterson, 2017; Hudec et al., 2019; Renaud & Perez, 2010). It is not just exposure that will make someone vulnerable. The real outcome of a hazard is dependent on social and institutional situations. That is why a more holistic approach is more effective than the one that is hazard-based (Hinkel, 2011; You et al., 2014).

The fundamental aspects of vulnerability are influenced by institutional capacity and quality of governance. Adaptive action is limited where policy coherence is less than ideal, mandates are not clearly defined, or the administrative capacity is limited (Adodoadji-Dogbe & Urban, 2023; Dogra et al., 2023; UNDRR, 2019). Inclusive governance, true decentralization, and stable policy structures can decrease vulnerability, however, in the long run. The process of rapid urbanization is even more complex. Growth of the informal settlements and poor urban infrastructure in most cities increases exposure to climates and restricts the options of adapting (Breil & Johnson, 2012; Rumbach & Kudva, 2011). It is thus necessary to incorporate climate adaptation together with land-use planning and the wider sustainable development objectives (John et al., 2025). Community-based adaptation (CBA) is a significant reaction to contextual vulnerability and particularly in marginalized environments. The mechanism behind CBA is the social capital development, collective action, and behavioral change at a community level (Adger, 2006; McLeod et al., 2015). It demonstrates one of the main principles that the top-down administrative strategies are not enough. Community engagement is a structurally required part of efficient vulnerability reduction.

5 Intervention (I): Strategic and Policymaking in Response to Climate Vulnerability

Interventions refer to the strategic and policy measures, which are calculated to respond to the identified context of vulnerabilities in the CIMO framework. They can be in many forms such as regulatory tools, institutional reforms, infrastructure investments, and community-based policies. They are acting on various levels of governance. Notably, they do not necessarily work. It relies on the ability to fit within the particular vulnerability context and on the mechanisms which they manage to activate (Boswell et al., 2012; Denyer et al., 2008; Serra et al., 2022). The formal climate adaptation governance is based on regulatory and legislative tools. An institutional and legal framework of adaptation is built by national adaptation plans, climate risk assessment conditions, new building regulations, and cross-sector environmental rules (Planos Gutiérrez, 2020; Shihab, K.K. et al., 2024; Smith & Lenhart, 1996). Such frameworks integrate adaptation into sectoral planning, establish binding standards, and distribute the responsibilities among governance tiers. Their performance is determined by the fidelity of implementation, political will, and consistency in the bigger governance landscape.

The second category of intervention is the institutional reforms. Structural drivers of vulnerability are met by cross-sectoral coordination mechanisms, the decentralization of decision-making, and enhanced sub-national planning capacities (Mortsch,

2011; OECD, 2020; Williamson et al., 2012). Such reforms are particularly significant in those systems of governance that are marked by fragmented powers and overlapping mandates. Another category of high impact is early warning systems and climate information services. These are the benefits that will result when communities and decision-makers have access to local and timely information about climate risks, which will help communities plan, reduce their exposure to hazards and make better-informed decisions at all levels of governance (Nikam et al., 2025). Examples of infrastructure-based interventions are flood barriers, drought-resistant water infrastructure, transport networks that are climate proof, and urban design that responds to heat (Breil & Johnson, 2012; Xie et al., 2023; You et al., 2014). These are especially significant in the context of fast urbanization. They are not durable and therefore are important long-term investments. Their effectiveness, however, relies on long-term maintenance and compliance with the overall planning goals. These formal interventions are supplemented by community-based adaptation (CBA) programs. Internally established adaptive capacity includes participation-based vulnerability assessment, livelihood diversification, ecosystem-focused interventions, and local capacity building programs (Bennett et al., 2016; Heltberg et al., 2009). They are particularly useful in areas where formal interventions of government have traditionally been limited.

6 Mechanism (M): The Way Interventions Decrease Vessels to Climate Change

The causal processes under which interventions produce adaptation outcomes are known as mechanisms. They do not only describe what occurs following an intervention, but how and why. This is a very important difference. The same intervention can be applied to two communities, yet the results can be completely different given that the mechanisms involved are different given different settings. CIMO is therefore analytically centered on mechanisms. They also associate interventions and outcomes, but they consider mediating variables of social, institutional, and environmental factors (Denyer et al., 2008; Goklany, 2007; Haase, 2022). This is one of the most valuable lessons in climate adaptation research because of its context-sensitivity. A program on infrastructure investment, early warn system, or livelihood diversification can be effective in this community and not in a different community. Variations in the ability to govern, social cohesion, and physical exposure all change –the functioning of interventions (Conway et al., 2019; Naylor et al., 2020). The CIMO method is the answer to this dilemma. It does not necessarily assume that outcomes are universally produced but rather focuses on the mediating processes by which it actually gets produced (Berrang-Ford et al., 2021; Denyer et al., 2008). One of the most basic forms of adaptive mechanism is capacity development. Most adaptation measures, be they policy oriented or localized in communities, operate by enhancing the capacity of individuals, communities, and institutions to learn and organize around and act on climate hazard. Governance preparedness is enhanced by investments in institutional fortification, the transfer of knowledge, and development of skills as well as administrative capacity (Hasan et al., 2023; Pundir et al., 2023). New information, technologies, and finances are also accessible due to these investments. The outcome is enhanced ability to adapt and a decrease in the exposure to climate-related risks (Ansari Mahabadi, 2024; Nalau et al., 2024).

A second important mechanism is learning. Climate change is unpredictable and nonlinear. This means that any strategy of adaptation that has been developed based on any static assumption cannot be held over time. Adaptive management systems should thus be learned through iteration. Social and institutional learning is created because of early warning systems, vulnerability assessments, pilot programs, and participatory planning processes. They allow the actors to update the strategies according to the new evidence and new results (Folke, 2006; McLaughlin, 2011). Policy innovation and lessening the threat of premature lock-in to interventions that become ineffective are also encouraged by learning mechanisms. A third important mechanism is institutional coordination. Climate vulnerability is cross-sectoral and multi-scalar in nature. It cuts across urban development, disaster risk reduction, water governance, agriculture, and even public health at the same time (Renaud & Perez, 2010; Serra et al., 2022; UNDRR, 2019). The key to adaptations is thus the coordination of action on and between these spheres. Coherence of policies is enhanced through clear institutional functions, horizontal coordination across the agencies, and vertical alignment of the governance levels (Adodoadji-Dogbe & Urban, 2023). Even well-constructed interventions may be significantly undermined when coordination cannot work or requires conflict. A fourth mechanism is behavioral change and social mobilization. Participatory governance, community programs, and inclusive decision-making procedures affect the way people see risk, get ready to hazards, and shared resources (Adger, 2006; Heltberg et al., 2009; Rumbach & Kudva, 2011). These are particularly critical in areas where the level of socio-economic vulnerability is elevated, and the scope of formal governance is constrained. They add to formal adaptation policies with the help of social capital, trust, and collective action capacity, making communities more resilient.

Combined, these four processes, which include capacity building, learning, coordination, and behavioral change, prove that the outcomes of adaptation are not an automatic result of interventions. They are created under multifaceted and context-specific dynamics, which occur with time. These processes are analytically apparent in the CIMO framework. It allows a deeper comprehension of the reasons why adaptation is successful in certain situations and failing in others (Berrang-Ford et al., 2021; Denyer et al., 2008).

7 Outcome (O): Adaptation and Resilience Outcomes

The CIMO model identifies outcomes as the visible and enduring impacts that are realized when adaptive mechanisms are effectively triggered by the means of proper interventions. They are not single and short-term outcomes. Instead, it is context-sensitive implication which emerges because of interaction between mechanisms and social, economic, institutional, and environmental systems (Folke, 2006; Shihab, K.K. et al., 2024; Zorita et al., 2025). In this, outcomes are not predetermined destinations of a process, but emergent properties of the process. One of the types of adaptation outcomes includes short-term risk reduction. The ability to lower the harm caused by the severe weather conditions and enhance the emergency response resources is not a negligible benefit. They are just a part of the picture. More notable and long-term effects are an increase in adaptive capacity, a decrease in structural vulnerability, and an improvement in the resilience of social-ecological systems (Ansari Mahabadi, 2024; Smith & Lenhart, 1996; Waiba & Dorji, 2024). These are the more long-term outcomes. They play a role in sustainable development paths that cannot be complemented by immediate reduction of risks. Effective adaptation should be hence not evaluated solely through the lens of short-term mitigations of risks, but also through the lens of systemic resilience and the general development benefits.

Adaptive capacity is never constant. It develops and leads to experiential learning, diversification of livelihood, social networking, and strengthening of institutions (Bennett et al., 2016). With these processes, systems get to be more prepared to handle uncertainty and future change. This is helped by education, knowledge sharing, and social learning. Institutions that incorporate reflexive learning in their governance systems are in a better position to develop resilience, promote informed decision-making, and prevent the lock-in effect of maladaptive pathways that lead to compromised long-term delivery (McLaughlin, 2011; Xie et al., 2023). The successful adaptation results have developed co-benefits as well. Like the case of climate risk embedded in sectoral planning, the benefits are not limited to reduction of risk. Integrating the need to address climate in agriculture, water, infrastructure, health, and urban development is an aspect that leads to more equitable and inclusive development, especially in unequal societies (Bank, 2021; Serra et al., 2022). It also serves to prevent the unintended effect of adapting actions that contribute to the further development of inequalities or creating new weaknesses. In the CIMO paradigm, the outcomes are not final. The cycle of adaptation can change the context with every cycle. Effective interventions enhance the institutions, equity of resources, and learning ability (Folke, 2006; Zhai & Lee, 2024). Such feedback processes are indicators of the nonlinear nature of climate adaptation. Resilience is not a place that has been arrived at. Learning, institutional change, and adaptive governance continuously produce and renew it. It is this process perspective that makes CIMO an effective lens to a view of climate adaptation as a complex process.

8 Climate Change Vulnerability and Adaptation Framework of CIMO

The CIMO framework has an analytical power in the way it relates its four components in one causal system. It demonstrates that there are no outcomes which are the result of isolated interventions. They appear because of the compatibility between situations of vulnerability, intervention aimed at them, and the processes that these interventions effectively trigger. This integrative perspective is critical. The complexity of climate adaptation cannot be represented by a linear cause and effect model (Berrang-Ford et al., 2021; Haase, 2022; Nalau et al., 2024).

Context (C) involves structural circumstances of vulnerability: socio-economic inequality, environmental vulnerability, governmental capacity, and development pathway. These circumstances define the way interventions are created and the experience that people go through (Lampis, 2013; Renaud & Perez, 2010). The interventions (I) must be contextual. Regulatory, institutional, infrastructural, and community-based interventions operate in different ways in accordance with the governance environment and realities of socio-economic environments they are in. Mechanisms (M) refer to avenues by which interventions produce change: there is capacity development, knowledge acquisition, coordination, and behavioral inspiration (Denyer et al., 2008; Folke, 2006). The outcomes (O) are the visible outcomes: the decrease in exposure and sensitivity, the increase in the adaptive capacity, the increase in resilience, and the movement toward sustainable development (Zorita et al., 2025).

Mechanism-based explanation is the analytical core of CIMO model. It provides a response to a rather simple yet significant question: why does this intervention have this effect? The transmission channels are capacity building, learning, coordination, and behavioral change where policy intent is brought to adaptive action (Denyer et al., 2008). In the case of the lack of such mechanisms, their inadequately designed or inappropriate to the contextual situation, the interventions fail, even when they are well-intended. This refers directly to one of the most frequent trends in climate adaptation maladaptive consequences of generic policy prescriptions that are not adjusted to local circumstance (Hinkel, 2011; Magnan et al., 2016; You et al., 2014). Context is based on outcomes. Institutional or community adaptation leads to stronger institutions, less inequality, and more competent learning systems (Bennett et al., 2016; Naylor et al., 2020). These types of feedback highlight the nonlinear nature of climate adaptation and support the usefulness of a systems-based approach to analysis and policy. A simplified summary of the integrated CIMO framework is provided in Table 2, indicating the conceptual position of each element in the process of adaptation.

Table 2 Integrated CIMO framework for climate change vulnerability and adaptation

Element	Core focus	Conceptual role in adaptation
Context (C)	Vulnerability settings	Determine exposure, sensitivity and adaptive capacity
Intervention (I)	Strategy and policy interventions	Addresses vulnerability drivers, by regulation, reform and community involvement
Mechanism (M)	Causal processes	Converts interventions into adaptive change that can be measured
Outcome (O)	Adaptation and resilience	Minimizes vulnerability and facilitates long-term sustainability pathways

Source Author's conceptualization based on the CIMO framework (Denyer et al., 2008) and climate change vulnerability and adaptation literature (Adger, 2006; UNDRR, 2019)

9 Policy and Strategic Implications

There are real-life implications of the CIMO framework on climate policy and strategic decision-making. These function on three levels which include context-sensitive policy design, multi-level governance, and rebranding adaptation as a continuous developmental process.

First, the framework elucidates that policies have to be context specific. The interaction between the community conditions, the sub-national capacities, the national policy settings, and the international commitment to climate determines vulnerability at the local level (Adodoadji-Dogbe & Urban, 2023; Nalau et al., 2024). Interventions that are technically advanced, without paying attention to local socio-economic realities or governance constraints, are not likely to be successful. The CIMO methodology thus requires systematic vulnerability assessment before intervention design. These indicators ought to be on exposure, sensitivity and adaptive capacity. It is also necessary that they maintain a vertical coherence between local adaptation plans, national policies, and international frameworks like the Paris agreement as well as SDGs (Bak, 2013; UNDRR, 2019; Waiba & Dorji, 2024). This is what renders the adaptation policies both locally and globally significant.

Second, the framework demonstrates that the policies need to operate not only at the institutional level but also within the institutional levels. Multi-level governance system does not automatically see shifts between knowledge and action. They must be facilitated purposely into organizational structures, intergovernmental levels of coordination, and linking responsibilities (Denyer et al., 2008; Mortsch, 2011; Serra et al., 2022). Policy responses consisting of horizontal coordination among the sectors and vertical coordination between governance levels are more likely to have lasting results compared to individual infrastructure investments or more strictly regulatory strategies. This corresponds with the general body of knowledge about the significance of institutional coherence of successful climate governance (OECD, 2020; Williamson et al., 2012; Xie et al., 2023).

Third, the CIMO framework requires a conceptual change in the way adaptation is conceptualized. It is not a single technical intervention. It is a systemic iterative process and developmental. When the accumulation of adaptation outcomes takes place, they can transform the conditions of vulnerability. They enhance institutional power, decrease the spatial and social inequalities, as well as enhance the social capital at community level (Folke, 2006; Zhai & Lee, 2024). These dynamics are related to the SDG agenda. Climate adaptation cuts across poverty eradication (SDG 1), food security (SDG 2), health (SDG 3), sustainable cities and communities (SDG 11), and climate action (SDG 13). By ensuring adaptation is included in overall development strategies, as opposed to it being a single environmental concern, coherence will be created between climate objectives and other sustainability priorities (Ansari Mahabadi, 2024; Serra et al., 2022). It also promotes the ability of governing in a flexible, learning-based manner that can adapt as conditions change.

10 Conclusion and Future Research Directions

This chapter has formulated a CIMO-based theoretical model of the study of climate change vulnerability and adaptation. The framework links context, intervention, mechanism, and outcome through an analytical structure in one. This is contrasted to much of the existing adaptation literature, which can be classified as being linear or output-oriented models which are not adequate to the complexity of the real adaptation processes. The CIMO model does not view context specificity and mechanism driven change as optional considerations of adaptation but as its essence.

The framework fills a longstanding gap in the research on adapting to adverse environments by incorporating vulnerability contexts, adaptive interventions, causal mechanisms, and resilience outcomes. Interventions have been over and over again considered to be universally transferable. The way and the reasons behind their performance in specific environments have not received adequate attention. This is corrected by CIMO that shifts between descriptive and explanatory causality. It defines the mediums under which socio-economic, institutional, environmental, and governance settings influence the manner in which mechanisms act and generate differentiated resilience outcomes amid settings. This way, it adds to the existing body of research on climate policy and sustainability by shedding light on the causal logic between the state of vulnerability and adaptive efficacy. The framework produces a practical diagnostic tool to the decision-makers. It aids with determining the state of vulnerability, choosing relevant interventions, and learn how the interventions promote long-term resilience to climate actions. This comes in very handy in situations where resources are limited and evidence is required to make evidence-based choices on priorities of adaptation investments.

The CIMO framework gives researchers a tool to base their empirical research on. The framework must be confirmed within various geographic, governance and sectoral contexts through future studies. Mixed-method designs are especially appropriate in this case. They are able to integrate the explanatory richness of a mechanism-based analysis with the comparative richness needed when generalizing their results in a policy-relevant manner. Particular focus must be drawn on under-researched locations, such as small island developing states, fast-urbanizing areas, and fragile states. The vulnerability in such situations is sharp, and the distance between generic policies and local realities is the most critical. The circumstances in which CIMO-identified mechanisms disintegrate, are inhibited or produce unintended effects should also be investigated in the future studies. This would be helpful in the learning of maladaptation and its structural reasons. It is critical to the designing of improved monitoring, learning, and adaptive management systems that contribute to the constant enhancement of climate adaptation policy as time goes on.

References

Adger, W. N. (2006). Vulnerability. *Global Environmental Change, 16*(3), 268–281. https://doi.org/10.1016/j.gloenvcha.2006.02.006

Adodoadji-Dogbe, C. D., & Urban, F. (2023). *Climate change, policy processes and local vulnerability*. Edward Elgar. https://doi.org/10.4337/9781800882119.00034

Ansari Mahabadi, S. (2024). Adapting to climate change through risk management. *IntechOpen*. https://doi.org/10.5772/intechopen.1005008

Bak, I. (2013). *Rethinking Climate Change Adaptation Policy: A Decision-Oriented Vulnerability Assessment*. Unpublished manuscript.

Bank, World. (2021). *Climate Change Action Plan 2021–2025: Supporting Green, Resilient, and Inclusive Development*. World Bank Group.

Bennett, N. J., Blythe, J., & Tyler, S. (2016). Communities and change in the anthropocene: Understanding social-ecological vulnerability and planning adaptations to multiple interacting exposures. *Regional Environmental Change, 16*(4), 907–926. https://doi.org/10.1007/s10113-015-0839-5

Berrang-Ford, L., Siders, A. R., Lesnikowski, A., et al. (2021). A systematic global stocktake of evidence on human adaptation to climate change. *Nature Climate Change, 11*(11), 989–1000.

Boswell, M. R., Greve, A. I., & Seale, T. L. (2012). Climate change adaptation strategies. *Island Press*. https://doi.org/10.5822/978-1-61091-201-3_6

Breil, M., & Johnson, K. (2012). Conceptualizing urban adaptation to climate change. *Social Science Research Network*. https://doi.org/10.2139/ssrn.2141453

Conway, D., Nicholls, R. J., Brown, S., et al. (2019). The need for bottom-up assessments of climate risks and adaptation in climate-sensitive regions. *Nature Climate Change, 9*(7), 503–511. https://doi.org/10.1038/s41558-019-0502-0

Cutter, S. L., Boruff, B. J., & Shirley, W. L. (2003). Social vulnerability to environmental hazards. *Social Science Quarterly, 84*(2), 242–261.

Denyer, D., Tranfield, D., & Van Aken, J. E. (2008). Developing design propositions through research synthesis. *Organization Studies, 29*(3), 393–413.

Dogra, P., Kaushik, A. K., Kalia, P., Kaushal, A. (2023). Influence of augmented reality on shopping behavior. *Management Decision, 61*(7), 2073–2098. World. https://doi.org/10.1108/MD-02-2022-0136

Downing, T. E., Patwardhan, A., Klein, R. J. T., et al. (2005). Assessing vulnerability for climate adaptation. In B. Lim & E. Spanger-Siegfried (Eds.), *Adaptation policy frameworks for climate change*. Cambridge University Press.

Fazey, I., Gamarra, J. G. P., Fischer, J., et al. (2010). Adaptation strategies for reducing vulnerability to future environmental change. *Frontiers in Ecology and the Environment, 8*(8), 414–422. https://doi.org/10.1890/080215

Folke, C. (2006). Resilience: The emergence of a perspective for social-ecological systems analyses. *Global Environmental Change, 16*(3), 253–267.

Goklany, I. M. (2007). Integrated strategies to reduce vulnerability and advance adaptation, mitigation, and sustainable development. *Mitigation and Adaptation Strategies for Global Change, 12*(5), 755–786. https://doi.org/10.1007/s11027-007-9098-1

Haase, E. (2022). The multilevel path to climate change adaptation. *RAE—Revista de Administração de Empresas, 62*(6). https://doi.org/10.1590/s0034-759020220609

Halofsky, J. E., & Peterson, D. L. (2017). *Climate Change Vulnerability and Adaptation in the Blue Mountains. USDA Forest Service.* Technical Report PNW-GTR-939. General.

Hasan, A., Tiwari, S., & Omrane, A. (2023). Development and validation of an online shopping scale adapted to specific cities of India. In *Digital Technologies for Smart Business, Economics and Education.* Springer.

Heltberg, R., Siegel, P. B., & Jorgensen, S. L. (2009). Addressing human vulnerability to climate change: Toward a 'no-regrets' approach. *Global Environmental Change, 19*(1), 89–99. https://doi.org/10.1016/j.gloenvcha.2008.11.003

Hinkel, J. (2011). Indicators of vulnerability and adaptive capacity. *Towards a Clarification of the Science–Policy Interface. Global Environmental Change, 21*(1), 198–208. https://doi.org/10.1016/j.gloenvcha.2010.08.002

Hudec, J. L., Halofsky, J. E., Peterson, D. L., & Ko, C. (2019). *Climate Change Vulnerability and Adaptation in Southwest Washington. USDA Forest Service.* Technical Report PNW-GTR-977. General.

Lampis, A. (2013). Vulnerabilidad y adaptación al Cambio Climático: Debates Acerca Del Concepto de Vulnerabilidad y Su Medición. *Cuadernos de Geografía: Revista Colombiana de Geografía, 22*(2), 37–49. https://doi.org/10.15446/rcdg.v22n2.37017

Magnan, A. K., Schipper, E. L. F., Burkett, M., et al. (2016). Addressing the risk of maladaptation to climate change. *Wires Climate Change, 7*(5), 646–665.

McLaughlin, P. (2011). Climate change, adaptation, and vulnerability: Reconceptualizing societal-environment interaction within a socially constructed adaptive landscape. *Organization & Environment, 24*(3), 269–291. https://doi.org/10.1177/1086026611419862

McLeod, E., Szuster, B. W., Tompkins, E. L., et al. (2015). Using expert knowledge to develop a vulnerability and adaptation framework and methodology for application in tropical island communities. *Coastal Management, 43*(4), 365–382. https://doi.org/10.1080/08920753.2015.1046803

Mortsch, L. D. (2011). Multiple dimensions of vulnerability and its influence on adaptation planning and decision making. In J. D. Ford & L. Berrang-Ford (Eds.), *Climate change adaptation in developed nations.* Springer. https://doi.org/10.1007/978-94-007-1770-1_5

Nalau, J., Gilmore, E., & Howden, M. (2024). Improving adaptation assessment in the IPCC. *Npj Climate Action, 3*(1), 76.

Naylor, A. W., Ford, J. D., Pearce, T., & Van Alstine, J. (2020). Conceptualizing climate vulnerability in complex adaptive systems. *One Earth, 2*(5), 444–454. https://doi.org/10.1016/j.oneear.2020.04.011

Nikam, R. J., Arunachalam, K. D., Mayani, S. V., et al. (2025). Evaluating climatic vulnerability in India: Moving from risk to resilience. *E3S Web of Conferences, 603, 04017.* https://doi.org/10.1051/e3sconf/202560304017

OECD. (2020). *OECD Environment Outlook to 2050: The Consequences of Inaction.* OECD Publishing.

Planos Gutiérrez, E. (2020). *Cambio Climático y La Gestión de La Adaptación.* Unpublished manuscript

Pundir, A., Tiwari, M., Venkatesan, K. G. S., Patil, K. S., Aher, H. R., & Hasan, A. (2023). Adaptation of sustainable business management for managing global success. *BioGecko, 12*(3), 3919–3926.

Renaud, F. G., & Perez, R. (2010). Climate change vulnerability and adaptation assessments. *Sustainability Science, 5*(2), 155–157. https://doi.org/10.1007/s11625-010-0114-0

Rumbach, A., & Kudva, N. (2011). Putting people at the center of climate change adaptation plans: A vulnerability approach. *Risk, Hazards & Crisis in Public Policy, 2*(4), 1–25. https://doi.org/10.2202/1944-4079.1096

Sarkodie, S. A., Ahmed, M. Y., & Owusu, P. A. (2022). Global adaptation readiness and income mitigate sectoral climate change vulnerabilities. *Humanities & Social Sciences Communications, 9*(1), 106. https://doi.org/10.1057/s41599-022-01130-7

Serra, V., Ledda, A., Ruiu, M., et al. (2022). Integrating adaptation to climate change into sustainable development policy and planning. *Sustainability, 14*(13), 7634. https://doi.org/10.3390/su14137634

Shihab, K. K., Vashisht, A., Hasan, A., & Joshi, A. (2024). Artificial intelligence and the metaverse. In The metaverse dilemma: Challenges and opportunities for business and society (Book chapter). Emerald Publishing Limited. https://doi.org/10.1108/978-1-83797-524-220241004

Shihab, K. K., Mir, F. A., Hasan, A., Agarwal, M., & Dixit, A. (2024). Pitch competitions–catalyzing innovation through alternative funding. In Alternative Finance (pp. 198–211). Routledge.

Smit, B., & Wandel, J. (2006). Adaptation, adaptive capacity and vulnerability. *Global Environmental Change, 16*(3), 282–292.

Smith, J. B., & Lenhart, S. S. (1996). Climate change adaptation policy options. *Climate Research, 6*(2), 193–201. https://doi.org/10.3354/cr006193

Soares, M. B., Gagnon, A. S., & Doherty, R. M. (2012). Conceptual elements of climate change vulnerability assessments: A review. *International Journal of Climate Change Strategies and Management, 4*(1), 6–35. https://doi.org/10.1108/17568691211200191

UNDRR. (2019). *Global Assessment Report on Disaster Risk Reduction 2019.* United Nations Office for Disaster Risk Reduction.

Waiba, T. N., & Dorji, Y. (2024). *From National to Local: A Comprehensive Vulnerability Assessment Framework for Climate Change Adaptation in Bhutan.* https://doi.org/10.21203/rs.3.rs-5037109/v1

Williamson, T., Campagna, M. A., & Ogden, A. (2012). *Adapting Sustainable Forest Management to Climate Change: A Framework for Assessing Vulnerability and Mainstreaming Adaptation into Decision Making.* Canadian Forest Service.

Xie, Y., Li, H., Liu, J., et al. (2023). The climate change impacts and responses index: Quantifying disparities and guiding policies for collective resilience. *International Journal of Sustainable Development & World Ecology, 31*(3), 272–286. https://doi.org/10.1080/13504509.2023.2268577

You, H., Connelly, E. B., Lambert, J. H., et al. (2014). Climate and other scenarios disrupt priorities in several management perspectives. *Environment Systems and Decisions, 34*(3), 476–490. https://doi.org/10.1007/s10669-014-9525-2

Zhai, L., & Lee, J. E. (2024). Investigating vulnerability, adaptation, and resilience: A comprehensive review within the context of climate change. *Atmosphere, 15*(4), 474. https://doi.org/10.3390/atmos15040474

Zorita, S., Milde, K., Peña Cerezo, N., et al. (2025). Extending from adaptation to resilience pathways: Perspectives from the conceptual framework to key insights. *Environmental Management.* https://doi.org/10.1007/s00267-025-02115-3

Part III
AI and Circular Economy Innovations

Chapter 12
Roadmap for AI-Powered Circular Value Chains for Carbon Neutrality and Corporate Sustainability in Emerging Economies

Sanjana Santra, Kishore Kumar, and Ankita Sharma

Abstract This chapter offers a strategic blueprint for using Artificial Intelligence (AI) to facilitate circular value chains for supporting carbon neutrality and corporate sustainability in emerging economies. As climate change and resource shortages call for moving from linear to circular ways, AI becomes a disruptive force to maximise material usage, promote transparency, and enhance closed-loop systems. With a focus on developing economies, the chapter delves into how AI can aid sustainable procurement, reverse supply chain flows, and waste minimisation across industries. Practical case studies identify the use of AI combined with Internet of Things (IoT), satellite imagery, and mobile platforms for tracking supply chains, emissions reduction, and making informal recycling networks formal. The chapter also discusses developing economy-specific barriers like data fragmentation, non-enforcement, and challenges in the informal sector and suggests how AI and digital infrastructure can help solve the issues. A step-by-step plan for companies to use AI-powered circular practices is discussed, from establishing sustainability objectives and testing cases to scaling effects through ecosystem partnerships. In summary, circular value chains enabled by AI present a potent synergy for realising environmental, social, and economic benefits, aligning emerging markets as leaders rather than followers of the global sustainability shift.

Keywords Circular value chains · Corporate sustainability · Carbon neutrality · Sustainable procurement · Resource efficiency

---

S. Santra · K. Kumar (✉)
BU Centre for Sustainable Development (BUCSD), School of Management, Bennett University, Noida, Uttar Pradesh, India
e-mail: Akishore001@gmail.com

S. Santra
e-mail: santrasanjana1995@gmail.com

A. Sharma
School of Management, IILM University, Greater Noida, India
e-mail: dr.ankitasharma912@gmail.com

K. Kumar et al. (eds.), *AI-Driven Circular Economy for Carbon Neutrality in Emerging Economies*, Environmental Footprints and Eco-design of Products and Processes,
https://doi.org/10.1007/978-3-032-30298-4_12

1 Introduction

Climate change and resource depletion are encouraging businesses to abandon conventional 'take-make-dispose' models in favour of circular economy approaches. In emerging economies, this is both a requirement and an opportunity (Goyal et al., 2018; Shukla et al., 2024). By rethinking value chains around reuse and recycling, companies can significantly reduce waste and carbon emissions. Indeed, circular solutions such as recycled inputs and prolonging product lifespan can help cut the carbon footprint of producing new materials (Crang et al., 2013; Johansen et al., 2022). Artificial intelligence (AI) is becoming an enabling force of real power in facilitating this change. Technology is crucial to realising this vision at scale, and AI can address complex optimisation issues quicker than ever. (Aldoseri et al., 2024). This chapter examines how corporations in developing economies can utilise AI to establish circular value chains, from green procurement and closed-loop production to AI-based reverse logistics, all in the service of carbon neutrality and long-term sustainability.

Developing economies have their own set of challenges and opportunities for circularity practices (Mishra et al., 2021). Industrialisation in these nations is accelerating at a breakneck pace, with increasing consumption and waste, but in many cases, with underdeveloped recycling methods (Cárcamo & Peñabaena-Niebles, 2022). Meanwhile, these economies can skip old systems by embracing 'digital power solutions, leveraging IoT, blockchain, and AI, to monitor materials and streamline processes for circularity. The stakes are high, as cities such as Delhi are suffering from massive landfills, with tens of millions of tonnes of waste piling, fuelling pollution and greenhouse gases (Batra et al., 2022). Businesses that adopt AI-facilitated circular value chains can reverse this tide, becoming carbon neutral while driving innovation and inclusive growth (Cheah et al., 2022). In the rest of this section, we move on to mainstream issues, from sourcing circularly with AI, to AI-enabled recycling systems, demonstrating each through real-world application examples across textiles, electronics, agriculture, construction and more. We also explain the impediments (data silos, illicit supply chains, poor enforcement) that firms will need to navigate, and lay out a vision for establishing AI-driven circular systems in developing countries.

1.1 Rethinking Value Chains: Circular Economy and Carbon Neutrality

To put this in perspective, it is essential to know what a circular value chain is and why it is vital for carbon neutrality. A circular economy is an economic model that substitutes the conventional 'end-of-life' idea with measures to decrease, reuse, recycle, and renew (Russell & Nasr, 2023). Rather than pulling new raw materials into each product and sending waste to landfills, a circular value chain keeps resources in circulation as long as possible, through methods such as product life extension, remanufacturing, and recycling waste as inputs (Delai & Alcantara, 2022). For businesses, this translates to rethinking every phase of the value chain: designing products to be long-lasting and repairable, sourcing sustainable and recycled inputs, employing production processes that create zero waste, and facilitating return loops to reclaim products after use. All these actions contribute directly to carbon neutrality. Through reusing material and 'encouraging recycling and reuse,' circular systems preclude the high emissions that come with the production of virgin materials. Producing something from recycled aluminium or plastic, say, uses a small fraction of the energy (and emissions) required to manufacture those materials in the first place (Shamsuddoha & Kashem, 2024). A circular economy not only significantly lowers waste production but also reduces the carbon impact, rendering it a central driver of business climate initiatives (Sönnichsen & Clement, 2020).

Rising economies have a great deal to benefit from by incorporating circular principles into their value chains from the outset. Such nations tend to experience severe pollution and resource shortage in tandem with growth. For example, India is at a very nascent stage of waste management infrastructure in place (Hossain et al., 2022a, 2022b). As a result, gigantic open dumps have been polluting the air and water (Soni et al., 2023). By making value chains circular, businesses can mitigate such environmental challenges while generating fresh value streams and employment opportunities. Notably, circular business models can stimulate economic growth by stimulating local recycling industries and innovation, at a similar carbon price as the previous linear models (Pieroni et al., 2021). This two-for-one advantage is why circular economy practice is regarded as key to achieving global climate ambitions (such as the Paris Agreement) on terms that are supportive of growth in emerging markets.

1.2 AI for Sustainable Procurement

Circular value chains begin with what companies purchase. Sustainable procurement involves procuring materials, components, and products that are environmentally friendly, recycled or recyclable, and produced ethically (Boruchowitch & Fritz, 2022; Kaith & Sachdeva, 2024). Yet, it is very difficult to find suppliers that have stringent environmental, social, and governance (ESG) standards in worldwide supply bases. That is where AI can make a huge difference. AI-powered procurement platforms can quickly scan enormous quantities of supplier information from environmental ratings and certifications to real-time news regarding labour or environmental processes and assist companies in making more sustainable sourcing choices (Camarena, 2022; Kumar et al., 2025; Prakash & Tiwari, 2025; Singh et al., 2025a, 2025b). For example, an AI-driven solution can accept a rigorous set of procurement specifications, such as sustainability parameters like low carbon content, recycled materials usage, non-toxic material usage, etc., and subsequently conduct internal and external searches among potential vendors who fit these (Bains et al., 2024; Elmousalami et al., 2025). MHP, a consulting firm, explains employing text-mining algorithms that scan internal supplier databases and internet news feeds to find suppliers that satisfy ESG (standardsmhp.com). Such a tool widens the pool of suppliers to include creative start-ups or local manufacturers with solid sustainability records that would otherwise remain under the radar (Agrawal, 2025; Joselin et al., 2025; Mohammadrezaei, 2024; Singh et al., 2024). By automatically ranking and filtering suppliers by sustainability data, AI enables procurement managers to look beyond cost and quality, introducing circular considerations into the decision process (Ittoo et al., 2016; Tripathi et al., 2022). This can, in its turn, stimulate demand for recycled content and components, as AI can bring to prominence vendors that supply, for example, recycled plastic resins or remanufactured goods.

AI can also enhance supply chain transparency, a vital component of sustainable sourcing (Pal, 2023). Supply chains of most emerging economies are opaque and multilayered, and this makes it hard for a corporation to know whether the raw material (e.g. cotton, palm oil, minerals) is produced using child labour or deforestation. Advanced analytics and satellite imagery are relevant here. Innovative companies have begun applying AI to satellite imagery to monitor their sourcing locations in real time. Unilever, for instance, uses a combination of geolocation information and AI analysis of satellite images to scan for deforestation in its palm oil supply chain in Southeast Asia (unilever.com). This 'eye in the sky' tool sends deforestation warnings if any of Unilever's suppliers or adjacent areas initiate forest clearing, enabling the company to step in and ensure a deforestation-free supply chain (Mukherjee et al., 2021; Singh et al., 2025b). They even estimate the carbon locked up in those forests and the climate impact if they were chopped down, linking procurement choices to carbon outcomes. Such monitoring with AI is especially useful in emerging economies where land-use policy enforcement may be weak. It effectively gives companies unprecedented visibility into the 'first mile' of their supply chain, including informal or smallholder sources that were out of reach before. With these facts in hand, procurement can favour suppliers that protect forests and reject those that do not, thus pushing sustainability upstream.

Case-AI for Sustainable Sourcing: In the textile and apparel sector, which is humongous in nations such as India and Bangladesh, AI is being leveraged to reduce material wastage at the design and sourcing level. One such cutting-edge example is start-up SXD (Shelly Xu Design), which developed an AI-powered design software to design zero-wastage garment patterns (Lu & Serafeim, 2023). Traditionally, garment pattern design leads to 15–20% of the cloth going as off-cuts (waste) onto the cutting room floor. SXD's AI technology can automatically calibrate and optimise pattern arrangements so that every inch of cloth gets utilised, minimising wastage by a whopping amount. The result is staggering, SXD's solution generated around 10× more material savings than conventional practice, equivalent to up to 69% less cloth being consumed by the same garmentaccenture.com. Since fabric making is a labour-intensive activity (textiles are one of the world's biggest industrial polluters), this saving also reduces emissions by an estimated ~ 80% for those garments (Visvanathan & Anbumozhi, 2018). By smartly scaling patterns, AI allows you to purchase less raw material for a production run yet achieve design goals (Wang et al., 2021). This AI-powered sustainable design ensures procurement is tightly integrated with circular objectives; you procure just what you need, and you select material that you know will be utilised to the fullest.

2 Barriers in Emerging Economies and How AI Helps Overcome Them

Moving towards AI-driven circular value chains is not problem-free, especially in the developing world. It is important to identify these challenges so that firms can address them fairly. Three significant challenges are worth mentioning here: data fragmentation, informal supply chain dominance, and lack of policy enforcement. All of these can slow down or make circular initiatives complex to implement. Fortunately, AI and digital strategies, along with facilitating policies and partnerships, can overcome many of these challenges.

1. *Data Fragmentation and Quality*:

Good quality data is the cornerstone of any analytics or AI solution. In most emerging market scenarios, however, little or intermittent data exists for supply chains or environmental performance. There may be numerous disconnected systems (or even paper-based data), and suppliers may never have measured waste or emissions. The disconnected nature means we cannot know where circular interventions may occur or be tracked. For instance, the construction industry has never had material consumption and waste data to hand, a report cites that lack of available data required to make informed decisions around waste management is a main issue (Burger et al., 2018), resulting in inefficiency and lost recycling potential (Zeng et al., 2017). Likewise, the major challenges are getting input data to AI both accurate and structured, and it takes much effort to clean up and consolidate erroneous, incomplete, or stale data sets from heterogeneous sources (Maharana et al., 2022; Xu et al., 2021).

AI alone can address some data issues. Machine learning can impute missing data by picking up patterns in what is available. For example, if a small supplier is not reporting their carbon footprint, an AI model can estimate it based on similar reportable suppliers, filling blanks for emissions tracking (Hossain et al., 2022a, 2022b). AI can also process and combine unstructured data, like parsing text out of supplier documents or extracting activity from satellite pictures, generating new points of data that are not in databases (Van Wassenhove, 2019). Effectively, AI can take several streams of semi-structured information and convert them into usable data. Companies must first spend money on creating a digital infrastructure: IoT sensors to create real-time data (e.g. on energy consumption, food waste bins fill level), (Srivastava et al., 2021) and cloud platforms to consolidate data across departments and partners (Kamble et al., 2022; Srivastava et al., 2018). With those in place, AI algorithms can identify anomalies (highlighting data errors) and provide a single, accurate picture of the truth. Digital twin models of supply chains also consolidate data, as all inputs and outputs are duplicated in a virtual model that AI can scan continuously for inefficiencies or waste (Saha et al., 2025).

2. *Inclusive Supply Chain Coverage*

As described, in most developing economies, a significant majority of value chain activities, especially end-of-life, are controlled by the informal sector, individuals, and small businesses outside formal company structures (Hu & Wen, 2015). For example, an estimated 80–90% of e-waste and plastic recycling in India is informally carried out by tens of thousands of waste pickers and backyard recyclers (accushred.net). Although such informal networks are highly entrepreneurial and have helped achieve relatively high recycling rates, their very informality is an issue (Laha, 2022; Srivastava et al., 2024). It creates data invisibility as their activities are not included in official corporate or municipal statistics. An Indian national study concluded that over 90% of India's e-scrap enters the informal sector for processing, in clusters like the Seelampur neighbourhood of Delhi with 25,000 informal workers (accushred.net). The government brought in extended producer responsibility (EPR) rules in 2016 (Gupta & Dash, 2023), mandating electronics makers to take e-waste responsibility, but no one has figured out how to integrate the Indian informal side successfully yet (Dutta & Goel, 2021). This is symptomatic of the broader issue of how to bring informal players into a circular economy system without hampering their livelihoods.

AI solutions can be designed to include and empower informal workers. With mobile apps and online marketplaces, as Recykal and others have implemented, informal collectors can be given a formal platform upon which to sell material, track their collections (Luthra et al., 2023), and receive good payment. This not only gives companies and authorities improved data as all transactions are tracked electronically, but it can also increase profit for waste pickers by eliminating middlemen and optimising prices. Another benefit AI provides is traceability, for instance, by using QR codes or low-cost RFID tags on products (Lukacs et al., 2025), which can be read by an informal recycler using a smartphone to mark recycling of the product. This model has been tested on PET bottles and electronics; it is a win–win in which business gets evidence towards their recycling commitments and informal workers potentially receive bonuses for being in the tracking system.

3. *Weak Policy Enforcement and Incentives:*

Enabling policies and enforcement systems is necessary to push corporate actions towards circularity. Even where good environmental laws are present in most emerging economies, their enforcement may fall behind because of a lack of resources or corruption (Mahmood et al., 2021). For instance, prohibitions on single-use plastics or requirements for waste segregation may be in place, but enforcement is uneven. In the business world, governments may ask companies to assume responsibility for the end-of-life of their products (such as India's EPR for electronics), but if companies perceive that enforcement is lacking, they might not comply in full, and business-as-usual linear behaviour will continue (De Abreu et al., 2021). Inadequate enforcement also implies that rogue players like illegal dumpers, polluting plants, undercut compliant companies, which can deter investment in circular solutions (Chandrappa & Das, 2024).

AI can enhance the monitoring and enforcement environment. Governments are starting to leverage remote sensing and artificial intelligence to identify environmental infractions, such as via satellite imagery or aerial vehicles, to identify unlawful dumping or incineration of waste (Divine et al., 2024). It is possible to train machine learning algorithms to identify characteristic indicators of, e.g. an unapproved landfill or an industrial outflow in a river. All of this can be done in real time with AI processing and raise alerts to send to inspectors to deal with. Such openness makes it more difficult for non-compliance to go unchecked. AI can also assist with monitoring compliance in supply chains by reviewing data and determining whether a company is hitting its recycling/reuse goals under legislation (Tariq, 2025).

Another policy-related area is the absence of common metrics and frameworks specific to emerging economies. Even with international frameworks such as GRI or the Science-Based Targets, local conditions can vary (e.g. greater need for informal sector inclusion, varying waste streams). Businesses might not know how to establish KPIs or how to measure circular development. This can be an action barrier. AI here is not a silver bullet, but data-driven methods can provide localised baselines and targets. Governments and industry associations may apply AI to simulate scenarios (such as the potential for CO_2 reduction if circular practices are implemented in, e.g. the textile industry) and establish evidence-based targets.

Cost and expertise are, in fact, generic obstacles to adopting AI as well. Several AI solutions demand an initial outlay and technical know-how that some emerging-market companies may not possess (Teece, 2025). But cloud computing and AI-as-a-service are decreasing this barrier, enabling companies to test without hefty infrastructure and as home-grown tech ecosystems come of age (India, e.g. has a robust IT industry with AI start-ups interested in local challenges), the knowledge barrier is also decreasing. Over time, we will see AI technologies become easier to use (no-code or low-code AI) so that sustainability teams can access them without necessarily requiring data scientists at each juncture (How et al., 2023; Simon, 2022).

In practice, emerging economies have additional challenges, dispersed data, big informal sectors, poor enforcement, but these are not insurmountable. AI and digital technologies, with clever policies and capacity development, can overcome many of the gaps. The following section presents a roadmap for businesses to implement AI-driven circular practices, bearing these challenges in mind and utilising partnerships to drive success.

3 Roadmap for AI-Derived Circular Value Chains

Initiating a circular economy shift with AI needs a well-defined strategy. Here we suggest a pragmatic roadmap or framework that firms (particularly in emerging markets) can follow to transition towards AI-driven circular value chains. The roadmap is a series of sequential steps and enabling actions, with key performance indicators (KPIs) to monitor progress. It also emphasises the need for partnerships, as no firm can create a circular ecosystem by itself.

1. *Establish a Circular Vision and Objectives*

All journeys begin with a destination. Businesses need to start by having a circular economy vision aligned with their carbon neutrality and sustainability ambitions. At the leadership level, this translates to a promise to redesign value chains and products for circularity, with clear objectives such as '100% recyclable or reusable packaging by 2025' or '50% recycled content in products by 2030' (Scrioşteanu & Criveanu, 2023). It is essential to link such goals to accepted frameworks, for instance, map them to applicable UN Sustainable Development Goals (such as Responsible Consumption and Production), apply GRI indicators (such as waste sent to landfill, materials re-used), and coordinate carbon-related goals for emissions. KPIs here are target-setting ones: % decrease in virgin material usage, % waste decrease, reduction of emissions (to help achieve net-zero), etc., that will steer the initiative.

2. *Evaluate the Value Chain and Determine Hotspots*

Second, carry out a baseline measurement of the existing value chain to identify where the largest linear problems are; these are the circular improvement hotspots. This may be done through a material flow analysis, potentially by using software and AI to model flows to identify where lots of waste or emissions take place. As an example, a consumer products company may learn that packaging contributes significantly to its waste profile, or that product returns and unsold stock are a significant problem at a fashion retailer. At this point, collect data from throughout the departments: procurement, manufacturing, logistics, product returns, and end-of-life. It is expected that data are not complete; the evaluation may employ estimates or industry averages wherever necessary, which can subsequently be improved with AI. Certain organisations employ lifecycle assessment (LCA) tools that are supplemented by AI to quickly estimate the impacts of products and see which phases cause the most waste and carbon. AI software can also be employed to search external sources (industry reports, research) to compare performance against peers. For example, an AI text-mining software could aggregate what other firms in your industry claim as their recycling rates or waste percentages. The result of this stage is a well-defined map of opportunities: e.g. 'Shifting to recycled plastic in product X can reduce 500 tons of CO_2'; 'Recycling and reusing product Y can save 20% on raw materials.'

3. *Establish Data Infrastructure and Digital Capabilities*

With priorities established, companies can make sure the data and technology foundation to enable AI solutions. This includes spending on data collection (sensors, IoT, ERP system upgrades) and integration (data lakes or platforms that integrate data silos). For instance, if one of the objectives is enhancing reverse logistics, you can use IoT trackers on returnable packaging or GPS on collection vehicles to collect data. If supplier sustainability is of importance, begin to implement a supplier data portal where suppliers enter key metrics (energy consumption, recycled material, etc.). Data cleaning and standardisation are crucial, perhaps by utilising an AI-driven master data management tool to consolidate duplicates and correct errors. Also, start building the digital competencies of your employees: upskill procurement teams to use AI analytics dashboards, upskill factory engineers to read AI predictive maintenance alerts, etc. During this phase, creating a small cross-functional 'Circular Innovation Team' with IT/data specialists and sustainability specialists can be useful to guide the tech rollout. KPIs here can be process-centric, count of data points that are being harvested, scores of data quality, or hours of training in new systems.

4. *Pilot AI Solutions in Key Areas*

Instead of implementing everything simultaneously, determine 1–3 high-impact use cases to pilot AI solutions at a reduced level. Piloting enables learning and iteration before expanding. For each initiative, set clear goals and metrics. Some examples of pilots:

- AI for Sustainable Procurement: Rank suppliers based on sustainability metrics using an AI platform. Maybe pick a single procurement category (e.g. packaging materials) and use AI to recommend substitute suppliers with higher recycled content or reduced carbon footprint. Metric: increase in selected suppliers' sustainability score, or decrease in supply chain CO_2 for that category.
- Smart Factory Waste Reduction: Install machine vision on a single production line to identify defects and optimise cutting patterns (such as the SXD method) in a test environment. Measurement: percentage scrap waste reduction, cost savings realised, and CO_2 reduction from reduced material usage.
- AI-Powered Reverse Logistics: Collaborate with a partner (perhaps a technology start-up) to implement a digital platform for back-end product take-back across one city or a single product line. For example, implement an AI-scheduling system for pickup of returned goods from consumers, or an incentive app that leverages AI to suggest the closest drop-off locations.

This is also an opportunity to do return on investment (ROI) calculations, which may entail more than outright cost savings but ultimate benefits such as regulatory compliance, brand equity, and carbon emissions reductions (with a financial payoff if there's carbon pricing or credits).

5. *Scale Up and Embed Circular Practices*

The initiatives that were successful or provided useful lessons can help to develop the scale up across the operations of the company. Scaling could involve taking an AI procurement tool to all product categories, deploying predictive waste reduction to all manufacturing plants, or scaling a take-back programme across a country. It is important here to embed the solutions into core business processes so that they are not isolated experiments. For instance, adding the AI supplier scoring into the formal supplier on-boarding procedure or bidding for the procurement system or putting the reverse logistics app in synergy with the corporation's primary customer app, to target additional users. Keeping track through KPIs: monitor corporate-level indicators now, such as total waste recycling rate, total secondary raw material utilised, total CO_2 emissions saved through circular activities, will hugely help. Most companies develop an internal dashboard of circular economy KPIs for leadership, typically using AI to predict whether they are likely to meet future targets. For example, if the objective is '50% recycled content by 2030,' an AI system may anticipate based on existing trends whether it would be achieved or not and if attempts must speed up.

6. *Collaborate Throughout the Ecosystem*

Building a circular value chain is an ecosystem endeavour. Companies should actively seek partnerships with various stakeholders: technology providers, suppliers, recyclers, start-ups, government agencies, and even competitors in some cases (pre-competitive collaboration for recycling infrastructure, for instance). Identify gaps that the company cannot fill alone. If there is a need for advanced AI algorithms, maybe partner with a local AI start-up or a university research lab. If the informal sector must be involved, work with NGOs or social enterprises that have networks of waste pickers (such as the network of waste picker cooperatives). As an example, entering partnership with a company such as Banyan Nation (an Indian company that includes informal plastic collectors in its system using technology) could provide your packaging with quality recycled plastic. Or participating in efforts such as the Africa Circular Economy Alliance or Platform for Accelerating the Circular Economy (PACE) can offer insights and co-development for AI solutions specific to emerging markets.

Government collaborations are also crucial, firms can test projects with city governments, e.g. smart waste collection technologies where data flows in both directions. Emerging economy governments can provide incentives or assistance for digital innovation that enhances waste management, such as access to municipal waste streams for recyclability, or economic incentives for technology that enhances recycling rates. One illustration of policy synergy is leveraging AI data to assist cities in implementing or enforcing laws on segregating waste, the firm achieves cleaner recyclable material, and the city achieves its landfill diversion goal.

7. *Measure, Report, and Refine*

Lastly, embrace a culture of ongoing improvement. Rigorously measure results and transparently report them. AI can significantly assist in measurement by taking data collection and analysis automatically (as with carbon tracking). Such firms as Unilever have also installed 'AI-powered carbon measurement systems' to obtain real-time product data. In the same way, an AI dashboard of circularity can display, say, live figures on waste recycled, water conserved, etc. Public disclosure according to standards such as GRI or CDP will not only illustrate advancement to stakeholders but also compel the firm to keep strong data procedures. Numerous companies now incorporate circular economy metrics into their annual integrated reports.

When reporting, qualitative insights are needed to be included as well, what did work, what did not, and what comes next. Reflective practice refines the strategy. Maybe the data indicates that one initiative reached a plateau, analyse why and iterate. Perhaps one AI model's performance degrades with time; plan regular reviews and model retraining as necessary.

Critical KPIs to monitor during this path will include:

- Material Circularity Indicators: For example, % recycled input materials, % of products designed for disassembly, material utilisation rate in manufacturing (how much of input ends up in final product vs scrap).
- Waste and Recycling Metrics: For example, total waste generated vs. waste recovered, landfill diversion rate (%), volume of product take-back (units or weight), recycling rate by material type.
- Carbon and Energy: For example, greenhouse gas emissions reduced due to circular activities (this can be estimated by AI models by referencing a baseline), energy conserved through utilisation of secondary materials or remanufacturing (because recycled aluminium takes ~ 95% less energy compared to primary, etc.).
- Economic Metrics: Cost avoided due to reuse of materials, revenue from secondary products (such as refurbished goods), and ROI on AI deployment.
- Social Indicators: Jobs created in reuse/recycling, rise in income for informal workers involved, training hours offered aligning with just transition principles is significant particularly in emerging economies.

Through the tracking of these, businesses can ascertain that the circular shift is paying dividends and remains on track. AI will be used not only to monitor these KPIs but also to extract insights (for instance, applying machine learning to determine what drivers most significantly contribute to an increase in recycling rate in an area consumer behaviour, incentives, frequency of collection). These insights are passed back into the subsequent cycle of strategy refinement.

8. *Scale Impact and Innovate*

Beyond the company's operations, the endgame is to influence the broader system. As the circular model proves its worth (both environmentally and financially), companies can start to advocate for broader adoption. This might mean open-sourcing some AI tools or sharing data that can benefit the whole industry (for instance, publishing anonymised data on material lifespans to help create industry benchmarks). It might also mean collaborating with policymakers to amplify reach, i.e. proving to a government that an AI-powered bottle return scheme is effective, and lobbying for it to be adopted city-wide or nationwide. Continuing innovation is also necessary; keep looking for next-generation technologies (e.g. high-end AI for recycling molecules, or applying machine learning in the circular product design stage to model recycled material performance, etc.). The space for AI and circular economy keeps changing, so keep adapting. Foster a culture of innovation within—maybe keep a 'digital circular innovation lab' which tests new concepts even after completing the primary transformation.

By using a blueprint such as this, a company can progress step by step towards an AI-optimised circular value chain. Begin with vision, validate ideas with data and pilots, and then scale up with the right partners and continue to refine. The process might take several years, but early success can build momentum. And the outcome is not merely compliance or positive PR, it is a more resilient, efficient, and future-proof business. Circular value chains minimise reliance on unstable raw material availability (a significant risk reducer), and AI introduces agility and forward thinking, together, which enable businesses to succeed in a resource-limited, carbon-limited future.

4 Conclusion

Emerging economies stand at a crossroads where fast growth can either seal a heritage of pollution and waste or turn towards a circular, sustainable future. Companies can play a key role in nudging the latter direction, and AI is the driver that can enable them to do so. As discussed in this chapter, AI-driven circular value chains present a way for firms to pursue ambitious sustainability targets such as carbon neutrality while simultaneously opening up economic possibilities. By applying AI to reimagine reverse logistics, production, design, and procurement, companies can extend the use of materials, extract greater value from resources, and reduce emissions and waste by a large margin.

The cases of India's recycling technology platforms to Brazil's precision farming, and South Africa's emerging circular construction projects indicate that this change has already begun. Business is leveraging AI on long-standing issues, whether optimising textile designs to reduce fabric waste or leveraging machine learning to convert e-waste into a valuable commodity, with striking success. As raw materials get pricier or harder to find and consumers and regulators increasingly call for sustainability, circular value chains will be the key to staying ahead.

The journey involves overcoming challenges specific to emerging markets, from bringing informal workers into the fold to enhancing data and governance. But as we have discussed, these challenges can be addressed with careful strategies and cooperation. Developing economies can leapfrog by constructing more circular systems from scratch, rather than emulating the wasteful paths that developed nations took in the last century. With the assistance of AI, a lack of legacy infrastructure becomes an opportunity to deploy. For example, a new city could implement an artificial intelligence-based waste management system before it ever constructs traditional landfills, thereby preventing the development of 'mountains of trash' in the first place.

For business leaders, the takeaway is that the circular economy and AI combined create a very potent synergy. AI brings the intelligence and predictive capability to run circular value chains efficiently, and circular models give AI projects concise, meaningful goals (such as minimising CO_2 or cost by reclaiming materials). Firms must approach these not as competing agendas (digital transformation versus sustainability) but as interwoven strands of one strategy. The roadmap presented here is a starting template, but each organisation's journey will be unique. The key is to begin piloting, begin partnering, and begin learning. Small wins will lead to more significant ones.

In summary, AI-driven circular value chains hold a promising pathway for corporations within emerging economies towards triple bottom line gains, environmental (carbon neutral, reduced pollution), social (green employment, inclusive growth), and economic (cost reduction, new markets). By rethinking value chains with circular principles and turbocharging them with AI, businesses can become frontiers of sustainable development.

References

Agrawal, D. (2025). A comprehensive review of investing and financing for sustainable tourism projects. *Environment, Sustainability, and Governance Insights, 1*(1), 53–70.

Aldoseri, A., Al-Khalifa, K. N., & Hamouda, A. M. (2024). AI-powered innovation in digital transformation: Key pillars and industry impact. *Sustainability, 16*(5), 1790.

Bains, A., Sridhar, K., Dhull, S. B., Chawla, P., Sharma, M., Sarangi, P. K., & Gupta, V. K. (2024). Circular bioeconomy in carbon footprint components of nonthermal processing technologies towards sustainable food system: A review. *Trends in Food Science & Technology*, 104520.

Batra, S., Srivastav, A. L., Chadha, P., & Madaan, N. (2022). Challenges of waste management in Delhi (India) and its scope of improvement to achieve cities' sustainability in developing nations: A review. *Cities, 121*, Article 103480.

Boruchowitch, F., & Fritz, M. M. (2022). Who in the firm can create sustainable value and for whom? A single case-study on sustainable procurement and supply chain stakeholders. *Journal of Cleaner Production, 363*, Article 132619.

Burger, C., Kalverkamp, M., & Pehlken, A. (2018). Decision making and software solutions with regard to waste management. *Journal of Cleaner Production, 205*, 210–225.

Camarena, S. (2022). Artificial Intelligence (AI) for sustainable institutional food systems: Implementation of AI tools for school nutrition program management in the United States of America. *Frontiers in Sustainable Food Systems, 6*, Article 743810.

Cárcamo, E. A. B., & Peñabaena-Niebles, R. (2022). Opportunities and challenges for the waste management in emerging and frontier countries through industrial symbiosis. *Journal of Cleaner Production, 363*, Article 132607.

Chandrappa, R., & Das, D. B. (2024). Waste Mafia. *Solid waste management: principles and practice* (pp. 657–697). Springer International Publishing.

Cheah, C. G., Chia, W. Y., Lai, S. F., Chew, K. W., Chia, S. R., & Show, P. L. (2022). Innovation designs of industry 4.0 based solid waste management: Machinery and digital circular economy. *Environmental Research, 213*, 113619.

Crang, M., Hughes, A., Gregson, N., Norris, L., & Ahamed, F. (2013). Rethinking governance and value in commodity chains through global recycling networks. *Transactions of the Institute of British Geographers, 38*(1), 12–24.

De Abreu, M. C. S., Webb, K., Araújo, F. S. M., & Cavalcante, J. P. L. (2021). From "business as usual" to tackling climate change: Exploring factors affecting low-carbon decision-making in the Canadian oil and gas sector. *Energy Policy, 148*, Article 111932.

Delai, I., & Alcantara, R. L. C. (2022). Circular value chain practices for developing resource value retention options. *Journal of Cleaner Production, 359*, Article 131925.

Divine, I. L. O. H., Aguma, C. P., & Olagunju, A. O. (2024). *Integrating AI enhanced remote sensing technologies with IOT networks for precision environmental monitoring and predicative ecosystem management.*

Dutta, D., & Goel, S. (2021). Understanding the gap between formal and informal e-waste recycling facilities in India. *Waste Management, 125*, 163–171.

Elmousalami, H., Maxy, M., Hui, F. K. P., & Aye, L. (2025). AI in automated sustainable construction engineering management. *Automation in Construction, 175*(10620), Article 106202.

Goyal, S., Esposito, M., & Kapoor, A. (2018). Circular economy business models in developing economies: Lessons from India on reduce, recycle, and reuse paradigms. *Thunderbird International Business Review, 60*(5), 729–740.

Gupta, D., & Dash, S. (2023). Challenges of implementing extended producer responsibility for plastic-waste management: Lessons from India. *Social Responsibility Journal, 19*(9), 1595–1612.

Hossain, R., Islam, M. T., Shanker, R., Khan, D., Locock, K. E. S., Ghose, A., et al. (2022). Plastic waste management in India: Challenges, opportunities, and roadmap for circular economy. *Sustainability, 14*(8), 4425.

Hossain, R., Islam, M. T., Shanker, R., Khan, D., Locock, K. E. S., Ghose, A., Schandl, H., Dhodapkar, R., & Sahajwalla, V. (2022). Plastic waste management in India: Challenges, opportunities, and roadmap for circular economy. *Sustainability, 14*(8), 4425.

How, M. L., Cheah, S. M., Chan, Y. J., Khor, A. C., & Say, E. M. P. (2023). Artificial intelligence for advancing sustainable development goals (SDGs): An inclusive democratized low-code approach. *The ethics of artificial intelligence for the sustainable development goals* (pp. 145–165). Springer International Publishing.

https://accushred.net/blog/factory-resets-101-the-most-overlooked-step-in-smart-device-disposal/

Hu, S., & Wen, Z. (2015). Why does the informal sector of end-of-life vehicle treatment thrive? A case study of China and lessons for developing countries in motorization process. *Resources, Conservation and Recycling, 95*, 91–99.

Ittoo, A., Nguyen, L. M., & van den Bosch, A. (2016). Text analytics in industry. *Computers in Industry, 78*(78), 96–107. https://doi.org/10.1016/J.COMPIND.2015.12.001

Johansen, M. R., Christensen, T. B., Ramos, T. M., & Syberg, K. (2022). A review of the plastic value chain from a circular economy perspective. *Journal of Environmental Management, 302*, Article 113975.

Joselin, D., Singh, A., & Tiwary, K. S. (2025). The Nexus between CSR practices and financial performance: Empirical evidence from the manufacturing sector in India. *Environment, Sustainability, and Governance Insights, 1*(1), 1–17.

Kaith, A., & Sachdeva, G. (2024). Bibliometric insights into green human resource management: A connection with business ethics and values. *PURUSHARTHA-A Journal of Management, Ethics and Spirituality, 17*(1), 51–63. https://doi.org/10.21844/16202117104

Kamble, S. S., Gunasekaran, A., Parekh, H., Mani, V., Belhadi, A., & Sharma, R. (2022). Digital twin for sustainable manufacturing supply chains: Current trends, future perspectives, and an implementation framework. *Technological Forecasting and Social Change, 176*, Article 121448.

Kumar, K., Singh, R., Choudhary, M., & Sharma, A. (2025). Climate change mitigation strategies in environmentally sensitive industries. *In Climate Change and Social Responsibility* (Vol. 25, pp. 185–204). Emerald Publishing Limited.

Laha, S. (2022). Governing the network: Trust in E-waste informality in India. *Geoforum, 134*, 1–12.
Lu, S., & Serafeim, G. (2023). Circular products and business models: Opportunities for entrepreneurs, managers, and investors. *Managers, and Investors (April 27, 2023).*
Lukacs, M., Toth, F., Horvath, R., Solymos, G., Alpár, B., Varga, P., et al. (2025). Advanced digital solutions for food traceability: Enhancing origin, quality, and safety through NIRS, RFID, Blockchain, and IoT. *Journal of Sensor and Actuator Networks, 14*(1), 21.
Luthra, A., Lowe, J., & Berkley, K. O. (2023). Dispossession by platformization: The rise of on-demand recycling in urban India. *Geoforum, 141*, Article 103731.
Maharana, K., Mondal, S., & Nemade, B. (2022). A review: Data pre-processing and data augmentation techniques. *Global Transitions Proceedings, 3*(1), 91–99.
Mahmood, H., Tanveer, M., & Furqan, M. (2021). Rule of law, corruption control, governance, and economic growth in managing renewable and nonrenewable energy consumption in South Asia. *International Journal of Environmental Research and Public Health, 18*(20), 10637.
Mishra, J. L., Chiwenga, K. D., & Ali, K. (2021). Collaboration as an enabler for circular economy: A case study of a developing country. *Management Decision, 59*(8), 1784–1800.
Mohammadrezaei, M. (2024). *Enhancing Sustainability Reporting Through Automated Text Analysis: A Systematic Review and Empirical Study in the Apparel Industry* (Doctoral dissertation, Université d'Ottawa/University of Ottawa).
Mukherjee, A., Panayotov, G., & Shon, J. (2021). Eye in the sky: Private satellites and government macro data. *Journal of Financial Economics, 141*(1), 234–254.
Pal, S. (2023). Integrating AI in sustainable supply chain management: A new paradigm for enhanced transparency and sustainability. *International Journal for Research in Applied Science and Engineering Technology (IJRASET), 11*, 2979–2984.
Pieroni, M. P., McAloone, T. C., & Pigosso, D. C. (2021). Circular economy business model innovation: Sectorial patterns within manufacturing companies. *Journal of Cleaner Production, 286*, Article 124921.
Prakash, A., & Tiwari, A. K. (2025). The role of institutional factors in achieving SDG 16: A thematic review of drivers of corporate anti-corruption action. *Environment, Sustainability, and Governance Insights, 1*(1), 34–52.
Russell, J. D., & Nasr, N. Z. (2023). Value-retained vs. impacts avoided: the differentiated contributions of remanufacturing, refurbishment, repair, and reuse within a circular economy. *Journal of Remanufacturing, 13*(1), 25–51.
Saha, A., Raut, R., Kumar, M., & Ghoshal, S. (2025). To explore the relationship between sustainability, digital technology, and sustainable development goals. *Business Strategy & Development, 8*(1), Article e70076.
Scrioşteanu, A., & Criveanu, M. M. (2023). Reverse logistics of packaging waste under the conditions of a sustainable circular economy at the level of the European Union States. *Sustainability, 15*(20), 14727.
Shamsuddoha, M., & Kashem, M. A. (2024). Zero plastic drive: A comprehensive review on unveiling innovative sustainable solutions for a circular plastics economy. *Sustainability, 16*(23), 10329.
Shukla, S. S., Jain, M., Jain, A., Bajpai, S., Mishra, A. K., & Pramanik, S. (2024). The circular economy: Changing the future of India. In *Green Transition Impacts on the Economy, Society, and Environment* (pp. 149–165). IGI Global.
Simon, P. (2022). *Low-code/no-code: Citizen developers and the surprising future of business applications.* Racket Publishing.
Singh, R., Kumar, K., & Khan, S. (2024). A comprehensive view of artificial intelligence (AI)–based technologies for sustainable development goals (SDGs). *Artificial Intelligence Enabled Management: An Emerging Economy Perspective, 183*, 9783111172408-012.
Singh, R., Sharma, A., & Kumar, K. (2025a). Navigating carbon neutrality: Strategies for sustainable future and environmental challenges. In *Climate Neutrality Through Smart Eco-Innovation and Environmental Sustainability* (pp. 69–80). Springer Nature Switzerland.

Singh, R., Tiwary, K. S., Kumar, K., & Kumar, A. (2025b). Empowering a sustainable future: Blockchain's role in economy, organizations, and finance towards smart circular economy. In *Blockchain Technologies for Smart Circular Economy and Organisational Sustainability* (pp. 271–283). Springer Nature Switzerland.

Soni, A., Das, P. K., & Kumar, P. (2023). A review on the municipal solid waste management status, challenges and potential for the future Indian cities. *Environment, Development and Sustainability, 25*(12), 13755–13803.

Sönnichsen, S. D., & Clement, J. (2020). Review of green and sustainable public procurement: Towards circular public procurement. *Journal of Cleaner Production, 245*, Article 118901.

Srivastava, S. K., Anshul, A., Pathak, P., & Bansal, J. (2018). Responsible consumption for curbing food wastage: An exploratory enquiry. *PURUSHARTHA-A Journal of Management, Ethics and Spirituality, 11*(2), 104–116.

Srivastava, S. K., Singh, S., & Srivastava, M. (2021). Food waste at household and social gatherings: Drivers and possible remedies. *PURUSHARTHA-A Journal of Management, Ethics and Spirituality, 14*(2), 117–133.

Srivastava, S. K., Sinha, A. K., & Kumar, S. (2024). Workplace spirituality empowered by spiritual leadership to promote organizational effectiveness: A bibliometric analysis. *PURUSHARTHA-A Journal of Management, Ethics and Spirituality, 17*(1), 1–22. https://doi.org/10.21844/16202117101

Tariq, M. U. (2025). Integrating advanced technologies for multi-stakeholder collaboration: Driving sustainability in supply chains through industry 4.0. In *Multi-Stakeholder Collaboration for Sustainable Supply Chain* (pp. 313–334). IGI Global Scientific Publishing.

Teece, D. J. (2025). The multinational enterprise, capabilities, and digitalization: Governance and growth with world disorder. *Journal of International Business Studies*, 1–16.

Tripathi, V., Singh, K., Parihar, S., & Srivastava, S. K. (2022). A bibliometric analysis on supply chain collaboration. *PURUSHARTHA-A Journal of Management, Ethics and Spirituality, 15*(2), 66–103. https://doi.org/10.21844/16202115205

Van Wassenhove, L. N. (2019). Sustainable innovation: Pushing the boundaries of traditional operations management. *Production and Operations Management, 28*(12), 2930–2945. https://doi.org/10.1111/poms.13114(Originalworkpublished2019)

Visvanathan, C., & Anbumozhi, V. (2018). Evolutionary acts and global economic transition: progress of the circular economy in ASEAN. *industry, 4*, 67–105.

Wang, L., Liu, Z., Liu, A., & Tao, F. (2021). Artificial intelligence in product lifecycle management. *The International Journal of Advanced Manufacturing Technology, 114*, 771–796.

Xu, W., Dainoff, M. J., Ge, L., & Gao, Z. (2021). From human-computer interaction to human-AI Interaction: New challenges and opportunities for enabling human-centered AI, *5*, 17. arXiv preprint arXiv:2105.05424.

Zeng, X., Wang, F., Li, J., & Gong, R. (2017). A simplified method to evaluate the recycling potential of e-waste. *Journal of Cleaner Production, 168*, 1518–1524.

Chapter 13
Beyond Today: Strategic Foresight in AI-Powered Circular Economies for Environmental Sustainability

Shubhra Mishra

Abstract This chapter is investigating how AI could assist in strategic management and strategic foresight in this fast-changing business environment. Conventional strategic planning approaches are usually too rigid or slow to address the multifaceted circumstances. This chapter explores ways in which AI can enhance foresight and support better long-term decision-making. The chapter reviewed the primary strategic foresight tools, horizon scanning, scenario planning, Delphi studies and trend analysis explaining the potential of AI to support each approach. AI solutions like machine learning, big data analytics and natural language processing allow companies to sort through vast amounts of information and spot trends early on, identifying the smallest signs of change. These features facilitates the creation of stronger and evidence-based future scenarios. Above all, the chapter emphasizes that AI integrated with strategic foresight has potential to make organizations more adaptive, robust and ready for future ambiguity.

Keywords Artificial intelligence · Strategic foresight · Circular economy · Environmental sustainability · Scenario planning · Horizon scanning

1 Introduction

Sustainability has become a significant element in today's economic decisions. The loss of biodiversity, global warming, pollutions and resources scarcity is the result of a linear economic system model based on an 'extract-produce-consume-dispose' logic that calls into question the long-term future of the planet (Upadhayay & Alqassimi, 2018). These 'straight line systems' have initiated jump starts in natural ecosystems, that leads to a production consumption cycle which is unsustainable. As the world's population continues to increase towards 10 billion by 2050 (WEF, 2021a, 2021b),

S. Mishra (✉)
Indian Institute of Technology Jodhpur, Jodhpur, India
e-mail: reachshubhra11@gmail.com

K. Kumar et al. (eds.), *AI-Driven Circular Economy for Carbon Neutrality in Emerging Economies*, Environmental Footprints and Eco-design of Products and Processes,
https://doi.org/10.1007/978-3-032-30298-4_13

so will the demands for resources. This will put further pressures on the environment and creating social and economic inequalities.

Within these challenges, the concept of the circular economy (CE) comes out to be a competent paradigm. Based on 'reduce, reuse, recycle' (the 3R principle) (Kirchherr et al., 2017, pg. 226, para 7), the concept of a CE adopts a regenerative concept of sustainable economic development. It aims to separate growth from resource mining, supporting innovation that makes economic mission consistent with ecological constraints (Ghisellini et al., 2016). Simultaneously, artificial intelligence (AI) is changing the ways we harness data. It improves performance and handle complexity that leads to tremendous opportunities to accelerate the shift to circularity (Danish & Senjyu, 2023). From live material tracking and smart waste sorting to predictive modelling of supply chains, Big Tech could become a powerful force for driving sustainability outcomes). Yet, despite the promise to AI-driven circular systems, a critical challenge remains: the lack of long-term proactive planning, in sustainability-oriented governance. This reactive mindset can lead to fragmented policy, inappropriate use of technology and missed chances to shape resilient systems. This is where Strategic Foresight (SF) becomes essential. It is a proactive anticipatory process that enable decision-makers to explore multiple possible futures, opportunities and threats. It uses tools like scenario planning, horizon scanning and backcasting to locate emerging trends, analyse long-term risks and design adaptive strategic designs (Vecchiato, 2012).

1.1 Circular Economy and Environmental Sustainability

The CE is an alternative model in which the 'take, make and dispose' linear model of economy. It is substituted by a regenerative system which aims to minimize environmental degradation, while enhancing the competence of resources (Benachio et al., 2020). It is based on three main pillars: **Reduce, which** includes the reduction of the waste resource input and waste generation within the production and consumption process (Kirchherr et al., 2017). **Reuse, that** refers to the duration of products and materials extended through repair, refurbishment and second-hand markets (Russell & Nasr, 2023). **Recycle** emphasizes on the retrieval of materials at the end of their use phase and their incorporation in new production cycles (Atif & Hassan, 2025; Srivastava et al., 2021, 2024). The environmental benefit of CE is enormous. It also has a high potential to reduce depletion of the resources, greenhouse gases, landfilling and to restore natural systems. For instance, The Ellen McArthur Foundation (2020) projects that through applying CE strategies, textile, food and plastic industry could lead to reduce around 45% of emission globally. Nevertheless, the transformation towards a circular economy is not free of obstacles. This includes: Economic obstacles, in which linear solutions tend to be more economical as environmental price is shifted, making circular investments less attractive (Ghisellini et al., 2016). Behavioural disinterests, which is really hard and a challenging issue to change consumer's patterns/it is hard to make producers think for circularity (Gomes et al.,

2022). Technological gaps includes lack of repairing and reverse logistics infrastructure continue to be the challenges due to which the underdeveloped nations are lagging behind (Grafström & Aasma, 2021). Policy mismatch includes some laws and regulations that still favour a linear business model or provide unclear guidance for circular standards (Wasserbaur et al., 2022). To tackle these problems, the CE must be sustained by intelligent technologies and active attitude. Here is where SF and AI can greatly facilitate CE transitions.

1.2 Contribution of AI to the Transition to Circular Economy

AI is being established as a powerful aid for circulate economy in different stages of the life of the products and resource (Fraga-Lamas et al., 2021). Machine learning, natural language processing and computer vision, e.g. can learn from a lot of data, find hidden patterns and take smart decisions in real time. These efficiencies are quite important for CE purposes. Key applications of AI technology in CE are covered here: **Waste Sorting**: AI-based robots can separates various categories of pollution and emission causing materials (plastics, metals, etc.) with a higher degree of efficiency thereby boosting recycling and reducing contamination. For example, for bio processing, technologies such as machine learning (ML) algorithm are able to isolate the pollutants such as stones and plastic from compost, solid digested and feed stock, which increase the plausible environmental risk (Guo et al., 2021). **Supply Chain Optimization**: AI algorithms can be used to improve circular supply chain management via demand management and constraints optimization concerning overproduction and efficient recycling/re-manufacturing of resources (Nayal et al., 2022; Tripathi et al., 2022). **Predictive Conservation**: In manufacturing and infrastructure, AI will predict when machinery will break down so it can be timely replaced, prolonging the life of the asset and reducing waste (Lee et al, 2020; Srivastava et al., 2018). For example, it employs sensors and data analytics to predict potential equipment problems in advance and thus allows preventive maintenance resulting in decreased downtime, waste and energy consumption (Zeb & Lodhi, 2025).

However, the introduction of AI in CE also has some drawbacks such as **Transparency and Bias**, while AI-powered algorithmic decision-making is often opaque, leading to trust questions or issues of fairness, accountability (particularly when AI is applied in public services, or environmental sensing) and other biases (Lepri et al, 2018). **Technological Determinism**: A hyper-dependence on AI could result in technocratic responses to our ecosystem's challenges that blind us to social, cultural, and ethical considerations of sustainability (Arogyaswamy, 2020; Kaith & Sachdeva, 2024). For example, Smart City Waste Collection Using AI, that improve garbage collection based on AI, while disregard of informal waste workers who lose their jobs making social inequality worse. **Data and Resource Intensity**: AI systems, whether learning, inference, or control, relies on extremely large data and high energy computation, which could outweigh the environment benefits if not well justified (Wu et al., 2022). Such systems are usually associated with the use of energy hungry datacentres, if not powered with clean energy and optimized for efficiency, their carbon emissions actually work against the ecological cause they are trying to promote (Nishant et al., 2020).

In order to steer through these gale-like forces, AI will require to be coupled with a diversity of open and flexible governance systems. This is where Strategic Foresight is key.

1.3 Strategic Foresight: Definition and Significance

Strategic Foresight (SF) 'is a systematic approach to learning and understanding possible futures and building shared visions and is aimed at guiding and enabling present day decisions' (Iden et al., 2017, pg. 8, para 5). Unlike classical planning that works with historical data or linear predictions, SF has no qualms about allowing the unknown and complex. It also incorporates different instruments and procedures to support long-term decision-making. The basic tenets of SF are addressed as follows: **Anticipation, which** involves recognizing and making sense of early signs for change, e.g. weak signals, emerging trends and opportunities (Rossel, 2012). **Future Scenarios, that** is related to developing scenarios in terms of best-case, worst-case, wildcard and implausibilities that are used to stress-test existing strategies (Burt & Nair, 2020). **Horizon Scanning, which** is a tool for systematically exploring trends and novelties in conjunction with a potential implication for sustainability pathways (Cuhls, 2020). **Backcasting** begins with a desirable future condition and works backwards to identify the steps, policies and procedures corresponding to the future condition we wish to achieve (Dreborg, 1996; Quist, 2016). By embedding SF into institutional activities, like city planning, industrial design methods and climate approaches, government and industry are preparing for uncertain futures, mitigating systemic risks and (ideally) channelling innovation into generative, distributed processes.

2 Strategic Foresight and Adaptive Sustainability-Oriented Governance

The traditional form of sustainability-oriented governance is reactive, meaning that problems are only addressed as they arise Kumar et al., 2025; Prakash & Tiwari, 2025; Singh et al., 2025a, 2025b). This reactive and backward approach is unsuited to the multi-causal, non-linear and uncertain dynamics of environmental and technological change. With SF, government can move from reactive to proactive governance, grounded in future rather than the past (Agrawal, 2025; Fuerth, 2009; Joselin et al.,

Table 1 SF-driven governance versus traditional governance approach in CE

Dimensions	SF-driven governance	Traditional governance
Time-horizon	Long-term (10 + years) (Ramos, 2014)	Short-term (1–5 years) (Ramos, 2014)
Approach decision logic	Proactive and anticipatory scenario based, adaptive planning (Hanssen et al., 2009)	Reactive linear planning (Hanssen et al., 2009)
Focus	Emerging signals, risks and opportunities (Ramos, 2014)	Past trends and present issues (Ramos, 2014)
Governance culture	Learning-oriented, innovative and flexible (Umbach, 2024)	Risk averse, bureaucratic (Umbach, 2024)
Stakeholder engagement	Participatory, multi-stakeholder (Hanssen et al., 2009)	Top-down, expert-driven (Hanssen et al., 2009)
Equity and justice lens	Centred through inclusive foresight methods (Ramos, 2014)	Often overlooked (Ramos, 2014)

2025; Singh et al., 2024). Table 1 shows the difference between SF-driven governance and traditional governance.

Through the use of tools such as horizon scanning, scenario planning and back-casting, institutions can look for weak signals and potential opportunities that are still outside narrow band of trends and act on these in the absence of clear confirmation (Cuhls, 2020; Singh et al., 2025b). For instance, SF can assist governments to foresee future resource scarcity and influence CE policies of encouraging resource substitution, recycling infrastructure and eco-design standards in advance (Mendoza et al., 2017). It is also about policy resilience, and flexibility. Through scenario-based exploration, policy-makers are able to assess the won/loss, and positive/negative impact of policies under a variety of different, possibly future, conditions. This responsive methodology is especially important for sustainability governance, that needs to be current to fluctuations in socioeconomic, ecological and technological dynamics. In addition, SF promotes interdisciplinary and cross-sector cooperation, an aspect that is of importance for circular economies, which interdependently interact in industrial ecosystems (Rittershaus et al., 2025). When SF is made part of governance, it promotes long-term visioning, alignment of organizational goals. And the early consulting with stakeholders including industry, governments, researchers and communities.

2.1 Enhancing Resilience Through Strategic Foresight

CE exists because of and from that which can absorb shocks and still continue to function and keep a structure (Kennedy & Linnenluecke, 2022)). SF further strengthens this ability by enabling institutions to plan ahead and address chronic issues such as global supply chain disruptions, climate change and socio-political shifts which can impact circular strategies (Mani & Goniewicz, 2023). For example, by modelling the future climate, SF can aid urban planning of climate-friendly waste and water infrastructure for continued circular resource flows in times of climate disasters. Likewise, companies may also deploy SF to make their supply chains more diversified or move their sourcing towards local or renewable resources to counteract geo-political and global price fluctuations.

SF also promotes the strengthening of institutions. Most companies face short-termism and are not structurally equipped to deal with deep uncertainty. This supports organizational learning and reflexivity at the institutional level, facilitating the adjustment of mental models as well as strategies and which helps to stay resilient in times of such volatility. In the case of developing countries, considering that sustainability governance is commonly constrained by capacity and financial constraints, SF can be used as an affordable high leverage instrument. If governments were to train policy planners and practitioners with SF tools, it could facilitate a growing institutionalization of future preparedness, increasing systemic resilience from the bottom up (Kimbell & Vesnić-Alujević, 2020).

2.2 *Strategic Foresight as a Governance Tool for Future-Proof Circular Economies*

International phenomena such as rising emissions, climate change, resource exhaustion and technological disruption are happening in an accelerating rate demanding governance that is proactive and responsive, not simply reactive (kaur et al., 2025). SF offers a forward-looking and system-based policy approach, which incorporates future-proofing into policy decisions (Kolios, 2024). It enables organizations, institutions and governments to respond to uncertainty well, align strategies with goals of long term sustainability, and develop more holistic, adaptive and resilient systems.

2.3 *Stakeholder Engagement and Participatory Planning*

An important concept in both sustainability and circularity is inclusion (Corona et al., 2019), making sure that no voice, particularly those at the fringes, is left behind when shaping the future. There is great credence to the argument that SF is a participative vehicle (Berkhout & Hertin, 2002) that enables such participation. SF methods such as inclusive visioning, future workshops and the Delphi tool can be employed to involve workers, stakeholders and communities in the co-creation of sustainable futures. Such inclusive conversations can bring up lived realities, local wisdom and ethical points of view that don't always get factored in to top-down planning processes. For example, SF within circular agricultural planning may not only encompass Agri-tech experts, but also smallholder farmers, women's cooperatives and climate-vulnerable communities, to ensure relevance and equity of solutions. This participatory SF method develops trust, social legitimacy and increases the likelihood for uptake and usage (Amanatidou & Guy, 2008). SF also offers a method for addressing challenging justice and ethical issues involved in CE transitions (Woensel, 2024). Who profits from the transition? Who pays for the disruption? By visually populating power distributions, distributional impacts and standard trade-offs in plausible futures, SF assists to mainstream principles of climate justice, equity, transparency and accountability into governance of CE.

3 Foresight Tools in Action: Scenario Planning, Horizon Scanning and Delphi Method

SF is embedded in a system of formalized instruments which support decision-making focused on identifying uncertainty, change and multiple futures (Vecchiato, 2012). The latter two, scenario planning and Delphi method, are particularly useful in promoting the transitions of CE. When accompanied and empowered by digital technologies and AI, they can empower a more fluent, democratic and accountable form of governance.

3.1 Scenario Planning in CE Transitions

Scenario planning is a popular foresight method that helps organizations to build and analyse several possible futures (Chermack, 2011). It allows decision-makers to get away from linear forecasting by accepting uncertainty and thinking in terms of a range of conclusions. Within the CE transitions, the scenario planning structure is generally applied: Pick a focused theme (like plastic waste in Indian cities) (Tonelli & Cristoni, 2018). Identify key drivers and uncertainties (including policy direction, consumer behaviour, technology breakthroughs) (Tonelli & Cristoni, 2018). Build in several alternatives such as:

(a) Best case: Policy and public values align to diminish single-use plastic by 80% (Bezold, 2010).
(b) Worst-case scenario: Economic reversals result in deregulation, worsening plastic pollution (Bezold, 2010).
(c) Possible future: Patchy progress made because of scattershot policy and uneven technological uptake.

Next will be, interpret implications for existing plans and build adaptable action plans that work across scenarios. 2 by 2 metrics can be used as scenarios templates (Rolland et al., 1998). Sustainable product design, investment decisions and long-term policy development could be informed by these scenarios by stress-testing ideas against a variety of possible futures. For example: A city planning team that uses scenario planning to help understand the city's future waste management needs with varying population growth and consumption patterns could put a stronger emphasis on prioritizing modular infrastructure and adaptive logistics systems.

3.2 Horizon Scanning for Emerging Trends and Technologies

Horizon scanning is the systematic identification of early signs of change also known as weak signals, that may evolve into significant trends, opportunities and threats (Rowe et al., 2017). It enables organizations to stay ahead of the curve by detecting disruptions before they fully materialize (Bengston et al., 2024). To use horizon planning in CE transitions, the following steps should be followed: First is, scanning diverse sources such as patents, news, social media and scientific literatures (Hines et al., 2019). Second, identifying irregularities, originalities and discontinuities that may indicate evolving shifts (Sharpe et al., 2016). Third is, clustering trends into broader themes (for instance, biodegradable materials, AI-based product lifecycle tracking) (Hines et al., 2019). Last is, prioritizing signals based on relevance and potential impact (Sharpe et al., 2016).

In environmental and CE context, horizon scanning can detect such as material innovation trends like enzyme-based recycling and bio-based packaging. Regulatory movements towards extended producer responsibility (Sharpe et al., 2016). Social shifts such as zero-waste lifestyle or green consumer coalitions. By embedding AI and Natural Language Processing (NLP) can enhance horizon scanning by automating signal detection from large unstructured data sources, accelerating SF processes.

3.3 Delphi Method for Expert-Driven Circular Futures

The Delphi methodology is a foresight technique that can be structured to collect and narrow expert opinion in a series of rounds (survey iterations) (Gallego & Bueno, 2014). This becomes particularly valuable when it is the uncertain and complex problem, and not any one actor, which possesses complete knowledge (Calleo et al., 2025). The process typically involves: Using multiple experts, with a variety of types such as scientists, government officials, industry and community members (Gordon, 1994). Multiple survey rounds of participants who answer questions on developments in future (e.g. by what date will AI have been fully integrated at scale into circular supply chains?) (Rejeb et al., 2022). Responses are shared and are anonymize after each round, so players can adjust their opinions (Rejeb et al., 2022). The aim is to come to convergence, or learn where consensus falls apart, so the institutions can build out robust scenarios and policy advice. For CE planning, the Delphi method can reveal expert consensus on key risks and game-changers like the time at which circular infrastructure might be ready or possible moral dilemmas concerning automation of waste systems (Försterling et al., 2023). Unlike conventional consultative methods, influence within the Delphi process is non-hierarchical and iterative, which encourages further thought and cross-sector synthesis (Liang & Segalas, 2025).

3.4 Integration of Tools in Strategic Governance Model

While each foresight tools offer unique strength and advantage, their true optimization and power lies in integration within strategic governance models. This model explains, through its use-case results, how the combination of scenario planning, horizon scanning and the Delphi method could converge to provide a holistic and participation-based foresight operational map for CE governance. Scenario planning contributes to the long-term view, horizon scanning to ongoing environmental scanning and agility, while the Delphi method brings expert depth and consensus. When combined, horizon scanning can be used as the basis for a scenario planning strategy to help with wise and rapid decision-making (Washida & Yahata, 2021). Delphi with expert review allows for redefinition and validation of outcomes (Linstone & Turoff, 1975). Delphi-based expert estimates may help to identify priorities for real-time horizon scanning and to guide scenario planning (Linstone & Turoff, 1975). It is likely to be used to facilitate real-time interfacing, supported by digital platforms and AI tools, between governments and organizations so that the collection and coding of foresight data are automated. It can support participative planning, by Delphi panels and scenario workshops. For instance, Singapore and Helsinki developed AI-augmented foresight systems that combine scenario planning with a digital trend monitoring dashboard for circular innovation in urban planning and resource management.

4 Case Studies: Strategic Foresight in Environmental Circular Innovation

SF enables long-term thinking in sustainability transitions, especially with combined with AI technologies in CE models. This section discusses two real-world cases from Ghana and the Netherlands, where foresight practices shaped AI-powered circular innovations. Their contrasting context, an emerging economy and a high-income urbanized state offer valuable comparative lessons for foresight driven environmental transitions.

4.1 Case Study 1: Ghana-AI-Powered Waste Mapping and Plastic Governance

Ghana produces approximately 1.1 million of plastic waste annually, with only about 9% formally recycled and the remaining largely mismanaged through land-filling, burning or open dumping (WEF, 2021a, 2021b). Urban establishment like Accra face growing environmental and health risks from plastic leakage, often exacerbated by informal waste management and inadequate infrastructure.

To tackle this crisis, Ghana has introduced the National Plastic Action Partnership (NPAP) in partnership with the World Economic Forum, UNDP and local partners. This effort utilized AI and geospatial prediction technologies to identify and track hot spots of plastic leakage. By applying machine learning to satellite photographs, developers produced detailed maps of informal dumps and plastic polluting zones across Accra and elsewhere. This information pour into live digital dashboards, which lawmakers could use to prioritize clean-up, map out the accumulation process and project future plastic movement. Predictive mapping also assisted in more efficiently targeting investments in recycling infrastructure and interventions at the community level.

NPAP also engaged in scenario planning for the future of CE in Ghana in 2020. Other scenarios included variables such as implementation of Extended Producer Responsibility (EPR) schemes, use of alternative materials (e.g. compostables), enhanced public awareness and education (Anokye et al., 2024). Backcasting methodologies were applied to develop roadmaps from defined future's states (e.g. 50% plastic recovery in 2030) back to proximal actions. Nevertheless, there are still obstacles to retain infrastructure investments and the integration of the informal waste economy in circular regimes.

4.2 Case Study 2: Netherlands-Inclusive Visioning and Smart Circulatory Systems

The Netherlands is one of the countries that has been most advanced in CE planning and has officially signed up to reach a full CE by 2050 (with 50% reduction in virgin materials application by 2030). Complicated, multi-material plastics are managed with difficulty, particularly in urban settings, even though recycling rates are already high. The country uses SF via participatory visioning institutionalized in city planning mechanisms such as 'Amsterdam Circular 2050'. The city organizes multi-stakeholder workshops with residents, NGOs, designers, scientists and waste companies, to co-create a future vision on sustainable material flows. In addition to visioning, the government carries out horizon scanning on new technologies and social trends (digital product passport, bio-plastics) and integrates this into a circular policy roadmaps.

Amsterdam tested a Smart Waste Management system in conjunction with their Smart City Initiative. The system includes: Bins with sensors that monitor the level of fill and the types of waste. AI-based garbage route optimization algorithms to decrease the frequency of collection and save fuel. Machine vision systems at Materials Recovery Facilities (MRFs) to enhance the auto-sorting of plastics. Real-time interfacing with waste to energy plants to optimize incineration schedule.

This is an entire, closed-loop, cyber-physical foresight ecosystem, with the scenario-led urban planning driving the tech-deployment and supported by AI-generated data for the next cycle. Through case examples we show how Ghana and Netherlands epitomize two successful, but context-specific ways of mainstreaming SF in circular plastics transitions. Ghana innovatively uses AI and real-time data to address the fundamental issues of dealing with urban waste. Netherlands is using participatory foresight and smart systems to improve its already advanced circular economy. Collectively, they are a blueprint for how technology, governance and SF can converge to confront the sustainability emergency.

4.3 Lessons for Political and Institutional Leaders

The learnings that political and institutional leaders can learn from the chapter are as follows: Institutions need to have operational intelligence along with SF. There can be a symbiosis between predictive AI tools and participatory visioning in coordinating long and short-term decision-making. The need to invest in futures literacy in all fields has become crucial. Essential condition-building, the capacity for scenario construction and use of digital data is critical to mainstream SF into governance frameworks. Focusing on both the informal settlement in Accra and the smart neighbourhoods of Amsterdam, circular innovation should not lead to digital or economic divides.

5 Challenges and Future Directions

Although SF has great potential in contributing to the development of sustainable and circular futures. However, the application of SF is more difficult in the developing country context within which the environmental systems are very complex. This section highlights the limitations, barriers and the challenges for systematically adopt foresight and its methodologies and the pathways for enhancing adaptive SF capabilities.

5.1 Limitations in Current SF Capacity in Emerging Economies

Transition economies often struggle to integrate SF into decision-making processes. These range from low institutional capacity, poor data infrastructure and no trained foresight experts in governments and companies. However, many future-oriented initiatives in the Global South are donor-driven or externally led with limited localization of process or commitment to implementation. Similarly, SF outcomes are commonly traditionalist in nature while remaining clustered within particular ministries or departments with minimal cross-sectoral coordination. For example, the experience in Ghana, shows that whereas international backing provided access to advanced AI-enabled waste mapping, the durable adoption of such innovations would always rely on local capacities and long-term investments. As AI is applied more and more widely in the context of foresight activities (for predictive analytics, scenario modelling and behavioural simulations), there is an increased danger of technological determinism. As a result, data-driven tools can be over-relied upon at the expense of the cultural, political and social aspects of change at scale. Too much reliance on AI could also sideline low-tech or community-based knowledge systems, particularly in a context of persistent digital divide. For example, use of smart waste systems in The Netherlands will increase efficiency, but may inadvertently leave behind vulnerable people who do not have access to digital solutions. It is important that AI enhance and not replace participatory and value-laden SF.

5.2 Strengthening Global Cooperation and Adaptive SF Capabilities

To address these limitations and enhance the value for SF, there is a pressing need for global cooperation and collaboration, through the development of adaptive foresight systems. The institutionalization of adaptive SF at the national and local level is also important. This includes: Incorporating Futures Thinking into Education and Public Administration Studies. Setting up specialized units for intelligence collection in ministries and commissions, with clear mandates and budgets. Promoting the creation of internal foresight labs by the private sector to focus on sustainable innovation. Advocating for hybrid foresight methods combining AI-assisted simulations with participatory tools scenario planning, visioning and environmental scanning.

6 Conclusion: A Call for Anticipatory, Ethical and Inclusive Futures

From the intersecting environmental, social and technological challenges the world is facing, the need to look beyond the immediate crisis has never been more urgent. In this chapter, we have discussed how SF, when sensibly embedded in CE actions, can help to create anti-fragile, just, and sustainable futures. Through scenarios from the reality of Ghana and the abstraction of The Netherlands, among other cases, it is clear that SF is not merely good to have; nor is it only for academics; it is a strategic need in the times of uncertainty and for speed in systemic transformation. AI technologies and data analytics when put together provide mighty ways to scan the horizons, co-create scenarios and dynamically adjust strategies on a real-time basis.

The example of Ghana showed how AI-enabled SF, paired with global collaborations, enabled the recognition of areas within and across which plastic waste accumulated and then pushed for national responsiveness. The Netherlands also demonstrated how participatory visioning and AI smart waste systems can work symbiotically to achieve inclusive urban CE. These cases serve to reinforce the fact that successful SF is about getting ready for SF, steering it and making it more democratic.

In order to make the best use of SF, this chapter proposes the following recommendations: Institutionalize SF practice in public governance systems has dedicated units, trained professionals and clear mandates linking short-term decisions to long-term consequences. Advocate for ethical and inclusive SF pathways that ensure the involvement of marginalized voices, informal sectors and digitally excluded communities in the design of circular transitions. Pair technology tools with indigenous wisdom. AI and predictive analytics must supplement, not supplant those participatory and social value and lived experience-based processes. Promote a global and intersectoral network for foresight, facilitating mutual learning, scalable innovation and collective actions to address global challenges for sustainability. Designing the circular systems of tomorrow ultimately starts with foresight today. The ability to envision diverse possible futures, opportunities, threats and emerging trends and to adopt inclusive ethical decisions is the key for succeeding CE transitions in the coming years. Through smart cities, regenerative industries or decentralized community systems, the future needs to be planned for, not just muscled through.

References

Agrawal, D. (2025). A comprehensive review of investing and financing for sustainable tourism projects. *Environment, Sustainability, and Governance Insights, 1*(1), 53–70.

Amanatidou, E., & Guy, K. (2008). Interpreting foresight process impacts: Steps towards the development of a framework conceptualising the dynamics of 'foresight systems.' *Technological Forecasting and Social Change, 75*(4), 539–557.

Anokye, K., Apea, O. B., Darko, A. O., & Kwabi, L. A. (2024). The future of waste management in Ghana: Assessing the feasibility and scalability of smart waste solutions amidst key adoption challenges. *Cleaner Waste Systems*, 100183.
Arogyaswamy, B. (2020). Big tech and societal sustainability: An ethical framework. *AI & Society, 35*(4), 829–840.
Atif, S., & Hassan, A. (2025). From bane to boon: How global fortune 500 companies disclose about using servitisation and industry 4.0 as circular economy strategy. *Corporate Social Responsibility and Environmental Management*.
Benachio, G. L. F., Freitas, M. D. C. D., & Tavares, S. F. (2020). Circular economy in the construction industry: A systematic literature review. *Journal of Cleaner Production, 260*, Article 121046.
Bengston, D. N., Mauno, T., & Hujala, T. (2024). Horizon scanning: A process for identifying emerging signals of change shaping the future of natural resources management. *Society & Natural Resources, 37*(7), 1108–1115.
Berkhout, F., & Hertin, J. (2002). Foresight futures scenarios: Developing and applying a participative strategic planning tool. *Greener Management International, 37*, 37–52.
Bezold, C. (2010). Lessons from using scenarios for strategic foresight. *Technological Forecasting and Social Change, 77*(9), 1513–1518.
Burt, G., & Nair, A. K. (2020). Rigidities of imagination in scenario planning: Strategic foresight through 'Unlearning.' *Technological Forecasting and Social Change, 153*, Article 119927.
Calleo, Y., Taylor, A., Pilla, F., & Di Zio, S. (2025). AI-assisted Real-Time Spatial Delphi: Integrating artificial intelligence models for advancing future scenarios analysis. *Quality & Quantity*, 1–33.
Chermack, T. J. (2011). *Scenario planning in organizations: how to create, use, and assess scenarios*. Berrett-Koehler Publishers.
Corona, B., Shen, L., Reike, D., Carreón, J. R., & Worrell, E. (2019). Towards sustainable development through the circular economy—A review and critical assessment on current circularity metrics. *Resources, Conservation and Recycling, 151*, Article 104498.
Cuhls, K. E. (2020). Horizon scanning in foresight-why horizon scanning is only a part of the game. *Futures & Foresight Science, 2*(1), Article e23.
Danish, M. S. S., & Senjyu, T. (2023). Shaping the future of sustainable energy through AI-enabled circular economy policies. *Circular Economy, 2*(2), Article 100040.
Dreborg, K. H. (1996). Essence of backcasting. *Futures, 28*(9), 813–828.
Försterling, G., Orth, R., & Gellert, B. (2023). Transition to a circular economy in Europe through new business models: Barriers, drivers, and policy making. *Sustainability, 15*(10), 8212.
Fraga-Lamas, P., Lopes, S. I., & Fernández-Caramés, T. M. (2021). Green IoT and edge AI as key technological enablers for a sustainable digital transition towards a smart circular economy: An industry 5.0 use case. *Sensors, 21*(17), 5745.
Fuerth, L. S. (2009). Foresight and anticipatory governance. *Foresight, 11*(4), 14–32.
Gallego, D., & Bueno, S. (2014). Exploring the application of the Delphi method as a forecasting tool in information systems and technologies research. *Technology Analysis & Strategic Management, 26*(9), 987–999.
Ghisellini, P., Cialani, C., & Ulgiati, S. (2016). A review on circular economy: The expected transition to a balanced interplay of environmental and economic systems. *Journal of Cleaner Production, 114*, 11–32.
Gomes, G. M., Moreira, N., & Ometto, A. R. (2022). Role of consumer mindsets, behaviour, and influencing factors in circular consumption systems: A systematic review. *Sustainable Production and Consumption, 32*, 1–14.
Gordon, T. J. (1994). The Delphi method. *Futures Research Methodology, 2*(3), 1–30.
Grafström, J., & Aasma, S. (2021). Breaking circular economy barriers. *Journal of Cleaner Production, 292*, Article 126002.
Guo, H. N., Wu, S. B., Tian, Y. J., Zhang, J., & Liu, H. T. (2021). Application of machine learning methods for the prediction of organic solid waste treatment and recycling processes: A review. *Bioresource Technology, 319*, Article 124114.

Hanssen, G. S., Johnstad, T., & Klausen, J. E. (2009). Regional foresight, modes of governance and democracy. *European Planning Studies, 17*(12), 1733–1750.
Hines, P., Yu, L. H., Guy, R. H., Brand, A., & Papaluca-Amati, M. (2019). Scanning the horizon: A systematic literature review of methodologies. *British Medical Journal Open, 9*(5), Article e026764.
https://www.ellenmacarthurfoundation.org/completing-the-picture
https://www.pbl.nl/en/topics/circular-economy
https://www.undp.org/ghana/press-releases/call-proposal-integrated-plastic-waste-management
https://www.weforum.org/press/2021/09/global-plastic-action-partnership-making-an-impact-in-fighting-plastic-pollution/
Iden, J., Methlie, L. B., & Christensen, G. E. (2017). The nature of strategic foresight research: A systematic literature review. *Technological Forecasting and Social Change, 116*, 87–97.
Joselin, D., Singh, A., & Tiwary, K. S. (2025). The Nexus between CSR practices and financial performance: Empirical Evidence from the manufacturing sector in India. *Environment, Sustainability, and Governance Insights, 1*(1), 1–17.
Kaith, A., & Sachdeva, G. (2024). Bibliometric insights into green human resource management: A connection with business ethics and values. *PURUSHARTHA-A Journal of Management, Ethics and Spirituality, 17*(1), 51–63. https://doi.org/10.21844/16202117104
Kaur, H., Reddy, K. K., Reddy, M. K., & Hanafiah, M. M. (2025). Collaborative approaches to navigating complex challenges and adapting to a dynamically changing world. In *Integration of AI, Quantum Computing, and Semiconductor Technology* (pp. 209–234). IGI Global.
Kennedy, S., & Linnenluecke, M. K. (2022). Circular economy and resilience: A research agenda. *Business Strategy and the Environment, 31*(6), 2754–2765.
Kimbell, L., & Vesnić-Alujević, L. (2020). After the toolkit: Anticipatory logics and the future of government. *Policy Design and Practice, 3*(2), 95–108.
Kirchherr, J., Reike, D., & Hekkert, M. (2017). Conceptualizing the circular economy: An analysis of 114 definitions. *Resources, Conservation and Recycling, 127*, 221–232.
Kolios, A. (2024). Retrofitting technologies for eco-friendly ship structures: A risk analysis perspective. *Journal of Marine Science and Engineering, 12*(4), 679.
Kumar, K., Singh, R., Choudhary, M., & Sharma, A. (2025). Climate change mitigation strategies in environmentally sensitive industries. In *Climate Change and Social Responsibility* (Vol. 25, pp. 185–204). Emerald Publishing Limited.
Lee, J., Ni, J., Singh, J., Jiang, B., Azamfar, M., & Feng, J. (2020). Intelligent maintenance systems and predictive manufacturing. *Journal of Manufacturing Science and Engineering, 142*(11), Article 110805.
Lepri, B., Oliver, N., Letouzé, E., Pentland, A., & Vinck, P. (2018). Fair, transparent, and accountable algorithmic decision-making processes: The premise, the proposed solutions, and the open challenges. *Philosophy & Technology, 31*(4), 611–627.
Liang, N., & Segalas, J. (2025). A deep and shallow sustainability intervention framework: A Taoist-inspired approach to systemic sustainability transitions. *Sustainability, 17*(11), 5170.
Linstone, H. A., & Turoff, M. (Eds.). (1975). *The Delphi method (Vol. 1975* (pp. 3–12). Addison-Wesley.
Mani, Z. A., & Goniewicz, K. (2023). Adapting disaster preparedness strategies to changing climate patterns in Saudi Arabia: A rapid review. *Sustainability, 15*(19), 14279.
Mendoza, J. M. F., Sharmina, M., Gallego-Schmid, A., Heyes, G., & Azapagic, A. (2017). Integrating backcasting and eco-design for the circular economy: The BECE framework. *Journal of Industrial Ecology, 21*(3), 526–544.
Nayal, K., Kumar, S., Raut, R. D., Queiroz, M. M., Priyadarshinee, P., & Narkhede, B. E. (2022). Supply chain firm performance in circular economy and digital era to achieve sustainable development goals. *Business Strategy and the Environment, 31*(3), 1058–1073.
Nishant, R., Kennedy, M., & Corbett, J. (2020). Artificial intelligence for sustainability: Challenges, opportunities, and a research agenda. *International Journal of Information Management, 53*, Article 102104.

Prakash, A., & Tiwari, A. K. (2025). The role of institutional factors in achieving SDG 16: A thematic review of drivers of corporate anti-corruption action. *Environment, Sustainability, and Governance Insights, 1*(1), 34–52.

Quist, J. (2016). Backcasting. In *Foresight in Organizations* (pp. 125–144). Routledge.

Ramos, J. M. (2014). Anticipatory governance: Traditions and trajectories for strategic design. *Journal of Futures Studies, 19*(1), 35–52.

Rejeb, A., Rejeb, K., Keogh, J. G., & Zailani, S. (2022). Barriers to blockchain adoption in the circular economy: A fuzzy Delphi and best-worst approach. *Sustainability, 14*(6), 3611.

Rittershaus, P., Aryan, V., & Renner, M. (2025). Systematic review on barriers, enablers, and interdependencies: sectoral challenges and opportunities towards adoption of circular practises in the mobile-phone industry. *Wiley Interdisciplinary Reviews: Energy and Environment, 14*(2), Article e70006.

Rolland, C., Ben Achour, C., Cauvet, C., Ralyté, J., Sutcliffe, A., Maiden, N., et al. (1998). A proposal for a scenario classification framework. *Requirements Engineering, 3*, 23–47.

Rossel, P. (2012). Early detection, warnings, weak signals and seeds of change: A turbulent domain of futures studies. *Futures, 44*(3), 229–239.

Rowe, E., Wright, G., & Derbyshire, J. (2017). Enhancing horizon scanning by utilizing pre-developed scenarios: Analysis of current practice and specification of a process improvement to aid the identification of important 'weak signals.' *Technological Forecasting and Social Change, 125*, 224–235.

Russell, J. D., & Nasr, N. Z. (2023). Value-retained vs. impacts avoided: the differentiated contributions of remanufacturing, refurbishment, repair, and reuse within a circular economy. *Journal of Remanufacturing, 13*(1), 25–51.

Sharpe, B., Hodgson, A., Leicester, G., Lyon, A., & Fazey, I. (2016). Three horizons: a pathways practice for transformation. *Ecology and Society, 21*(2).

Singh, R., Sharma, A., & Kumar, K. (2025a). Navigating carbon neutrality: Strategies for sustainable future and environmental challenges. In *Climate Neutrality Through Smart Eco-Innovation and Environmental Sustainability* (pp. 69–80). Springer Nature Switzerland.

Singh, R., Tiwary, K. S., Kumar, K., & Kumar, A. (2025b). Empowering a sustainable future: Blockchain's role in economy, organizations, and finance towards smart circular economy. In *Blockchain Technologies for Smart Circular Economy and Organisational Sustainability* (pp. 271–283). Springer Nature Switzerland.

Singh, R., Kumar, K., & Khan, S. (2024). A comprehensive view of artificial intelligence (AI)–based technologies for sustainable development goals (SDGs). *Artificial Intelligence Enabled Management: An Emerging Economy Perspective, 183*, 9783111172408–012.

Srivastava, S. K., Sinha, A. K., & Kumar, S. (2024). Workplace spirituality empowered by spiritual leadership to promote organizational effectiveness: A bibliometric analysis. *PURUSHARTHA-A journal of Management, Ethics and Spirituality, 17*(1), 1–22. https://doi.org/10.21844/162021 17101

Srivastava, S. K., Anshul, A., Pathak, P., & Bansal, J. (2018). Responsible consumption for curbing food wastage: An exploratory enquiry. *PURUSHARTHA-A Journal of Management, Ethics and Spirituality, 11*(2), 104–116.

Srivastava, S. K., Singh, S., & Srivastava, M. (2021). Food waste at household and social gatherings: Drivers and possible remedies. *PURUSHARTHA-A Journal of Management, Ethics and Spirituality, 14*(2), 117–133.

Tonelli, M., & Cristoni, N. (2018). *Strategic management and the circular economy*. Routledge.

Tripathi, V., Singh, K., Parihar, S., & Srivastava, S. K. (2022). A bibliometric analysis on supply chain collaboration. *PURUSHARTHA-A journal of Management, Ethics and Spirituality, 15*(2), 66–103. https://doi.org/10.21844/16202115205

Umbach, G. (2024). Futures in EU governance: Anticipatory governance, strategic foresight and EU Better Regulation. *European Law Journal, 30*(3), 409–421.

Upadhayay, S., & Alqassimi, O. (2018). Transition from linear to circular economy. *Westcliff International Journal of Applied Research, 2*(2), 62–74.

Van Woensel, L. (2024). Foresight in EU policy-making: Purpose, mindsets and methods. *European Law Journal, 30*(3), 361–381.

Vecchiato, R. (2012). Environmental uncertainty, foresight and strategic decision making: An integrated study. *Technological Forecasting and Social Change, 79*(3), 436–447.

Washida, Y., & Yahata, A. (2021). Predictive value of horizon scanning for future scenarios. *Foresight, 23*(1), 17–32.

Wasserbaur, R., Sakao, T., & Milios, L. (2022). Interactions of governmental policies and business models for a circular economy: A systematic literature review. *Journal of Cleaner Production, 337*, Article 130329.

World Economic Forum. (2021a). *Global Plastic Action Partnership: Making an impact in fighting plastic pollution.* World Economic Forum.

World Economic Forum. (2021b). *Global Risks Report 2021.* World Economic Forum.

Wu, C. J., Raghavendra, R., Gupta, U., Acun, B., Ardalani, N., Maeng, K., et al. (2022). Sustainable AI: Environmental implications, challenges and opportunities. *Proceedings of Machine Learning and Systems, 4*, 795–813.

Zeb, S., & Lodhi, S. K. (2025). AI for predictive maintenance: Reducing downtime and enhancing efficiency. *Enrichment: Journal of Multidisciplinary Research and Development, 3*(1), 135–150.

Chapter 14
If 'AI Waste-Sorting Robots Are the Future of Circular Economy,' How Are Emerging Economies Responding?

Sanjana Santra, Kishore Kumar, and Rubee Singh

Abstract The problem of waste management has remained a challenge in achieving environmental sustainability and circular economy goals. Especially in the context of emerging economies, this has been a lasting problem for decades, requiring academic and industrial focus on acquiring effective solutions. This chapter argues that with the advent of artificial intelligence (AI) and machine learning (ML) mechanisms, this problem can have a significant chance of getting solved with the innovative technological disruption. While AI technologies are leveraged across all aspects of creating interventions in all the pillars of Sustainable Development Goals (SDGs), developing economies are still suffering from inadequate technological infrastructure and complex cultural settings. The chapter investigates the efforts of the leading waste management companies that have adopted AI in this regard to create a lasting impact. With a cross-analysis approach of multiple cases from India, the chapter explores how the developing nations are responding to the problem of waste with technological solutions. The discussion also presents a comparative analysis between the developed and developing nations, differentiating the approaches taken by both economies, while stating the gap in progress and sustainable outcomes. The study may help future scholars in analysing this problem in depth, focusing on isolated cases, creating an impact regionally and mapping the impact on the SDGs, while also helping policymakers and climate entrepreneurs understand the importance of integrating AI for creating circular and sustainable impact that spreads across the environment, social and economic development of an emerging economy.

S. Santra (✉) · K. Kumar
School of Management, Bennett University, Greater Noida, Uttar Pradesh, India
e-mail: santrasanjana1995@gmail.com

K. Kumar
e-mail: akishore001@gmail.com

R. Singh
Institute of Business Management, GLA University, Mathura, India
e-mail: dr.rubeerajput.delhi@gmail.com

K. Kumar et al. (eds.), *AI-Driven Circular Economy for Carbon Neutrality in Emerging Economies*, Environmental Footprints and Eco-design of Products and Processes,
https://doi.org/10.1007/978-3-032-30298-4_14

Keywords Artificial intelligence · Circular economy · Sustainable development goals · Emerging economies · Waste management

1 Introduction

In today's time, emerging economies face a multitude of problems in meeting the circular economy goals. One of the major issues in this regard has remained the problem of waste management (Adami & Schiavon, 2021; Prakash & Tiwari, 2025). Even though the developed economies have quite advanced methodologies to segregate the solid and liquid waste, the developing nations are still striving to achieve the required efficiency level to manage the waste generated (Mmereki et al., 2016). Waste is a material that is produced from individual home units to organizational and industrial levels. The type of waste becomes different in composition and nature across all the levels, and it is often difficult to segregate the waste into definite categories to take adequate measures to eliminate it from the environment (Barr et al., 2013). In the vision of meeting Sustainable Development Goals (SDGs), this problem of waste has become a major barrier, especially for the emerging nations (Kumar et al., 2025; Rodić & Wilson, 2017; Singh et al., 2025a, 2025b). With increasing modernization efforts, more waste is being produced. In nature, there is no waste generated that does not contribute to the next cycle of creation. However, in man-made industrial processes, the waste is a problem that needs an extra step to eliminate (King et al., 2006). Most of the time, this extra step becomes a burden that not only requires extra cost but is also time-consuming. Also, without awareness and a sense of responsibility of the people towards planetary health, the work becomes more daunting (Joselin et al., 2025; Wabnitz et al., 2020). In India, this problem is very prominent, as many places, especially in urban and semi-urban areas, have an increased need for resources, utilities, food and living infrastructure; the waste production goes out of hand (Banerjee et al., 2022; Srivastava et al., 2018). From a behavioural perspective, many frameworks such as the Theory of Planned Behaviour (Ajzen, 1991), Thomas and Rogers' (1998) Diffusion of Innovations Theory, Technology Acceptance Model have been used to understand how the adoption of AI will be perceived by organizations and society. The perceived usefulness of integrating AI in waste management is measured by reduced waste volumes, improved recycling cycle rates and lowered operational costs (Aydin & Yildirim, 2021; Singh et al., 2025b; Srivastava et al., 2021, 2024). Even though the waste management technologies have been proven to be very effective, with the integration of AI, the impact on the outcome has increased in a significant manner. The developing economies are working very actively in this regard, with highly advanced machine learning and leveraging AI in the existing technological infrastructure (Agrawal, 2025; Mhlanga, 2021). Multiple organizations are leading the market to provide eco-friendly solutions to the waste management problem. Not only does it segregate the waste efficiently to reduce the landfills, but it also improves the economic development of the area, increases human safety in the process and better staffing strategies in the organization (Golzary et al., 2024; Nanda & Berruti, 2021).

The impact of increasing AI integration in waste management has been shown to improve circularity in the environment by developed countries; however, it has remained an increasing concern for developing nations (Shekdar, 2009; Yadav et al., 2023). A rapid urbanization or industrialization creates a massive disruption in the emerging societies, which becomes a challenge later to manage the waste production due to the rapid increase in migration, temporary habitats and industrial wastes (Adnan et al., 2022; Singh et al., 2024). As it may already look complex with the challenges, the environmental effects, coupled with socioeconomic barriers, make it more problematic. In this chapter, these aspects have been explored through business cases from India. Five cases on organizations that have realized the need of integrating AI in waste management systems of developing nations have been purposively taken and exemplified to understand what are the strategies that may lead us to achieve the circular economy goals in the emerging contexts. Circular economic traits have been seen to emerge with the advent of Industry 4.0 from a service-based circular business model (CBM), better resource management strategies and autonomous refurbishment systems (Nascimento et al., 2019). However, in developing contexts, it is not without its challenges. One of these challenges remains in the waste management, as the countries still suffer from heaping wastelands, landfills and polluted spaces in random places in the urban spaces (Awuah & Abdulai, 2022). Even though circular business models have been taken into consideration, there has been a lack of effort in managing the generated waste from the industries (Oghazi & Mostaghel, 2018). Not only the factory waste, but the lack of awareness in the general public also creates a significant impact. The sustainable strategies of waste segregation in households have been shared by the government and advised to keep separate dustbins to throw dry and liquid wastes separately (Trushna et al., 2024). Even after several awareness campaigns and announcements, the behavioural aspects of the people in these areas did not seem to improve in a manner that can create a noticeable difference (Kaith & Sachdeva, 2024; Wan & Du, 2022). It often becomes difficult for households to understand the different waste materials to categorize them well and discard them in the prescribed way (Nell et al., 2022; Yusoff et al., 2024). With more knowledge-sharing mechanisms and creating definite guides, it has improved with more awareness in sustainability and circularity principles (Sun et al., 2025), however, in the larger spaces where urbanization has rapidly taken place or the landfills outside the urban spaces where industrial wastes are discarded or the slum areas and open spaces (Siddiqua et al., 2022), where creating the awareness about the waste management has become difficult to achieve. In this gap, AI makes it easier to handle these complexities with ease. With AI's advanced identification technologies, this segregation of waste as per the composition and material becomes very quick and efficient (Lakhouit, 2025). Many companies have applied high-end methodologies in these identification parameters with AI monitoring and then implementing machine learning in the process of segregation (Shreyas Madhav et al., 2022). To understand this process and its implication, we have identified five organizations who are operating in India. We aim to understand their technological infrastructure, technical methods and outcomes of the process with impact assessment. The implications of the study lie in different conceptual and practical aspects. Through the analysis of multiple cases, the study aims to bring focus to the ongoing sustainability practices in the developing nations. The problem of waste management has been explored by many scholars in regional contexts, and the challenges are explored, measured and reported. This study discusses the emerging companies that are trying to achieve the circularity goals in the socioeconomically nuanced and exclusive developing contexts. The study may help future researchers develop full case studies on either of the organizations and report measurable impact. The study may also be beneficial for social and environmental entrepreneurs in getting important insights into the problem and designing actionable strategies to tackle the problem of waste management.

The chapter is organized in the following manner. Followed by the current introduction section, we present an overview of the role of AI in waste management and how the developed countries have undertaken the goal of creating circularity through high-end AI mechanisms. In the third section, we discuss the five selected cases from India, with cross-case analysis on the problem statement. A conclusion to the study comes afterwards in section four, with a statement on the implications and future scope of the study in section five.

2 Role of AI in Waste Management: What's Happening in the Global North?

In this section, we have discussed the strategies and implications of AI, adopted by organizations from the developed economies, who are working towards creating a waste-free, clean environment. Amidst the rising concern of waste generation in the world, these companies are working towards this cause to not only eliminate waste but also towards creating a circular value chain through recycling the waste for a more inclusive, circular economic society. In the developed countries, many entrepreneurial ventures have come up with innovative ideas of recycling waste in various industries and benefit the societal needs through innovative products. Most of these companies use AI and machine learning in these processes. For example, AMP Robotics in the USA has designed robotic sorters that can easily segregate waste as per its type and then transfers to recycling facilities. Their technology includes machine learning with computer vision that understands the difference in the compositions of plastics, metals and paper. It improves the sorting accuracy and recovery rates. Similarly, Greyparrot Waste Intelligence from the UK is doing a wonderful job by not only using AI in waste segregation but also states that there is a huge amount of uncaptured value in the waste streams. Greyparrot tracks the material, mass balance, food-grade, financial value and carbon footprint of every waste production in real time. The AI used in the process also produces instant customizable dashboards displaying live insights on the waste streams, spotting trends and issues as they arise. The segregation system also reduces the amount of valuable material lost to residue. Their GreySync technology functions as the brain of the facilities for a fully automated, responsive and efficient operation. This is how they ensure a 100% visibility into waste composition to unlock a new and more intelligent waste management system.

The title of the study, 'AI Waste-Sorting Robots Are the Future of the Circular Economy,' is taken from the vision statement of the company, Recycleye. It is a Germany-based company that has been leading the market with its AI-sorting robots equipped with computer-vision technology. The company has emphasized on circular economy principles as the core mission. Recycling waste removes the waste from the cycle and maximizes the reusable components. Their computer-vision technology enables the computers to design algorithms and techniques that help the machines analyse visual data to recognize patterns and detect features. The different elements of the waste, such as paper, aluminium and different plastics, are easily detected by the computer vision by scanning the waste stream according to a variety of factors. Like GreySync, Recycleye Vision technology differentiates even the packaged and non-packaged materials, food and non-food-grade materials according to their shape and colour. This enhances Material Recovery Facilities (MRFs), that sorts the waste within. A specific timeframe for increasing recycling output. They exclusively use AI Robotics to integrate technologies such as Natural Language Processing (NLP), Sensor Technology and Machine Learning. This blend between computer vision and robotics enhances the MRFs that help the company cover more household and commercial waste. The company claims that while only computer vision alone is very useful in the waste recycling process, integrating AI increases its usefulness even further within MRFs. Many MRFs are now adopting these solutions for boosting outputs, financial benefits and contributing towards an effective circular economy. Recycleye also claims that their technology is not only a very effective solution in creating a circular economy but also can improve workplace safety and minimize operations cost. Recycleye has also collaborated with VALTRIS in Belgium, Cireco in Scotland, recovering aluminium cans, rolling PET left on the residual line in MRFs with AI-powered robotic pickers, and Panda Ireland, especially in the automation on the aluminium line.

One of the pioneer organizations in this domain is ZenRobotics 4.0, which has realized this need of integrating advanced technologies in managing waste for sorting since 2007. They have the right robotic solutions for every need of an MRF. Their robots have served across countries and reached their circular economy goals. Their technology comes in three categories of services: Standalone Solution, Retrofitted Solution and Greenfield Solution. From building a fully automated and independent waste-sorting station to integrating waste-sorting robots into the existing facilities of an organization or designing a new large-scale facility around the same robotics, the company has led the path to circular economy solutions in a very effective manner. Very similar to GreySync and Recycleye Vision, their ZenBrain integrates AI inside both heavy-picker and fast-picker sorting robots. This enables the company to handle all kinds of waste materials with utmost efficiency. With a more AI-intensive approach, the company Glacier has reimagined the recycling work to a new level.

With China's ban on importing most of the foreign recyclables, the USA has faced increasing challenge on processing more waste than before, making it only about a third of total waste products recycled (Brooks et al., 2018). The company has emphasized on keeping communities clean by creating cleaner waste streams and better jobs by using AI. The work of sorting waste is often a very dangerous job, and human involvement in sorting biohazard products often causes serious injury. MRFs have faced repeated incidents of fire hazards due to lithium-ion batteries, with millions of dollars in revenue lost and damage repair costs. These incidents have also caused the MRFs to be understaffed. Glacier brought AI into this gap to create a safer environment. Using real-time analytics for better recovery rates, resulting in fewer materials in landfills. With automation, the company has been able to perform the largest single-site deployment in Seattle. Partnering with Recology King County (RKC), Glacier successfully conducted the project to reduce waste and support a more sustainable solution. The project not only helped in reducing waste to promote circularity in the environment but also enriched the community engagement and workforce training, facilitating a social change.

The above discussion has articulated the role of AI as a driving force in creating a circular economy. The developed economies are taking a lead in waste management to bring more circularity. Japan and South Korea have also adopted AI in their waste management strategies with good outcomes. Studies done in these countries have explored this domain and found that the amount of waste that is sent to the landfills can be reduced and also reducing methane emissions to contribute towards SDG 13 (Ram & Bracci, 2024; Wilson et al., 2024). It helps in supporting the environmental sustainability of the area and the economic development of the people. In this aspect, to understand the public adoption of AI in waste management, many behavioural frameworks have been applied to understand the factors in depth, as it is needed to understand how individual attitudes will matter in AI-based solutions for circular economy initiatives.

3 Cross-Case Analysis from India: How the Developing Nations Are Working Towards a Waste-Free Future

In a developing country like India, identifying waste generation patterns and measuring their impact on the environment is a daunting task. The previous studies have shown that this pattern varies across states, regions and cultures. The average waste generation in India is approximately 160,000 tons of municipal solid waste per day, with a per capita waste generation rate of 0.49–0.62 kg per person per day in 2018 (Central Pollution Control Board [CPCB], 2021). To tackle the municipal solid waste, Shahab and Anjum (2022) proposed a multipath convolutional neural network (mp-CNN) to detect the waste dumps as per evaluation parameters of detection, classification and localization. Many studies have also identified the problem of plastic waste as an alarming challenge that needs to be addressed with urgent remedies. In other developing countries, such as Nigeria, this is a huge problem due to the insufficient waste management infrastructure. Similarly, in India, the waste management infrastructure is often poor in many places. Even though the growing awareness in the general public showed some positive results, the problem still persists in almost every region. Identifying this issue, many companies have come up with AI-driven approaches to manage waste reduction with enhanced efficiency and effectiveness. India's unique challenges remain in the barriers such as lack of technological infrastructure, data scarcity, unplanned urbanization, rapid industrialization and lack of awareness, among others. Addressing these challenges requires a holistic approach and technologically efficient approach that may go beyond the impact of such barriers and can provide actionable solutions. We have discussed a few such cases here to understand the strategies, challenges and outcomes created by the leading companies that have integrated AI in waste management systems to contribute to the sustainable development of India.

To begin with, we bring Recykal, which is driving India's circular economy through high-end technology. Their approach has empowered every stakeholder in the waste management ecosystem, which is still not managed in a significant capacity. Recykal is a cloud-based waste management company operating in Hyderabad, employing Extended Producer Responsibility (EPR) through market expansion, economies of scale and reduction in uncertainty (Bhadra & Mishra, 2021). Recykal's vision of 'sustainable circularity' brings the much-needed solutions to India's waste problem with smart technologies. The company provides a platform to offer buying and selling options for recyclable materials such as plastics, metals and e-wastes. The company collects about 10,000 to 15,000 kg of waste per day (Bhadra & Mishra, 2021). Their primary aim is to help the fast-moving consumer goods, along with electronic brands, in bringing consumer awareness to convert the informal waste yards to recycling measures. Owned by Rapidue Technologies, Recykal provides the Uzed App for individual household consumers, along with schools and colleges. The app uses reward points that can be redeemed to purchases items through the app. There are multiple solutions for every stakeholder through providing online tools and a marketplace to help industries as well as households in participating into sustainable circular initiatives. However, in the current discussion of AI integration in their services, they are employing high-end online tools at their MRFs. Recykal has used Google AI's CircularNet to detect the types of waste and categorize accordingly. Recykal now operates in multiple states in India, providing an online marketplace for buying and selling metal scraps, plastic scraps, recyclates (premium recycled plastic) and several e-waste categories of goods. In India, the segregation of waste is still done by hand today, and it causes serious human injuries, health adversities and low efficiency in managing recyclables. Materials often end up in the wrong bins, residues are not cleaned off, liquid waste leaks and damages the environment or ends up in the waterbodies, accessed by certain communities. With AI coming into this space, these problems can be diverted to a great extent. The initial segregation of the waste itself is the first step to tackle the problem of human involvement in the process. Very similar to Recycleye Vision or Greyparrot, if the entire system can be shifted to the AI-driven processes, then the mechanism becomes easier to handle and more efficient in categorizing. Google AI technology, CircularNet, solves this problem from the basic ground level, so that human interventions in handling the waste manually come down to a negligible level. The SmarkSkan at Recykal focuses on categorizing this waste with accuracy and helps the company deliver transparency and actionable insights to all stakeholders. The company states that by integrating AI into their systems, they have been able to scale their capacity, increasing profitability for waste management centres and reforming India's waste management ecosystems. Going towards the discussion on managing a more robust and different kinds of waste materials, the Indian company, Ishitva, has quite made a stir in the domain of circular economy and sustainability with their advanced robotic systems for sorting and recycling varied waste materials. The sole mission of the company is to enable AI in the sorting solution for a profitable circular economy. Ishitva's mission is to automate sorting, eliminating contamination and maximizing resource recovery at every scale. Ishitva's AI-driven sorting solution detects a wide range of contamination in PET bales with a high-purity production and impurity-free output. Not only PET-sorting solutions, but they have also segmented their AI training in segregating mixed plastics, polymers, textiles, municipal soil waste, food waste and metals with complex e-waste. They have created quite a disruption in this segment with their AI-powered Air-sorter, autonomous MRF called Sanjivani, Flake analyser, Flake Sorter and PET1. All these tools are fully operated by AI with machine learning techniques, providing innovative solutions to recycle every kind of waste into defined segregated categories and ensuring actionable solutions. This company is one of the leading organizations that has high-end technological solutions, seen in companies from developed countries, as we have previously discussed. It is setting a benchmark in creating a 'take-make-dispose' model with an AI-powered solution for closing the circular economy loop.

Another waste management company based in Hyderabad is Banyan Sustainable Waste Management Private Limited, or Banyan Nation. This company works mainly on plastic recycling and works towards managing informal waste collectors, providing dignity to them. This company is also EPR-compliant and has a circular, responsible supply chain (Tripathi et al., 2022). Banyan Nation states that their strategies are especially aimed towards advancing the circular economy through developing inclusive and responsible approaches. Their plastic washing technology effectively eliminates contaminants and volatile compounds from waste. Banyan's data intelligence has democratized waste management across the country effectively by providing high-quality recycled high-density polyethene and recycled polypropylene. For this process, their AI-based sorting and washing system is leading in the manufacturing market in India. The company was established two years before the Paris Agreement in 2015, and its vision is aligned with the environmental sustainability goals, being one of India's first vertically integrated plastics recycling companies. As we previously argued, the challenges created by rapid urbanization the Banyan Nation to take up this segment to automate urban recycling operations. The company is one of the first of its kind in India to digitalize the waste management process, even before the advent of AI and created a significant nationwide impact. The Telangana state government in India has signed a memorandum of understanding to take up their technology across multiple municipalities to tackle the municipal solid waste. As part of their smart cities mission, the company has deployed on-field bin sensors, analytical visualization, landfill measurement toolkits and GPS truck routers. As a part of their programmes, Banyan Sustainable Waste Management was incubated in Hyderabad at the T-Hub foundation. Even though the Banyan nation has not fully transitioned into AI-driven technologies, it is in the transitioning phase, adopting Indian-based AI trainings and IoT solutions. Their vision to contribute to Swachh Bharat India can progress more if AI can be integrated more into their operations. They are solving the problem of the waste management crisis by targeting the problem of irresponsible dumping of used plastics. Major AI function is being adopted in their process optimization, and with the use of Intel IoT analytics, they have designed the Jan Unnati Tablet to streamline and automate the urban recycling operations. In creating cleaner cities, one company that consciously integrated AI for merging government interests and business profitability is Ecowrap, based in Rajasthan, India. With the government's collaboration, their AI tracks entities with source-segregation ratings, optimizing city-wide waste management with the help of machine learning, and tracks city-wide waste generation patterns. They help businesses in obtaining quarterly waste audit compliant reporting access AI-generated sustainability ratings. Their mission is to create India's cities with zero dumping policies and reinforce responsibilities for waste processing through AI. Ecowrap analyses and forecasts waste generation patterns in a city and then works towards providing a local solution with data-driven strategies.

Along with Banyan Nation, another company operating solely on plastic waste management in South Asia is Cleanhub, which has applied plastic tracking very efficiently with high-end tracking systems. Their track and trace software is a detailed, efficient tool that not only helps in collecting waste from households, but also accurately measures the amount of waste collected and tracks the plastic elements in it. The software helps in tracking the entire system till the final recovery. Their app is enriched with data logging options and processes that seamlessly help the collection teams on the ground. Their services work in five consecutive steps: to collect the waste by scanning QR codes on the front door, then measure the waste through a small scale, and then, with their advanced AI models, assess the weight distribution. Next, the crew report the weight with values cross-checked with data accuracy. The app aggregates the collected waste from all the households. After the collection is complete with no manual handling, the waste reaches their MRF, where they cross-compare its volume. The reports are checked multiple times across this timeline and then recorded in order to understand what the sources of the wastes are apart from the households. After the waste reaches the facility, with AI-integrated machines, the crew sorts out the recyclables and non-recyclables. The recyclables are reused, and the non-recyclables are sent to produce fuels. At this point, they register the waste again to monitor its mass balance. The data is monitored minutely for any discrepancies and inaccuracies. This is how the data generates quality insights into the waste generation patterns from the households. Their centralized tool helps in simplifying mass balance reporting, like ISCC plus that removes the possibility of error-prone spreadsheets. In this way, the organization ensures data transparency that is free of any anomalies. While sending the trash to the facilities, the waste-carrying trucks go over weighing bridges to register the weight again. This is how it is monitored how much waste is still left at the location. This rigorous weighing and monitoring process is what makes the company's technology exclusive and audit-proof. In Kerala, India, Cleanhub has worked very efficiently in protecting the coastal communities by reducing plastic waste from the ocean. The communities, such as Konad, Chemnad and Kasargod, show that local actions can create a huge impact on the overall goal of sustainability of a country. These communities faced major challenges in waste dumping, trash burning, and other issues that threatened the local biodiversity. With Cleanhub's help, the waste management here has become much easier, organized and balanced. The Cleanhub hubs in these areas sort, bale and store the waste efficiently with data accuracy, faster recycling and monitored transparency. Cleanhub has not only helped in cleaning the coasts and the ocean, but also developed a local community through creating jobs, building resilient communities, and rebuilding environmental sustainability. Their collection partner, Green Worms have worked across many municipalities in South India, and materialized Cleanhub's goal of a circular economy with innovative practices.

Apart from these organizational cases, many isolated initiatives and start-ups have taken up AI-driven solutions to tackle the waste problem for sustainable development in India. Few honourable mentions are Trashcon, a technology-first company with circularity in its mission, Green Nerds solutions, providing innovative solutions for the green cause, and Bin-e, smart waste bins that automatically sort waste with AI-based monitoring and reporting. Bin-e has emerged from Poland but has already partnered with many MRFs from developing countries to provide green solutions at the ground level. In Delhi, India, Gem Enviro Management is also creating a significant impact by creating sustainable solutions for waste management. In Delhi. Waste is a growing concern that has created a challenge to keep the city clean and pollution-free. In this space, Gemrecycle has provided innovative solutions with ESG consulting, including EPR, fulfilment of all kinds of dry waste, ESG consulting, BRSR reporting, project advisory and management. Even though in the developing countries the progress of these initiatives is slower in creating a significant impact on the environment, society and economy, with the cultural nuances, this current growth is not negligible. While the developed nations have done quite well in managing the problem with very high-end technological solutions, the developing counterparts are not behind. Given the population strength, complex nature of the societies and cultural issues in collective communities, more impact needs to be created in the emerging economies. With the integration of AI in this process, it can disrupt with a bolder impact in future, given these organizations' mission and vision towards creating a sustainable circular economy.

4 Conclusion

In this chapter, we have presented a comparative exploration of the AI-driven waste management strategies and solutions adopted by the developed and developing countries. With a cross-case analysis between the leading MRF organizations producing insightful data and reports on the patterns of waste generation and the effective solution provided through a machine learning mechanism integrated with AI, the chapter has highlighted the implications of AI for creating a circular economy in both economic contexts. While the cases from the global north provide very high-end solutions in tackling the issues, the emerging economies are still struggling; however, not very behind in its realization or adoption of emerging technologies. Several key findings have emerged from our discussion, among which the robotic sorting systems with AI algorithms for the automated sorting process, predicting waste generation patterns with household behaviour in a developing context, is one of the major takeaways. While India focuses on scalable and adaptable AI solutions for waste segregation, minimizing human intervention and providing dignity to the waste collector, the organizations also provide effective ESG consulting and help businesses provide better BRSR reporting options. As we conclude our discussion on a high hope for the impact spreading across other developing contexts, the call for adopting AI for every MRF with a scalable approach remains an imperative in the immediate future for achieving a more sustainable and inclusive socioeconomic circularity model.

5 Implications of the Study

The study provides important insights for the key stakeholders in the government as well as the business corporations. The takeaways will not only help policymakers in understanding the need of integrating AI for urban waste management solutions but also to design the smart cities initiatives with the circularity perspectives in check. It may also help the social and climate entrepreneurs creating sustainable impact regionally with local solutions for understanding waste generation patterns and creating awareness accordingly. The theoretical implication of the study calls for more academic research in this regard, exploring the minute technological analysis of the particular mechanisms that may create a major difference in the environment, biodiversity and the people. The cases can be undertaken in isolation in future studies and analysed deeply with empirical case analysis, with organizational insights into the MRF operations and creating impact assessment reports accordingly. In this sense, the present discussion shows several possibilities and pathways to diversify the problem into multiple contextual studies based on the researcher's area of interest and discipline, as it covers all the pillars of the Sustainable Development Goals. After all, the problem of waste management cannot be taken as an isolated issue but a complex problem that shifts with social, geographical and cultural differences;

however, as this chapter suggests, the goal of circularity in all these aspects can be quite achievable with AI integration, creating a sustainable disruption and change irrespective of the regional differences.

References

Adami, L., & Schiavon, M. (2021). From circular economy to circular ecology: a review on the solution of environmental problems through circular waste management approaches. *Sustainability, 13*(2), 925.

Adnan, M., Xiao, B., Xiao, P., Zhao, P., & Bibi, S. (2022). Heavy metal, waste, COVID-19, and rapid industrialization in this modern era—Fit for sustainable future. *Sustainability, 14*(8), 4746.

Agrawal, D. (2025). A comprehensive review of investing and financing for sustainable tourism projects. *Environment, Sustainability, and Governance Insights, 1*(1), 53–70.

Ajzen, I. (1991). The theory of planned behavior. *Organizational Behavior and Human Decision Processes, 50*(2), 179–211.

Awuah, K. G. B., & Abdulai, R. T. (2022). Urban land and development management in a challenged developing world: An overview of new reflections. *Land, 11*(1), 129.

Aydin, A. E., & Yildirim, P. (2021). Understanding food waste behavior: The role of morals, habits and knowledge. *Journal of Cleaner Production, 280*, 124250.

Banerjee, A., Sarkar, A., & Ghosal, S. (2022). Challenges and strategies of water supply and wastewater management in coastal urban and semi-urban areas. In *Current directions in water scarcity research* (Vol. 6, pp. 425–446). Elsevier.

Barr, S., Guilbert, S., Metcalfe, A., Riley, M., Robinson, G. M., & Tudor, T. L. (2013). Beyond recycling: An integrated approach for understanding municipal waste management. *Applied Geography, 39*, 67–77.

Bhadra, U., & Mishra, P. P. (2021). Extended producer responsibility in India: Evidence from Recykal, Hyderabad. *Journal of Urban Management, 10*(4), 430–439.

Brooks, A. L., Wang, S., & Jambeck, J. R. (2018). The Chinese import ban and its impact on global plastic waste trade. *Science Advances, 4*(6), eaat0131. https://doi.org/10.1126/sciadv.aat0131

Central Pollution Control Board (CPCB). (2021). *Annual report on implementation of Solid Waste Management Rules, 2016*. Ministry of Environment, Forest and Climate Change, Government of India. Available at: https://cpcb.nic.in/uploads/MSW/MSW_AnnualReport_2021-22.pdf

Davis, F. D. (1989). Perceived usefulness, perceived ease of use, and user acceptance of information technology. *MIS Quarterly, 13*(3), 319–340.

Golzary, A., Nematollahi, H., & Tuysserkani, M. (2024). Assessment and pathways for improving municipal solid waste management in rapidly urbanizing Tehran, Iran. *Clean Technologies and Environmental Policy, 26*(6), 1901–1917.

Joselin, D., Singh, A., & Tiwary, K. S. (2025). The nexus between CSR practices and financial performance: Empirical evidence from the manufacturing sector in India. *Environment, Sustainability, and Governance Insights, 1*(1), 1–17.

Kaith, A., & Sachdeva, G. (2024). Bibliometric insights into green human resource management: A connection with business ethics and values. *PURUSHARTHA-A Journal of Management, Ethics and Spirituality, 17*(1), 51–63. https://doi.org/10.21844/16202117104

King, A. M., Burgess, S. C., Ijomah, W., & McMahon, C. A. (2006). Reducing waste: Repair, recondition, remanufacture or recycle? *Sustainable Development, 14*(4), 257–267.

Kumar, K., Singh, R., Choudhary, M., & Sharma, A. (2025). Climate change mitigation strategies in environmentally sensitive industries. In *Climate change and social responsibility* (Vol. 25, pp. 185–204). Emerald Publishing Limited.

Lakhouit, A. (2025). Revolutionizing urban solid waste management with AI and IoT: A review of smart solutions for waste collection, sorting, and recycling. *Results in Engineering*, 104018.

Mhlanga, D. (2021). Artificial intelligence in the industry 4.0, and its impact on poverty, innovation, infrastructure development, and the sustainable development goals: Lessons from emerging economies? *Sustainability, 13*(11), 5788.
Mmereki, D., Baldwin, A., & Li, B. (2016). A comparative analysis of solid waste management in developed, developing and lesser developed countries. *Environmental Technology Reviews, 5*(1), 120–141.
Nanda, S., & Berruti, F. (2021). Municipal solid waste management and landfilling technologies: A review. *Environmental Chemistry Letters, 19*(2), 1433–1456.
Nascimento, D. L. M., Alencastro, V., Quelhas, O. L. G., Caiado, R. G. G., Garza-Reyes, J. A., Rocha-Lona, L., & Tortorella, G. (2019). Exploring Industry 4.0 technologies to enable circular economy practices in a manufacturing context: A business model proposal. *Journal of Manufacturing Technology Management, 30*(3), 607–627.
Nell, C., Schenck, C., Blaauw, D., Grobler, L., & Viljoen, K. (2022). A three-pronged approach to waste composition determination. *Journal of Environmental Management, 303*, 114203.
Oghazi, P., & Mostaghel, R. (2018). Circular business model challenges and lessons learned—An industrial perspective. *Sustainability, 10*(3), 739.
Prakash, A., & Tiwari, A. K. (2025). The role of institutional factors in achieving SDG 16: A thematic review of drivers of corporate anti-corruption action. *Environment, Sustainability, and Governance Insights, 1*(1), 34–52.
Ram, M., & Bracci, E. (2024). Waste management, waste indicators and the relationship with sustainable development goals (SDGs): A systematic literature review. *Sustainability, 16*(19), 8486.
Ramachandra, T. V., Bharath, H. A., Kulkarni, G., & Han, S. S. (2018). Municipal solid waste: Generation, composition and GHG emissions in Bangalore, India. *Renewable and Sustainable Energy Reviews, 82*, 1122–1136.
Rodić, L., & Wilson, D. C. (2017). Resolving governance issues to achieve priority sustainable development goals related to solid waste management in developing countries. *Sustainability, 9*(3), 404.
Shahab, S., & Anjum, M. (2022). Solid waste management scenario in India and illegal dump detection using deep learning: An AI approach towards the sustainable waste management. *Sustainability, 14*(23), 15896.
Shekdar, A. V. (2009). Sustainable solid waste management: An integrated approach for Asian countries. *Waste Management, 29*(4), 1438–1448.
Shreyas Madhav, A. V., Rajaraman, R., Harini, S., & Kiliroor, C. C. (2022). Application of artificial intelligence to enhance collection of E-waste: A potential solution for household WEEE collection and segregation in India. *Waste Management & Research, 40*(7), 1047–1053.
Siddiqua, A., Hahladakis, J. N., & Al-Attiya, W. A. K. (2022). An overview of the environmental pollution and health effects associated with waste landfilling and open dumping. *Environmental Science and Pollution Research, 29*(39), 58514–58536.
Singh, R., Kumar, K., & Khan, S. (2024). A comprehensive view of artificial intelligence (AI)–based technologies for sustainable development goals (SDGs). *Artificial Intelligence Enabled Management: An Emerging Economy Perspective, 183*, 9783111172408–012.
Singh, R., Sharma, A., & Kumar, K. (2025a). Navigating carbon neutrality: Strategies for sustainable future and environmental challenges. In *Climate Neutrality Through Smart Eco-Innovation and Environmental Sustainability* (pp. 69–80). Springer Nature Switzerland.
Singh, R., Tiwary, K. S., Kumar, K., & Kumar, A. (2025b). Empowering a sustainable future: Blockchain's role in economy, organizations, and finance towards smart circular economy. In *Blockchain Technologies for Smart Circular Economy and Organisational Sustainability* (pp. 271–283). Springer Nature Switzerland.
Srivastava, S. K., Anshul, A., Pathak, P., & Bansal, J. (2018). Responsible consumption for curbing food wastage: An exploratory enquiry. *PURUSHARTHA-A Journal of Management, Ethics and Spirituality, 11*(2), 104–116.

Srivastava, S. K., Singh, S., & Srivastava, M. (2021). Food waste at household and social gatherings: Drivers and possible remedies. *PURUSHARTHA-A Journal of Management, Ethics and Spirituality, 14*(2), 117–133.

Srivastava, S. K., Sinha, A. K., & Kumar, S. (2024). Workplace spirituality empowered by spiritual leadership to promote organizational effectiveness: A bibliometric analysis. *PURUSHARTHA-A journal of Management, Ethics and Spirituality, 17*(1), 1–22. https://doi.org/10.21844/16202117101

Sun, Y., Shahzad, M., Ali, A., & Razzaq, A. (2025). Synergistic effect of green knowledge sharing and green creative climate for circular economy practices: role of artificial intelligence information quality. *Journal of Manufacturing Technology Management.*

Thomas, E., & Rogers, B. E. M. (1998). Diffusion of innovations theory and work-site AIDS programs. *Journal of Health Communication, 3*(1), 17–28.

Tripathi, V., Singh, K., Parihar, S., & Srivastava, S. K. (2022). A bibliometric analysis on supply chain collaboration. *PURUSHARTHA-A Journal of Management, Ethics and Spirituality, 15*(2), 66–103. https://doi.org/10.21844/16202115205

Trushna, T., Krishnan, K., Soni, R., Singh, S., Kalyanasundaram, M., Annerstedt, K. S., et al. (2024). Interventions to promote household waste segregation: A systematic review. *Heliyon, 10*(2).

Wabnitz, K. J., Gabrysch, S., Guinto, R., Haines, A., Herrmann, M., Howard, C., et al. (2020). A pledge for planetary health to unite health professionals in the Anthropocene. *The Lancet, 396*(10261), 1471–1473.

Wan, Q., & Du, W. (2022). Social capital, environmental knowledge, and pro-environmental behavior. *International Journal of Environmental Research and Public Health, 19*(3), 1443.

Wilson, D. C., Paul, J., Ramola, A., & Filho, C. S. (2024). Unlocking the significant worldwide potential of better waste and resource management for climate mitigation: With particular focus on the Global South. *Waste Management & Research, 42*(10), 860–872.

Yadav, H., Soni, U., & Kumar, G. (2023). Analysing challenges to smart waste management for a sustainable circular economy in developing countries: a fuzzy DEMATEL study. *Smart and Sustainable Built Environment, 12*(2), 361–384.

Yusoff, N. I. B. M., Godsell, J., & Woolley, E. (2024). Towards zero waste: A comprehensive framework for categorizing household food waste. *Sustainable Production and Consumption, 48*, 1–13.

Chapter 15
Integrating Generative AI for Sustainable and Resilient Supply Chains

Vivek Gupta and Ravi Shankar

Abstract The increasing volatility of global supply chains, driven by climate change, geopolitical disruptions, and regulatory pressures, necessitates the integration of advanced digital capabilities with sustainability and resilience strategies. This study develops a theory-driven conceptual framework explaining how Generative Artificial Intelligence (GenAI) capabilities enable sustainable and resilient supply chains through dynamic capability development. Drawing upon the Resource-Based View (RBV), Dynamic Capability Theory (DCT), the Triple Bottom Line (TBL) perspective, and Supply Chain Resilience Theory, the framework positions GenAI as a higher-order digital capability that enhances scenario generation, carbon-aware network design, risk intelligence, and adaptive reconfiguration. The model proposes that GenAI capabilities positively influence sustainability practices, which mediate the development of supply chain resilience and ultimately contribute to sustainable competitive advantage, with environmental turbulence acting as a moderating condition. By shifting the focus from predictive analytics to generative strategic design, the study reframes sustainability as a structural enabler of resilience. Theoretically, the framework integrates GenAI, sustainability, and resilience literatures into a unified model of dynamic capability development. Practically, it offers managers a strategic roadmap for embedding Generative AI into supply chain design to achieve carbon efficiency, disruption preparedness, and long-term competitive advantage.

Keywords Generative AI · Sustainable · Resilient · Digital transformation · Supply chain resilience

---

V. Gupta (✉)
Bharti School of Telecommunication Technology and Management, Indian Institute of Technology, New Delhi, India
e-mail: bsz238164@iitd.ac.in

R. Shankar
Department of Management Studies, Indian Institute of Technology, New Delhi, India
e-mail: ravi1@dms.iitd.ac.in

K. Kumar et al. (eds.), *AI-Driven Circular Economy for Carbon Neutrality in Emerging Economies*, Environmental Footprints and Eco-design of Products and Processes,
https://doi.org/10.1007/978-3-032-30298-4_15

1 Introduction

Ensuring the stabilization of the global supply chain is crucial to the world's sustainable development (Qin et al., 2023). Sustainable development satisfies the current needs without impairing the ability of future generations to meet demands (Niemets et al., 2021), where stabilizing the global supply chain is an important aspect (Qadeer et al., 2022). United Nations (UN) Sustainable Development Goals (SDGs) are considered as essential benchmarks for the growth of industry infrastructure, and innovation, with SDG-9 focusing on the new technologies and Artificial Intelligence (AI) specific areas (Shankar & Gupta, 2024). Generative Artificial Intelligence (GenAI) is emerging as a transformative force in supply chain resilience, offering new ways to enhance decision-making, automate operations, and improve adaptability to disruptions (Boone et al., 2025). Moreover, businesses that align their strategies with global sustainability frameworks such as the UN Sustainable Development Goals (SDGs) and ESG (Environmental, Social, Governance) criteria gain access to green financing, tax incentives, and government grants. These financial advantages further support long-term profitability and business growth.

Firms face increasing pressure to reduce carbon emissions while remaining resilient to geopolitical, climate, and demand shocks. Existing AI applications are largely predictive. Generative AI (GenAI) enables scenario generation, adaptive network redesign, and sustainability–resilience trade-off optimization, but its strategic integration is underexplored. The global supply chain is characterised by complexity and vulnerability (Shen, 2021), that is, each node is a 'vulnerable point', and the collapse of a node could adversely affect the upstream and downstream of supply chains and other links (Jomthanachai, 2022). With the increasing emissions of greenhouse gases, the occurrences of extreme climate, including drought, flood, heat wave and cold injury, are significantly frequent (Qin et al., 2022). On the one hand, the frequent extreme weather could damage production and transportation infrastructures (Nobre et al., 2019), inevitably weakening the supply capacity and disrupting the global supply chain.

Recent literature with respect to GenAI in supply chain area has received a multiple attention among professionals and academicians. GenAI can dramatically change the entire SCOR model (Plan, Source, Make, Deliver, Return). In previous research, analytics has served as the important topic of discussion, that how generative algorithms will change decision-support systems to make them work more successfully. For example, Siciliani et al. (2023) emphasize AI's strategic impact in procurement, focusing on natural language processing and robotic process automation as opposed to generative frameworks. Jackson et al. (2024a, 2024b) proposed a rational framework to guide practitioners and researchers on AI and GenAI applications in operations and SCM, emphasizing improved decision-making, process efficiency, strategic investment, and skill development. GenAI, offers distinct capabilities such as contextual learning, content generation, and scenario simulation (Aggarwal et al., 2025). Although GenAI applications have been extensively studied in Supply chain, such as advancements in predictive analytics (Kuo-Wei, 2025), sustainable SC (Pamucar et al., 2025), and AI-driven compliance monitoring (Fosso-Wamba et al., 2024) but their application in sustainability activities is still not very well developed. Presently, there is a scarcity of research that emphasizes theory-driven rationalizations. This may emerge from clarity issues in the theoretical frameworks; hence, our study seeks to address this gap and offer a conceptual framework for understanding how Generative AI will enhance resilience and sustainable supply chain. This purpose of this study is to examine how GenAI transforms sustainability and resilient outcomes in the supply chain. The research objective are as follows:

RO1: To develop a theory-driven framework explaining how Generative AI enables sustainable and resilient supply chains.

RO2: To propose a structured conceptual framework linking Generative AI capabilities, sustainability dimensions, and supply chain resilience outcomes.

This research contributes to a number of theoretical areas. It goes forward with digitizing the first transformation literature through a conceptualization of Generative AI (GenAI) as a higher-order dynamic capability which increases firms sense making, seizing, and reconfiguring supply chain resources under uncertainty. Second, it fills gaps in research areas by synthesizing past streams of research. The theory of resilience, sustainable supply chain management, and artificial intelligence are all transformed into a unified framework. This paper transforms the prevailing vision of AI in operations from anticipatory analytics to generative strategic design, encompassing scenario creation, adaptive configuration, and proactive decision intelligence. Lastly, it redefines sustainability not as an objective of compliance but as an enabler of resiliency on a strategic level, showing how long-term and adaptive capacity, which is enhanced by environmentally and socially responsible practices competitive advantage.

2 Background and Literature Review

Recent literature suggests that Artificial Intelligence (AI) plays very significant role in supply chains improvements. Nevertheless it also explains how AI can help different companies in improving their flexibility and faster response. Risks coming in supply chains are considered as growing concern. AI or GenAI looks at information like news, social media, and weather to predict early signs of problems shortages or delays (Jackson et al., 2024a, 2024b). Few researchers mentioned that AI helps in finding the best suppliers and making contracts faster (Gaddala, 2022). Another area where AI helps is in making supply chains more sustainable and reliable. AI tools help companies reduce waste and use resources carefully (Maghsodi et al., 2025). SCM focuses on variables such as speed, cost, and reliability, whereas sustainable SCM is extended to include the efforts addressing environmental and social factors (Hahn et al., 2025).

2.1 Generative AI (GenAI) in Supply Chains

Supply chain management (SCM) means handling the flow of goods, money and information from one place to another. Today, this process is very complicated because of many reasons like global trade, changing customer needs, and risks such as delays or shortages (Fosso Wamba et al., 2023). Artificial intelligence assists in computer learning and smart decision-making. A Generative AI (GenAI) is highly practical since it has the ability to generate new content. For example, it is able to write text, create images or create predictions. The skills can assist the companies in planning better decisions, identify risks at an early stage, and enhance their business (Culot et al., 2024; Goswami et al., 2025). Generative AI uses models like Variational Autoencoders (VAEs), Generative Adversarial Networks (GANs), and large language models (LLMs) like GPT. These models can do many things for the supply chain, such as forecasting demand, making smarter buying plans, or automating tasks (Touvron et al., 2023). The ability of these models to generate new data similar to what they were trained on makes them extremely useful in simulating various supply chain scenarios, optimizing decisions and automating communications. Among the various AI technologies, GenAI stands out due to its ability to autonomously generate context-relevant content and is particularly transformative for supply chains, offering capabilities beyond traditional AI systems (Grimes et al., 2023).

Today, many companies use AI to improve their supply chains. Such tools allow preventing delay, cutting expenses, and enhancing customer satisfaction (Brintrup et al., 2024). They also help in planning risks, stock management, and business decision-making that is more rapid (Jackson et al., 2024a, 2024b). Also, AI helps in combining multi-tier supply chains and offer a holistic perspective that assists businesses to operate under complexities and make sound decisions. Advanced AI algorithms facilitate improved future market changes and supplier actions, improving supply chain resilience.

2.2 Sustainable Supply Chain Management

Organizations across several sectors are making significant investments in sustainable supply chain management (SSCM), concentrating on integrating social and environmental factors into procurement, logistics, and production. Supply Chain Sustainability offers a systematic method to incorporate environmental, social, and economic sustainability into their supply chain activities (Ansari & Kant, 2017). This framework comprises numerous essential elements that assist firms in minimizing their carbon impact, improving social responsibility, and securing long-term economic sustainability. A Supply Chain Sustainability (SCS) framework focuses on harmonizing industrial operations with social well-being while ensuring economic viability and environmentally friendly. A major focus is reducing the carbon footprint through transportation optimization techniques, energy-efficient process, and sustainable procurement. Green Supply Chain Management (SCM) integrates eco-friendly practices, including sustainable purchasing, efficient manufacturing, optimized distribution, and reverse logistics. The three sustainability pillars—social responsibility (People), economic viability (Profit), and environmental protection (Planet)—ensure a balanced approach. Sustainability starts initially from every stage of supply chain starting from sourcing to production, warehousing and distribution and recycling. A truly sustainable supply chain operates with minimal harm to natural and social systems while maintaining long-term profitability. The sustainable market framework aims to prevent overproduction by reducing surplus supply and minimizing reverse supply through repair, recycling, and remanufacturing, fostering a responsible and efficient supply chain. This framework ensures that businesses not only reduce their ecological footprint but also create value by adopting ethical and efficient supply chain practices.

2.3 Supply Chain Resilience

The capability to overcome and react to disruptions is called supply chain resilience (SCR). A robust and flexible supply chain is necessary as a way of sustaining the economy due to the fact that it allows firms to withstand pandemics, climate change, geopolitical shocks. revolution, and other external factors. The world events like the COVID-19 pandemic demonstrated the vulnerability of the supply systems, which are based on nations. The continuity of businesses under disruptions can be ensured through local suppliers and Acquiring supplies at various places. Also, companies can anticipate risks and resort to precautionary actions by investing in. such digital supply chain technologies as real-time data monitoring, IoT tracking, and AI-powered predictive analytics. GenAI-based demand forecasting, e.g. reduces costs and enhances efficiency by avoiding overstocking or shortage of supplies. Companies can also integrate climate adaptation strategies, including investing in low-carbon logistics, using substitute consumption of transport systems such as rail or electric trains and increasing infrastructure infrastructures climate-sensitive locations. Focusing on efficiency, agility, and strategic decision-making, the GenAI technology is transforming the horizons of demand forecasting, operations, industrial, inventory control, and logistical resiliency (Mohamed, 2023). Over time, these actions not just make them resilient but also reduce operating costs and carbon emissions.

2.4 Identified Research Gap

AI traditional methods have already been used in supply chain management to success, optimizing demand forecast and logistics rule-based decision automation (Attaran, 2020; Min, 2019). However, GenAI presents new dimensions by not only assessing pre-defined scenarios, it also creates scenarios and appropriate strategies, allowing for more proactive and adaptive responses to disruptions (Naz et al., 2022; Zamani et al., 2023). For example, the ability to model alternate supplier networks, dynamically reconfigure inventory strategies and reinforce risk mitigation models might make it a critical part of creating an always-on, GenAI-powered supply chain resilience (Belhadi et al., 2024). A remarkably, given that academia is considering more and more AI in the SCM literature, there are still research gaps regarding what exactly GenAI implies for the SC resilience (Dubey et al., 2024; Lin and Zhu (2025). GenAI among the game-changing AI technologies, GenAI emerges as one of the most revolutionary technologies, capable to autonomously generate context-relevant artefacts and being transformative to supply chains by exceeding traditional AI systems' abilities (Grimes et al., 2023). Here we endeavour to address this gap by reviewing the potential opportunities and challenge of using GenAI for Supply Chain Resilience.

Although the number of studies on AI, sustainability and supply chain resilience is growing the literature remains incoherent and does not provide a unified theoretical explanation that connects Generative AI capabilities with sustainability followed by supply chain resilience. Most research emphasis has been placed on operational use and predictive analytics, with virtually no strategic level theorizing rooted in theories such as the resource-base view or dynamic capability theory. As a result, GenAI has not been well situated as a higher-order capability that combines sustainability and resilience to create long-term competitive advantage.

3 Theoretical Foundation

3.1 Resource-Based View (RBV)

The Resource-Based View (RBV) postulates that firms could achieve sustainable competitive advantage by developing resources that are valued, rare, unique, and non-substitutable (Barney, 1991). Gen AI is rapidly becoming a core capability in supply chain as key activities, such as inventory prediction, supplier assessment, and demand forecasting, are being redesigned by exploiting big data, learning algorithms, and contextual intelligence embedded in daily decision-making (Chen et al., 2022). Implementing of Generative AI into normal operations of supply chain transforms into a business's growth. AI-driven supplier intelligence solutions can monitor procurement part, vendor selection, and compliance in real time. Moreover, predictive technologies using AI or GenAI will enhance inventory management. These capabilities will transform GenAI from a mere technological tool into a strategic asset. From an RBV lens, GenAI-enabled supply chain demonstrates valued, rare, unique, and non-substitutable attributes, boosting competitiveness in supply chain (Wamba-Taguimdje et al., 2020). The RBV suggests that digital supply chain capabilities, driven by GenAI serve as a strategic resource that provides a competitive advantage. Furthermore, the Dynamic Capabilities Theory highlights the significance of constantly evolving these capabilities to maintain and improve ESG performance in a dynamic market environment (Wang & Zhang et al., 2025).

3.2 *Dynamic Capability Theory (DCT)*

The dynamic capabilities theory helps understand how companies grow and adapt in environments that are dynamic and technologically driven. DCT suggests that organizations must develop the ability to integrate, construct, and reconfigure internal and external competencies to adapt to the dynamic environment (Teece et al., 1997). GenAI represents a frontier technology with transformative potential for supply chain sensing, learning, and orchestration, core dynamic capabilities as defined by Teece (2007). By developing GenAI capabilities, businesses can ensure to adopt new opportunities and challenges associated with it. This necessitates a continuous learning and development, as well as a readiness to allocate new resources and tools accordingly. Nonetheless, the prospective advantages are considerable, including enhanced efficiency, improved decision-making, and a more competitive stance in the market (Dubey et al., 2024). Jackson et al. (2024a, 2024b) develop a capability-based framework linking Gen AI capabilities to supply chain responsiveness and adaptability, emphasizing its role in enabling flexible reconfiguration of processes and networks. Hence, cultivating Generative AI capabilities is crucial for every organization aiming to remain pertinent and competitive in the contemporary corporate landscape. Organizations may sustained growth and secure long-term success by adopting a strategic approach to AI development.

3.3 *Triple Bottom Line*

The TBL is used to evaluate sustainability. The TBL is considered to primarily consist of three sustainability aspects: economic, social, and environmental aspects, which are essential for a firm's prolonged prosperity (Elkington, 1998). When in today's context, modernizing technology has emerged as an enabler for countries to develop sustainably, on an industry level, modern and green technology innovations lead to enhanced performance and better terms of work (Khan et al., 2023). The Triple Bottom Line (TBL) concept serves as a fundamental perspective for comprehending sustainability in supply chains by highlighting the concurrent achievement of environmental, social, and economic performance. Within the framework of this research, TBL functions as the sustainability mechanism via which Generative AI (GenAI) capabilities provide resilience results. We need to ensure accessibility to such modern technologies to achieve a global standardization. GenAI can facilitate the early stages of sustainable business innovation.

3.4 Resilience Theory

Resilience theory conceptualizes supply chains as adaptive systems capable of anticipating, absorbing, responding to, and recovering from disruptions (Christopher & Peck, 2004; Ponomarov & Holcomb, 2009). Rather than merely restoring pre-disruption performance, resilience emphasizes adaptability, flexibility, and reconfiguration under uncertainty (Teece, 2007). In supply chain contexts, resilience is reflected through agility, redundancy, and recovery capability, enabling firms to sustain operations amid environmental turbulence (Sheffi & Rice, 2005). Within this study, resilience is viewed as an outcome of sustainability-driven adaptive capabilities supported by Generative AI integration. Supply chain resilience literature emphasizes that adaptive capabilities—such as agility, flexibility, and redundancy—help firms absorb and recover from disruptions. Technological capabilities, including AI-powered tools, contribute to resilience when embedded within organizational routines and dynamic processes rather than treated as isolated operational solutions (Haddad et al., 2026).

4 Conceptual Framework Development

In this section, a theory-based conceptual framework is developed as to how capabilities of GenAI allow sustainable and resilient supply chains in the form of dynamic capability development. Considering Resource-Based View (RBV), Dynamic Capability Theory (DCT), Triple Bottom Line (TBL), and Supply Chain Resilience Theory, the framework positions GenAI as a `higher-order digital capability that transforms sustainability practices into resilience outcomes, ultimately leading to sustainable competitive advantage.

> "The proposed framework follows a sequential logic: GenAI Capabilities → Sustainability Practices → Supply Chain Resilience → Sustainable Competitive Advantage, with environmental turbulence acting as a boundary condition that strengthens these relationships".

4.1 Generative AI Capabilities as Higher-Order Dynamic Capabilities

Generative artificial intelligence, characterized by its extraordinary learning and self-improvement capabilities (Victor et al., 2023), is profoundly transforming all sectors (Kanitz et al., 2023), particularly in supply chain management (Wu et al., 2022), where it exhibits immense potential and influence. GenAI and various digital supply chain technology, are significantly contributing towards sustainable development goals (SDGs). Generative AI, such as ChatGPT, in digital supply chains can enhance innovation, collaboration, and, ultimately, ESG performance (Wang & Zhang, 2025).

The integration of GenAI and AI has initiated a revolution in business operations, automating processes to ensure profitability (Carayannis et al., 2024) and support a wide range of supply chain functions, including logistics, procurement, sustainability, and forecasting. By integrating GenAI in supply chain operations, businesses can increase strategic emphasis on demand versus supply optimization. There is a growing emphasis on achieving sustainable development goals (SDGs), prompting supply chain experts to adopt new technologies (Holgado et al., 2024). GenAI helps with sustainability and well-informed decision-making by automating compliance in document creation. GenAI and sustainable supply chains in manufacturing lead to innovative, environmentally friendly products and services, creating sustainable business opportunities and enhancing SSCP (Malhotra & Ramalingam, 2023).

Preposition 1: *Generative AI capabilities positively influence the development of sustainable supply chain practices.*

4.2 Sustainability as a Strategic Mediating Mechanism

Sustainable supply chain has gained considerable attention in the recent past owing to the growing environmental concerns and regulatory pressures (Schilling & Seuring, 2024). Supply chain sustainability is concerned with reducing its environmental impact and preserving efficiency and profitability. Companies are more and more willing to use green technologies to minimize carbon footprints, increase energy efficiency, and adhere to international environmental policies. The sustainability initiatives have traditionally been considered as cost-constraining or compliance driven measures. Nevertheless, GenAI integration has been achieved intelligent design and adaptive optimization make repositions sustainability (Rahman et al., 2026). Carbon footprint calculators powered by AI assist companies in tracking and monitoring minimize their effect on the environment. It helps in adherence to carbon laws and enhances corporate social responsibility (CSR). GenAI and AI-based route optimization assistants saving fuel and emissions through finding the most efficient ways of delivery. It cuts as important as it reduces transportation expenses and carbon emissions (Singh et al., 2026). GenAI can be used environmentally to support carbon-conscious routing, inventory planning that is resource-efficient and minimizing production scheduling. Socially, the traceability is improved through AI Regulatory compliance, ethical sourcing, and transparency (Yadav, 2025). Economically, adaptive network optimization can enhance the cost-effectiveness and keeps the environment standards. In this way, sustainability will be incorporated as part of the logic of operational decisions instead of imposed externally. Critically, the sustainability is a mediating process between GenAI and resilience. Sustainable practices, e.g. supplier diversification, localized sourcing, and circular material flows make structures less vulnerable and enhance recovery ability during disruptions. It is not that a greener supply chain is merely environmentally friendly but structurally sound also. Generative AI is able to quickly produce and assess novel resolutions, including environmentally friendly packaging styles and energy saving measures (Rahman et al., 2026).

Preposition 2: *Sustainability practices mediate the relationship between Generative AI capabilities and supply chain resilience*

4.3 Sustainability-Driven Supply Chain Resilience

The issues of sustainability have remained eminent in the chain supply operations. The interdependence between supply chains resilience and sustainability indicates a change of considering these ideas as separate goals towards the acknowledgment of them as mutually supporting capacities that support the viability of operations in the long term. The GenAI-driven technologies can also increase the resilience within supply chains because automation and digitalization can optimally be utilized ultimately, the resources and minimize waste and increase transparency and traceability resilience improvement (Ahmed et al., 2025). It is automating its operations like demand forecasting procurement and customized service, and GenAI boosts efficiency, saves money, and enhances resilience (Nitsche et al., 2021). The use of GenAI in supply chains is a revolutionary change in the way companies use sustainability, innovation, and build organizational resilience (Dubey et al., 2024; Storey et al., 2025). GenAI facilitates the design of adaptive manufacturing systems that enhance flexibility and responsiveness to market shifts and consumer demands (Ghobakhloo et al., 2024).

Preposition 3: *Sustainability-enabled supply chains positively influence overall supply chain resilience*

4.4 From Resilience to Sustainable Competitive Advantage

The integration of advanced technologies, and artificial intelligence (AI) or GenAI, is revolutionizing supply chain management, and enhancing performance through improved visibility, traceability, and decision-making (Nandi et al., 2023). Grounded in dynamic capabilities theory, Lin and Zhu (2025) examine the interconnections among generative AI capabilities, competitive advantage, and service innovation under environmental dynamism. Ruangkanjanases et al. (2025) examine the implementation pathway of Generative AI Capabilities (GAIC) in manufacturing organisations, focusing on the roles of organisational creativity and green innovation ambidexterity in achieving a green competitive advantage. Sustainability is now a major factor in long-term profitability and competitive advantage, not only a moral decision. Businesses that integrate green practices into their supply chains benefit from improved investor and customer trust as well as regulatory backing, which raises profits and lowers expenses. Businesses that make investments in biodegradable materials, ecological packaging, and energy-efficient production techniques, e.g. draw in eco-aware customers who value eco-friendly brands.

Preposition 4: *Supply chain resilience derived from Generative AI-enabled sustainability contributes to sustainable competitive advantage.*

4.5 Moderating Role of Environmental Turbulence

The business environment is always changing and change creates risk. Managing the risk of the unpredictable future is a challenge which demands resilience the ability to survive, adapt and grow when faced with the rough change (Pettit, 2008). The success of GenAI incorporation is depending on the degree of environmental turbulence, such as climate uncertainty, political instability, regulatory changes, and fluctuations in demand. The degree of change in the environmental is usually defined in terms of external environment within which a firm is operating dynamism (Gupta et al., 2023). Environmental dynamism is often related to increased volatility, uncertainty, and ambiguity in the decision-making (Azadegan et al., 2013), which may interfere with resource setups, destabilize routines and slow down strategic execution. The utility of scenario generation, adaptive, is high in turbulent environments of high degree of turbulence.

The optimization and carbon-conscious design is emphasized. From a positive viewpoint, dynamic environments usually subject firms to fresh openings and swiftly changing customer demand, leading to the greater need on better sensing and seizing capabilities according to DCT (Fosso Wamba et al., 2026). Generative AI in this regard capabilities augment sensing as it helps firms to discern undiscovered customer needs, recognizing latent ones market patterns, and foresee the change in technology with the help of enormous amounts of data, natural process of language, and multimodal pattern recognition (Sedkaoui & Benaichouba, 2024).

Companies in predictable environments can have growth in benefits in an incremental manner, and those with exposure to challenges can have growth in compensation through the incremental approach high uncertainty achieve the significant benefit of GenAI resiliency. Environmental turbulence thus enhances the association between GenAI capabilities and resilience outcomes.

Preposition 5: *Environmental turbulence moderates the relationship between Generative AI capabilities and supply chain resilience*

The suggested conceptual framework makes Generative AI one of the backbone digital dynamics abilities that facilitates greener practices, which subsequently empower resilience and motivation sustainable competitive advantage. The incorporation of the development of AI capabilities and the framework offers a consistent explanation of how based on sustainability and resilience theory digital intelligence transforms supply chain design in time of environmental and geopolitical uncertainty. This conceptualization advances the understanding of supply chains as adaptive, sustainability-driven systems supported by generative digital intelligence rather than static efficiency-oriented networks.

5 Managerial and Policy Implications

The rapid rise of Generative Artificial Intelligence (GenAI) as a game-changer in both the business world and the academic world has been made possible by advances in technology in the fields of big data, processing power, and machine learning algorithms. The conceptual prepositions developed provide actionable insights for policymakers and businesses to mitigate supply chain risks and enhance resilience. These insights come in the form of suggestions that can help you use GenAI processes to make better decisions. Generative artificial intelligence is a relatively new thing that has confused researchers who are trying to figure out what it could mean for the future. The findings of this study may aid organizations in formulating strategies to efficiently oversee the implementation of the technology. Businesses need to make sure that their sustainability strategy is in line with their current level of AI maturity in order to get the most out of GenAI in the supply chain. This will make sure they get the most benefits. Companies that have reached a mature stage in the development of AI should strategically use GenAI to create new products and services that are more deeply embedded in their existing ones.

6 Conclusion

This study develops a theory-driven conceptual framework explaining how Generative Artificial Intelligence (GenAI) can enable sustainable and resilient supply chains through dynamic capability development. With the combination of the Resource-Based View, Dynamic the Triple Bottom Line approach, Supply Chain Resilience Theory, and Capability Theory, the framework orients GenAI as a more advanced strategic ability as opposed to just an operational technology. The model shows that GenAI capabilities—such as scenario generation, carbon-conscious network design, and adaptive reconfiguration—strengthen sustainability practices that subsequently promote supply chain resilience and contribute to it competitive advantage that is sustainable. The research contributes to the existing body of knowledge by filling the gap divided streams of research on AI, sustainability, and resilience into a single theoretical one structure. It changes the prevailing emphasis on predictive analytics to generative strategic design and reinvigorates sustainability as an organizational facilitator of resilience, as opposed to a compliance-based goal. By doing it, the framework emphasizes the strategic value of embedding generative intelligence in supply chain architecture to be proactive in managing environmental unpredictability and the uncertainty of the longer term. All in all, this study highlights that future of supply chain management is in the nexus of digital intelligence, sustainability integration, and adaptive resilience. In order to enhance more research in this field, theory-based research may be helpful important with regard to broadening the boundaries of the existing study.

References

Aggarwal, M., Verma, S., Kaur, N., Kaur, J., Rani, S., & Jolly, A. (2025). AI-IoT-blockchain synergy: A next-gen framework for smart cities. In *Innovations in Computing* (pp. 567–573). CRC Press.

Ahmed, Q., Sumbal, M. S., & Lee, C. K. (2025). Exploring the role of generative AI to enhance knowledge management capabilities for improved supply chain resilience in large-scale initiatives. *International Journal of Engineering Business Management, 17*, 18479790251399880.

Ansari, Z. N., & Kant, R. (2017). Exploring the framework development status for sustainability in supply chain management: A systematic literature synthesis and future research directions. *Business Strategy and the Environment, 26*(7), 873–892.

Attaran, M. (2020, July). Digital technology enablers and their implications for supply chain management. In *Supply chain forum: An international journal* (Vol. 21, No. 3, pp. 158–172). Taylor & Francis.

Azadegan, A., Patel, P. C., Zangoueinezhad, A., & Linderman, K. (2013). The effect of environmental complexity and environmental dynamism on lean practices. *Journal of Operations Management, 31*(4), 193–212.

Barney, J. (1991). Firm resources and sustained competitive advantage. *Journal of Management, 17*(1), 99–120. https://doi.org/10.1177/014920639101700108

Belhadi, A., Mani, V., Kamble, S. S., Khan, S. A. R., & Verma, S. (2024). Artificial intelligence-driven innovation for enhancing supply chain resilience and performance under the effect of supply chain dynamism: An empirical investigation. *Annals of Operations Research, 333*(2), 627–652.

Boone, T., Fahimnia, B., Ganeshan, R., Herold, D. M., & Sanders, N. R. (2025). Generative AI: Opportunities, challenges, and research directions for supply chain resilience. *Transportation Research Part e: Logistics and Transportation Review, 199*, Article 104135.

Brintrup, A., Kosasih, E., Schaffer, P., Zheng, G., Demirel, G., & MacCarthy, B. L. (2024). Digital supply chain surveillance using artificial intelligence: Definitions, opportunities and risks. *International Journal of Production Research, 62*(13), 4674–4695.

Carayannis, E. G., Dumitrescu, R., Falkowski, T., & Zota, N. R. (2024). Empowering SMEs "harnessing the potential of gen AI for resilience and competitiveness." *IEEE Transactions on Engineering Management, 71*, 14754–14774.

Chen, B., Simchi-Levi, D., Wang, Y., & Zhou, Y. (2022). Dynamic pricing and inventory control with fixed ordering cost and incomplete demand information. *Management Science, 68*(8), 5684–5703.

Christopher, M., & Peck, H. (2004). *Building the resilient supply chain.*

Culot, G., Podrecca, M., & Nassimbeni, G. (2024). Artificial intelligence in supply chain management: A systematic literature review of empirical studies and research directions. *Computers in Industry, 162*, Article 104132.

Dubey, R., Gunasekaran, A., & Papadopoulos, T. (2024). Benchmarking operations and supply chain management practices using Generative AI: Towards a theoretical framework. *Transportation Research Part e: Logistics and Transportation Review, 189*, Article 103689.

Elkington, J. (1998). Partnerships from cannibals with forks: The triple bottom line of 21st-century business. *Environmental Quality Management, 8*(1), 37–51.

Fosso Wamba, S., Guthrie, C., Queiroz, M. M., & Minner, S. (2023). ChatGPT and generative artificial intelligence: an exploratory study of key benefits and challenges in operations and supply chain management. *International Journal of Production Research*, 1–21.

Fosso Wamba, S., Guthrie, C., Queiroz, M. M., & Minner, S. (2024). ChatGPT and generative artificial intelligence: An exploratory study of key benefits and challenges in operations and supply chain management. *International Journal of Production Research, 62*(16), 5676–5696.

Fosso-Wamba, S., Guthrie, C., Queiroz, M. M., & Oyedijo, A. (2026). Building AI-enabled capabilities for improved environmental and manufacturing performance: evidence from the US and the UK. *International Journal of Production Research, 64*(2), 545–564.

Gaddala, V. S. (2022). Prompt engineering in supply chain enterprise data. *IRE Journals.*

Ghobakhloo, M., Fathi, M., Iranmanesh, M., Vilkas, M., Grybauskas, A., & Amran, A. (2024). Generative artificial intelligence in manufacturing: opportunities for actualizing Industry 5.0 sustainability goals. *Journal of Manufacturing Technology Management, 35*(9), 94–121.

Goswami, S. S., Mondal, S., Sarkar, S., Gupta, K. K., Sahoo, S. K., & Halder, R. (2025). Artificial intelligence-enabled supply chain management: Unlocking new opportunities and challenges. In *Artificial Intelligence and Applications* (Vol. 3, No. 1, pp. 110–121).

Grimes, M., Von Krogh, G., Feuerriegel, S., Rink, F., & Gruber, M. (2023). From scarcity to abundance: Scholars and scholarship in an age of generative artificial intelligence. *Academy of Management Journal, 66*(6), 1617–1624.

Gupta, S., Modgil, S., Choi, T. M., Kumar, A., & Antony, J. (2023). Influences of artificial intelligence and blockchain technology on financial resilience of supply chains. *International Journal of Production Economics, 261*, Article 108868.

Haddad, H., Jum'a, L., Alkalha, Z., & Madanat, H. (2026). AI-Powered tools for supply chain resilience: A dynamic capabilities perspective from jordanian manufacturing firms. *Logistics, 10*(1), 24.

Hahn, R., Hahn, R., Land, A., & Gattiker, T. (2025). Individual behavior in sustainable supply chain management: A systematic literature review. *Journal of Purchasing and Supply Management*, 101037.

Holgado, M., Blome, C., Schleper, M. C., & Subramanian, N. (2024). Brilliance in resilience: Operations and supply chain management's role in achieving a sustainable future. *International Journal of Operations & Production Management, 44*(5), 877–899.

Jackson, I., Ivanov, D., Dolgui, A., & Namdar, J. (2024a). Generative artificial intelligence in supply chain and operations management: A capability-based framework for analysis and implementation. *International Journal of Production Research, 62*(17), 6120–6145.

Jackson, T., et al. (2024b). AI in Logistics. McKinsey Reports.

Jomthanachai, S. (2022). *Development of performance evaluation and prediction models in logistics service and supply networks towards a sustainable and robust supply chain* (Doctoral dissertation).

Kanitz, R., Gonzalez, K., Briker, R., & Straatmann, T. (2023). Augmenting organizational change and strategy activities: Leveraging generative artificial intelligence. *The Journal of Applied Behavioral Science, 59*(3), 345–363.

Khan, S. A. R., Yu, Z., & Farooq, K. (2023). Green capabilities, green purchasing, and triple bottom line performance: Leading toward environmental sustainability. *Business Strategy and the Environment, 32*(4), 2022–2034.

Kuo-Wei, L. (2025). An integrated framework for Gen AI-assisted management learning: Insights from Kolb's learning cycle theory and knowledge types perspectives. *The International Journal of Management Education, 23*(2), Article 101164.

Lin, B., & Zhu, Y. (2025). Does AI elevate corporate ESG performance? A supply chain perspective. Business *Strategy and the Environment, 34*(1), 586–597.

Maghsoudi, M., Noorbakhsh, A., Khanizadeh, S., Shokouhyar, S., & Shokoohyar, S. (2025). Mapping the emergence and maturation of sustainable supply chain technologies: a patent-based assessment of technology life cycles. *Annals of Operations Research*, 1–32.

Malhotra, G., & Ramalingam, M. (2023). Perceived anthropomorphism and purchase intention using artificial intelligence technology: Examining the moderated effect of trust. *Journal of Enterprise Information Management*

Min, H. (2019). Blockchain technology for enhancing supply chain resilience. *Business Horizons, 62*(1), 35–45.

Mohamed, O. A. M. (2023). How generative AI transforming supply chain operations and efficiency?.

Nandi, S., Gonela, V., & Awudu, I. (2023). A resource-based and institutional theory-driven model of large-scale biomass-based bioethanol supply chains: An emerging economy policy perspective. *Biomass and Bioenergy, 174*, Article 106813.

Naz, F., Agrawal, R., Kumar, A., Gunasekaran, A., Majumdar, A., & Luthra, S. (2022). Reviewing the applications of artificial intelligence in sustainable supply chains: Exploring research propositions for future directions. *Business Strategy and the Environment, 31*(5), 2400–2423.

Niemets, K., Kravchenko, K., Kandyba, Y., Kobylin, P., & Morar, C. (2021). World cities in terms of the sustainable development concept. *Geography and Sustainability, 2*(4), 304–311.

Nitsche, A.-M., Schumann, C.-A., Franczyk, B., & Reuther, K. (2021). Artificial intelligence inspired supply chain collaboration: A design-science research and system dynamics approach. In *2021 IEEE International Conference on Engineering, Technology and Innovation (ICE/ITMC)*, pp. 1–9.

Nobre, G. G., Muis, S., Veldkamp, T. I., & Ward, P. J. (2019). Achieving the reduction of disaster risk by better predicting impacts of El Niño and La Niña. *Progress in Disaster Science, 2*, Article 100022.

Pamučar, D., Yazdani, M., Tarighi Shaayesteh, M., Pavur, R., & Simic, V. (2025, June). Sustainable performance analysis of remanufacturing suppliers in supply chains: circular economy perspective. In *Supply chain forum: An international journal* (pp. 1–30). Taylor & Francis.

Pettit, T. J. (2008). *Supply chain resilience: Development of a conceptual framework, an assessment tool and an implementation process* (No. AFITCI090032).

Ponomarov, S. Y., & Holcomb, M. C. (2009). Understanding the concept of supply chain resilience. *The International Journal of Logistics Management, 20*(1), 124–143.

Qadeer, A., Anis, M., Ajmal, Z., Kirsten, K. L., Usman, M., Khosa, R. R., Zhao, X., et al. (2022). Sustainable development goals under threat? Multidimensional impact of COVID-19 on our planet and society outweigh short term global pollution reduction. *Sustainable Cities and Society, 83*, Article 103962.

Qin, M., Su, C. W., Umar, M., Lobonţ, O. R., & Manta, A. G. (2023). Are climate and geopolitics the challenges to sustainable development? Novel evidence from the global supply chain. *Economic Analysis and Policy, 77*, 748–763.

Qin, M., Su, C. W., Zhong, Y., Song, Y., & Lobonţ, O. R. (2022). Sustainable finance and renewable energy: Promoters of carbon neutrality in the United States. *Journal of Environmental Management, 324*, Article 116390.

Rahman, S. M., Raihan, A., & Abudalfa, S. (2026). Generative artificial intelligence for environmental assessment: A new paradigm for sustainability analysis. *Environmental Management, 76*(3), 93.

Ruangkanjanases, A., Chen, S. C., Sivarak, O., & Khan, A. (2025). Achieving green competitive advantage through generative AI: The mediating roles of organisational creativity and green innovation ambidexterity in manufacturing. *International Journal of Logistics Research and Applications*, 1–21.

Schilling, L., & Seuring, S. (2024). Linking the digital and sustainable transformation with supply chain practices. *International Journal of Production Research, 62*(3), 949–973.

Sedkaoui, S., & Benaichouba, R. (2024). Generative AI as a transformative force for innovation: A review of opportunities, applications and challenges. *European Journal of Innovation Management*.

Shankar, R., & Gupta, L. (2024). An integrated AI framework for managing organizational risk and climate change concerns in B2B market. *Industrial Marketing Management, 117*, 173–187.

Sheffi, Y., & Rice Jr, J. B. (2005). A supply chain view of the resilient enterprise. *MIT Sloan management review*.

Shen, J. (2021). Uncertain two-echelon green supply chain models based on revenue sharing contract. *International Journal of Machine Learning and Cybernetics, 12*(7), 2059–2068.

Siciliani, L., Taccardi, V., Basile, P., Di Ciano, M., & Lops, P. (2023). AI-based decision support system for public procurement. *Information Systems, 119*, Article 102284.

Singh, V., Hughes, L., Albashrawi, M. A., Jeon, I., & Dwivedi, Y. K. (2026). Generative artificial intelligence (GenAI) in procurement and supply chain management: Applications, opportunities and challenges. *Journal of Systems and Information Technology, 28*(1), 123–144.

Storey, V. C., Yue, W. T., Zhao, J. L., & Lukyanenko, R. (2025). Generative artificial intelligence: Evolving technology, growing societal impact, and opportunities for information systems research. *Information Systems Frontiers*, 1–22.

Teece, D. J., Pisano, G., & Shuen, A. (1997). Dynamic capabilities and strategic management. *Strategic Management Journal, 18*(7), 509–533.

Teece, D. J. (2007). Explicating dynamic capabilities: the nature and microfoundations of (sustainable) enterprise performance. *Strategic management journal, 28*(13), 1319–1350.

Touvron, H., Lavril, T., Izacard, G., Martinet, X., Lachaux, M. A., Lacroix, T., & Lample, G. et al. (2023). Llama: Open and efficient foundation language models. arXiv:2302.13971.

Victor, B. G., Kubiak, S., Angell, B., & Perron, B. E. (2023). Time to move beyond the ASWB licensing exams: Can generative artificial intelligence offer a way forward for social work? *Research on Social Work Practice, 33*(5), 511–517.

Wang, S., & Zhang, H. (2025). Promoting sustainable development goals through generative artificial intelligence in the digital supply chain: Insights from Chinese tourism SMEs. *Sustainable Development, 33*(1), 1231–1248.

Wamba-Taguimdje, S.-L., Fosso Wamba, S., Kala Kamdjoug, J. R., & Tchatchouang Wanko, C. E. (2020). Influence of artificial intelligence (AI) on firm performance: The business value of AI-based transformation projects. *Business Process Management Journal, 26*(7), 1893–1924. https://doi.org/10.1108/BPMJ-10-2019-0411

Wu, L., Sun, L., Chang, Q., Zhang, D., & Qi, P. (2022). How do digitalization capabilities enable open innovation in manufacturing enterprises? A multiple case study based on resource integration perspective. *Technological Forecasting and Social Change, 184*, Article 122019.

Yadav, A. (2025). AI-driven enforcement of human rights in global supply chains: enhancing transparency, accountability, and corporate compliance. *International Review of Law, Computers & Technology*, 1–16.

Zamani, E. D., Smyth, C., Gupta, S., & Dennehy, D. (2023). Artificial intelligence and big data analytics for supply chain resilience: A systematic literature review. *Annals of Operations Research, 327*(2), 605–632.

Chapter 16
The Greta Effect and AI-Driven Circular Practices in Sustainable Hospitality: A Pathway to Guest Engagement and Innovation

Praveen Srivastava, Shelly Srivastava, and Gautam Shandilya

Abstract This study assesses the influences of a moral climate IQ, viewed as the perceived intensity of climate activism and social awareness, specifically by younger generations who have been referred to as experiencing the "Greta Effect", upon pro-environmental behaviours at hotels. Specifically, it will identify whether AI-enabled technologies that are viewed as useful and easy to use are aligned with the principles of the circular economy (CE) and increase the degree of involvement by guests in the environmentally responsible service offerings of hotels. This study will utilize social impact theory and sustainability-oriented innovation (SOI) to create an energy-efficient conceptual framework based upon the technology acceptance model (TAM) that identifies how the use of pragmatic, user friendly AI-based interventions can motivate hotel guests to participate in circular behaviour such as waste recycling, reducing energy consumption, and reusing facilities; however, the framework will also include the environmental beliefs of guests and the level of trust they have in their hotel's sustainability policy. The aim of current study is to investigate the impact of Greta Effect on perceived usefulness of AI-based green solutions. Combining the concepts of circular economy, digital innovation, and environmental activism, this study provides a new, customer-focused way to promote sustainability in the hotel industry using both social influence and acceptance of technology as a means to explore the linkages between the two.

Keywords Circular economy practices · Greta effect · Guest engagement · Technology acceptance model (TAM) · Sustainability-Oriented innovation (SOI)

P. Srivastava · S. Srivastava · G. Shandilya (✉)
Department of Management Studies, BIT Mesra, Ranchi, India
e-mail: rajshandy@gmail.com

P. Srivastava
e-mail: praveenbit@gmail.com

S. Srivastava
e-mail: shelly@bitmesra.ac.in

K. Kumar et al. (eds.), *AI-Driven Circular Economy for Carbon Neutrality in Emerging Economies*, Environmental Footprints and Eco-design of Products and Processes,
https://doi.org/10.1007/978-3-032-30298-4_16

1 Introduction

There are several trends in customer preferences for the use of green environmentally friendly practices (Jana et al., 2025). Additionally, growing public awareness of climate change issues and the increasing visibility of influential environmentalists, including Greta Thunberg and her "Greta Effect" (the impact on global awareness and environmentalism that has come from her activism), have also contributed to these changes. More specifically, an ever-increasing segment of travellers and eco-aware hotel guests are now pushing hotel companies towards creating environmentally conscious sustainable business models and strategies that are both transparent and circular, as well as align with the values of those customers and the larger global sustainability goals.

At the same time, technology has provided new opportunities for the hotel industry through the integration of digital technologies, particularly artificial intelligence (AI), which will enable a number of sustainable operational efficiencies within the hotel industry (Zahidi et al., 2024). For example, from using smart thermostats and AI-based waste monitoring to enabling automation of guest communication and providing behaviour prompts, AI is now becoming increasingly used to support circular economy (CE) projects in hotels and resorts. However, the success of these technologies will depend not only on their technological capabilities but also on how they are perceived by guests, i.e. their practicality and ease of use. Perceptions of these factors will determine whether guests are prepared to take part in environmentally beneficial behaviours including waste separation, towel reuse programmes, and energy conservation (Han & Hyun, 2018).

This chapter examines the interrelationship of social influence, the Greta Effect, perceptions of AI-based technologies, technology acceptance, and circular economy adoption in the hotel sector. Conceptual framework depicts the interaction of customer with AI-driven CE initiatives which can result in sustainability-oriented innovation (SOI) and guest engagement with green services (GEGS) by extending technology acceptance model by incorporating construct like environmental concern, trust, and social influence. Additionally, perception of ease of involvement and trust in sustainability promise of hotel can moderate the relationship between innovation and action.

The aim is to provide insight into how technology and social influence interact to impact sustainable customer behaviour. By doing this, it provides practical advice for technology designers, sustainability managers, and hotel managers looking to use artificial intelligence and the Greta Effect to promote circularity and creativity in service encounters.

2 Literature Review

2.1 The Greta Effect and Social Influence in Sustainability

Inspired by Greta Thunberg the "Greta Effect" describes the explosion of environmental knowledge and activity (Ainger & Fanetti, 2024). Particularly among Millennials and Gen Z customers, research in environmental psychology and marketing has shown that such social impact greatly impacts public views and conduct (Mochla & Tsourvakas, 2024). Social influence is a strong motivator for green behaviour adoption in the hotel setting as visitors are driven by peer approbation, celebrity sponsorships, and activist-led narratives to help sustainable businesses (Ayodele et al., 2017).

Social influence as the perceived society forces that change individual conduct, especially via the influence of close friends and important people (Ajzen, 1991). In the framework of sustainability, like in "The Greta Effect", where young activists and public personalities inspire eco-consciousness, social influence becomes essential to grasp behavioural changes. Without legislative enforcement, they shape consumer behaviour by means of environmentally friendly items, energy conservation, or trash reduction. Policymakers and marketers use these standards to support good environmental practices and deter negative behaviours. Studies confirm that using social influence is a calculated approach to promote sustainable behaviour and increase our knowledge of human responses to environmental challenges (Melnyk et al., 2022). Emphasising two key perceptions, its utility (PU) and its ease of use (PEOU) (Hu et al., 1999), traditional approaches like the technology acceptance model (TAM) assess how consumers embrace technology. These opinions show how easily consumers can interact with a technology and if they think it can help them to achieve their objectives. A lot of study has been done over the years on these aspects and suggested approaches to improve TAM, either by including outside factors or combining it with other models.

Recent artificial intelligence (AI) and mobile learning research repeatedly show that PU-AI and PEOU-AI greatly affect how people develop intentions towards utilising AI tools (AlGerafi et al., 2023). When users believe AI is really useful in simplifying processes or enhancing results, PU-AI typically shows to be a powerful predictor of use behaviour. If the AI is included into everyday task, its acceptance increases (Al-kfairy et al., 2024; Rout et al., 2024). The Greta Effect shows how a well-known personality can inspire general behavioural change. When powerful voices are incorporated into popular conversation support artificial intelligence, it often improves PU-AI by presenting the technology as both vital and required. Likewise, broad acceptance by peers or leaders might help AI seem more friendly and less scary (Krauter, 2024), hence enhancing PEOU-AI.

Thus, growing AI acceptability today is not just about enhancing systems themselves but also about deliberately using social narratives and trusted influencers to favourably shape how beneficial and simple these technologies seem. Promoting larger and more sustained interaction with AI technology depends on an awareness of this junction of usability, usefulness, and social impact (Rajendran et al., 2024). The attitudes and opinions of peers may help to define the thoughts and intentions people have towards using a technology (Alhwaiti, 2023). Social influence theory holds that people are more likely to engage in environmentally conscious conduct when it starts to be socially acceptable (Cialdini et al., 1990). The Greta Effect shows this by turning environmental care from a specialist interest into a mainstream moral expectation, therefore directly influencing consumer decision-making in travel and hospitality (Srivastava et al., 2024).

The social influence of The Greta Effect is that it may greatly affect the behaviour of consumer (Seyfi & Hall, 2023). Guest in hotel may include in various practices of circular economy, like reusing the linen (Gopalakrishna et al., 2025) if they are influenced by social influences supporting sustainability (Zaki & Farrag, 2024).

Hence, the following hypotheses are proposed:

H_{1a}: *The Greta Effect positively influences perceived usefulness of AI (PU-AI).*
H_{1b}: *The Greta Effect positively influences perceived ease of use of AI (PEOU-AI).*
H_{1c}: *The Greta Effect positively influences circular practices adoption (CPA).*

2.2 AI in Hospitality and Technology Acceptance

Driven by advancements in personalising, resource optimisation, and guest service automation, artificial intelligence (AI) has quickly become a transforming tool in hotel operations. Smart hotel management systems, predictive analytics for guest preferences, and sustainability solutions including energy-efficient systems and smart waste management now centre artificial intelligence technology (Ivanov & Webster, 2019).

The technology acceptance model (TAM) is extensively used to grasp personal adoption of new technologies (Davis, 1989). Two main impressions, perceived usefulness (PU) and perceived ease of use (PEOU), define user acceptability. In the sustainability space, visitors are more inclined to support and take part in environmentally friendly projects when they see AI apps as helpful to their comfort and simple to interact with (Morosan & DeFranco, 2016). Therefore, the success of AI-driven circular practices relies not only on their technical design but also on how visitors perceive and convey them (Bashynska, 2023). Users' adoption of the technology will determine whether AI-based sustainability projects are implemented successfully (Rane et al., 2024). Guests are more likely to support circular economy initiatives if they find AI-enabled green solutions such as automated energy systems, digital communication on recycling, or AI-assisted personalising of green choices as useful and simple to use. When artificial intelligence technologies are purposefully included into hotel operations they are seen to improve operational performance while matching with environmental aims (Bakshi & Singh, 2024).

The apparent value of artificial intelligence in the hotel industry spans many angles. Particularly in handling sustainability measurements, artificial intelligence promotes responsibility, transparency, and trust at the organisational level (Kulkov et al., 2024). Economically, via best use of resources, it helps lower running costs, increase turnaround times, and foster resilience. Strategically, artificial intelligence is seen as a driver of ethical, transparent, circular business models, hence generating long-term benefit for the environment as well as the company (Shaik et al., 2024). From a usability standpoint (PEOU-AI), hotel managers like that AI systems enabling circular practices usually need little effort to run and learn from. Routine chores such as inventory monitoring, demand forecasting, or guest experience optimization often find simplification by these systems (Avula et al., 2024). AI technologies promote clarity, enabling quicker decision-making, and help to handle exceptions, thus decreasing both physical and cognitive burden and so lowering mistakes and smoother operations. Given past research showing that simple and user-friendly technologies are more likely to be welcomed and deployed throughout hotel teams (Najafi, 2024), this simplicity of use is very vital.

Building on past studies and the findings, the paper suggests the following hypotheses:

H_{2a}: *Perceived usefulness of AI (PU-AI) positively influences circular practices adoption (CPA).*

H_{2b}: *Perceived ease of use of AI (PEOU-AI) positively influences circular practices adoption (CPA).*

2.2.1 *Circular Economy Practices in Hospitality*

Principles of the circular economy (CE) stress resource efficiency, waste minimisation, and lifecycle thinking, all of which are progressively finding application in hotel operations (Geissdoerfer et al., 2017). CE measures that advance sustainability while improving operational efficiency include composting, computerised check-ins, linen reuse programs, and refillable facilities.

When visitors are aware of their presence and find them simple to interact with, research has shown that they are more inclined to participate in circular practices (Sørensen & Bærenholdt, 2020). Therefore, determining behavioural acceptance depends mostly on perceived simplicity of involvement and perceived value of CE projects. Furthermore, hotels that open their CE initiatives usually inspire more trust, which strengthens guest eagerness to help (Seyfi et al., 2025). Driving sustainability-oriented innovation (SOI) in the hotel industry is increasingly considered as depending on matching corporate actions with social and environmental aims. Beyond basic environmental practices, the sector has embraced more complete transformations, that is, reinterpretation of services, processes, and organisational structures to provide environmental and social value alongside commercial returns (Adams et al., 2016; Klewitz & Hansen, 2014).

In the hotel industry, SOI promotes operational reforms including environmentally friendly service design, waste reduction, and resource-efficient policies. This covers changing hotel, resort, and restaurant energy use, trash management, and supplier interaction. Crucially, innovation in this area often transcends the boundaries of the particular property and calls for cooperation with outside partners to create more environmentally friendly service ecosystems (Adams et al., 2016). Offering a framework that stresses resource reduction, reuse, and recycling, vital in an industry renowned for high consumption, the circular economy (CE) enhances SOI. CE facilitates the creation of restoring hospitality models, therefore enabling companies to move from conventional linear operations to more robust and sustainable systems (Bocken et al., 2016).

Adopting isolated circular methods, therefore, could not bring about any significant transformation. True impact occurs when CE is combined with more general SOI plans to support ongoing innovation and development (Rattalino, 2018). Many hotels start their change with sustainability projects like zero-waste programs or green certifications, which eventually develop into overall innovative models. In this regard, CE provides a basis for reconsidering the whole service model and attaining long-term sustainable development (Brown et al., 2019; Lüdeke-Freund et al., 2019). By encouraging a culture of eco-efficiency, experimentation, and ongoing development, the acceptance of circular practices may propel more general sustainability-oriented innovation (SOI). Hotels that include their visitors in circular activities are more likely to investigate new business models and service improvements that fit sustainability targets (Florido et al., 2019).

The following hypothesis is established to experimentally test the conceptual framework generated in this work:

> *H_3: Circular practices adoption (CPA) positively influences sustainability-oriented innovation (SOI).*

2.3 Sustainability-Oriented Innovation (SOI)

Sustainability-oriented innovation is the creation and use of fresh ideas, technologies, or processes that lower environmental impact while producing social and financial benefit (Adams et al., 2016). In hospitality, SOI refers to design, guest services, and operational changes that advance sustainability. By allowing data-driven choices, predictive maintenance, and guest-personalized sustainability recommendations, artificial intelligence may act as a catalyst for SOI. Many studies have shown the connection between CE adoption and innovation; circular practices seem to be a portal to more general organisational transformation (Kristensen et al., 2021). Supported by digital technology and driven by environmental ideals, these developments may set businesses apart and increase visitor interaction.

Customised green experiences or AI-driven eco-feedback systems are two creative sustainability ideas that improve visitor experience in addition to lowering environmental effect. Guests' involvement is probably going to rise if they see a hotel's green policies as original and real (Jana et al., 2025; Merli et al., 2019). Since, hotel service innovation is dynamic and people-centric, it varies from industrial innovation. It serves as a kind of social technology depending on adaptable schedules and shared responsibilities between employees and visitors (Nelson & Sampat, 2001). One important difference in hospitality is the active participation of guests in co-creating their experiences, which may favourably affect sustainability by means of green practices (Warren & Coghlan, 2016). Notwithstanding this potential, many lodging companies still run under strict management systems that restrict guest interaction (Chathoth et al., 2014), and smart solutions that enable customised, environmentally friendly experiences are yet developing (Warren, 2018).

Particularly in terms of promoting sustainable guest behaviour, service companies differ in their ability to innovate. Adopting their knowledge of consumer behaviour, creative suppliers provide environmentally friendly options including incentives for adopting recyclable products. However, many small and medium-sized lodging companies (SMAPs) follow conventional models with limited staff expertise and conservative business practices that stifle sustainable innovation (Coles et al., 2014). By purposefully aligning all elements of service people, processes, and products that is, with sustainability goals, sustainability-oriented innovation (SOI) presents a paradigm for transcending these constraints (Adams et al., 2016). Although SOI has been used generally in SMEs (Klewitz & Hansen, 2014), its inclusion into SMAPs is still under development. Importantly, SOI calls for active worker, guest, supplier, and community stakeholder cooperation rather than just using green technology. Since these changes visitors from passive consumers into participants in sustainable hospitality, guest engagement with green services (GEGS) becomes an essential component of SOI (Juvan et al., 2018).

Based on the research on sustainability and creativity, we formulate the following hypothesis to investigate the suggested links:

> *H_4: Sustainability-oriented innovation (SOI) positively influences guest engagement with green services (GEGS).*

2.4 Guest Engagement with Green Services (GEGS)

Engagement of guests in a brand's products is a multi-dimensional notion including emotional, cognitive, and behavioural participation. Guest engagement with green services (GEGS) in the context of sustainable hospitality is the degree to which visitors support, interact with, and advocate for environmentally friendly hotel operations. Engaged visitors are more likely to share their experiences, engage in circular behaviours, and grow loyalty to green businesses (Mansoor et al., 2025). Many antecedents shape engagement: awareness, social influence, technological acceptability, and trust (Beldad & Hegner, 2018). When these elements come together, visitors are more likely to participate actively in the environmental purpose of the hotel, therefore changing from passive beneficiaries to co-creative players in sustainability.

Using circular methods not only increases environmental responsibility in the hotel but also raises guest and host pleasure, thereby fostering more unforgettable and significant events. Shifting from conventional linear models to circular ones, with an eye towards recycling, reusing, and decreasing, the hotel sector may be transformed (Tavares de Carvalho et al., 2024). Still, many businesses, especially small hotels mostly follow these guidelines in a restricted, hierarchical fashion, with decrease most often occurring. By impacting guest expectations and helping the adoption of digital and sustainable technologies in hospitality (Zaki & Farrag, 2024), the circular economy (CE) is progressively reshining tourism trends. It presents a viable substitute for conventional growth models by matching business practices with environmental and social objectives. For example, agritourism shows how visitor conduct could change by direct, ethical interaction with circular concepts (Tavares de Carvalho et al., 2024).

Hotel companies that want to effectively embrace CE have to carefully combine these ideas so that they improve sustainability without upsetting livelihoods, cultural legacy, or local communities (Jones & Wynn, 2019). Guest engagement with green services (GEGS) becomes crucial in this process as visitors actively help and influence sustainable operations by their decisions and involvement. As companies answer to changing customer expectations and handle environmental issues, the incorporation of circular processes into hotel operations becomes even more important (Geissdoerfer et al., 2017). Approaches like resource efficiency, waste minimisation, and closed-loop supply chains under circular practices adoption (CPA) provide the operational foundation for sustainable service delivery (Kirchherr et al., 2017). But companies have to use sustainability-oriented innovation (SOI) if they are to transform these internal processes into significant consumer experiences. Soi is the term used to describe innovations that clearly include social and environmental factors into goods, services, and processes (Adams et al., 2016), therefore integrating the internal sustainability efforts with external stakeholder value.

SOI transforms operational circularity into creative guest-facing green services acting as a mediator. CPA improves the environmental performance of the company, but it is via SOI that these ideas are recreated and presented as interesting, value-added experiences to consumers (Evans et al., 2017). Products of SOI that clearly and tangibly show customers circular behaviours include eco-designed hotel rooms, gamified sustainability programs, or immersive green customer journeys (Cheng, 2016). Without such creativity, the effect of circular practices might be limited to back-end operations, therefore restricting their influence on visitor view and involvement. Recent empirical research implies that guests' impression of real, creative sustainability activities (Han & Hsu, 2009) significantly affects their interaction with green services (GEGS). GEGS includes cognitive, emotional, and behavioural reactions that include taking part in green initiatives, spreading good word-of-mouth, and developing loyalty for eco-friendly companies. Using SOI, companies improve openness, perceived value, and emotional resonance by including sustainability into consumer touchpoints (Klewitz & Hansen, 2014). Therefore, by building a consistent and compelling sustainability story, SOI not only improves the effectiveness of circular processes but also raises guest involvement degrees. Therefore, the mediating function of SOI is crucial; without it, the link between CPA and GEGS may remain weak or uncertain, thereby stressing the necessity of innovative sustainability measures in the hotel industry.

Active participation in green activities helps visitors to connect more with the sustainability vision of the hotel, therefore improving their emotional and behavioural involvement (Yu et al., 2024). This involvement is probably showing up as frequent visits, good word-of-mouth, and active support of environmental projects.

This paper suggests the following assumptions based on the examined literature and the theoretical links among the constructs. First of all, it is anticipated that implementing circular practices would help to increase visitor interaction with environmentally friendly services. Moreover, the research suggests that in this link sustainability-oriented innovation acts as a mediator.

H_5: Circular practices adoption (CPA) positively influences guest engagement with green services (GEGS).
H_6: Sustainability-oriented innovation (SOI) mediates the relationship between circular practices adoption (CPA) and guest engagement with green services (GEGS).

Conceptual Framework

Figure 1 shows the conceptual framework used in the study. The three mentioned hypotheses together provide a model that combines environmental, behavioural, and technical elements to explain sustainable involvement in the hotel sector. Key antecedents of circular practice adoption, the model places social impact (Greta Effect) and technological acceptance (PU/PEOU of AI), which in turn encourages sustainability-oriented innovation and guest involvement. For the hotel industry, the combination of digital transformation, circular economy thinking, and climate advocacy offers both possibilities and problems. While the Greta Effect increases moral drive and societal pressure, artificial intelligence provides the instruments to enable and customise environmentally friendly conduct. If implementation plans take user impressions, trust, and convenience into account, together they may inspire visitor involvement and creativity. This part offers a conceptual framework combining these ideas to represent the road towards improved sustainability results in the hotel. The approach for empirical testing these links will be covered in the next part.

Fig. 1 Conceptual framework

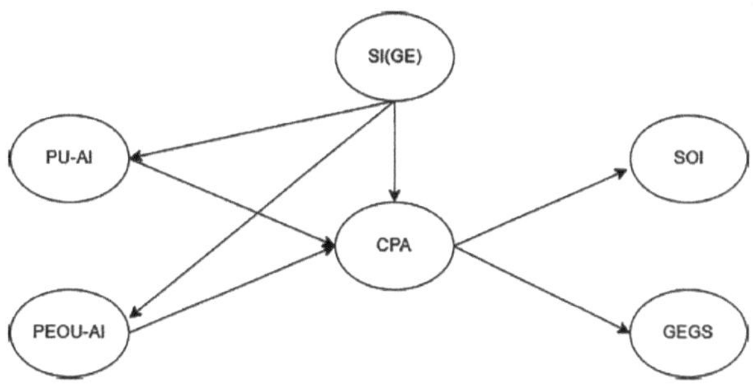

3 Research Methodology

This study adopts a quantitative, cross-sectional survey design. A structured questionnaire was administered online to examine how the Greta Effect, perceived usefulness and ease of use of AI, and circular-practice adoption drive sustainability-oriented innovation (SOI) and guest engagement with green services (GEGS) in the upscale hotel segment.

The target population comprised domestic leisure or business travellers who, within the past 12 months, had stayed in at least one 3- to 5-star hotel. Given the absence of a publicly available guest list that meets these criteria, a non-probability purposive sampling approach was applied. The survey link was disseminated through social media, LinkedIn hospitality groups, Facebook, WhatsApp, and sustainable-travel communities to maximise contact with qualified respondents. All scale items were adapted from validated instruments in tourism and information-systems research (Al Halbusi et al., 2022; Chen & Tung, 2014; Davis, 1989). Each item was carefully re-worded to fit a hotel service context and reviewed by three subject-matter experts (one hotel operations manager and two hospitality academics) for face and content validity. Minor lexical adjustments were made to ensure clarity and cultural neutrality. All constructs were operationalised on a five-point Likert scale (1 = "strongly disagree", 5 = "strongly agree"). Appendix A lists the final items.

To detect inattentive responding, the questionnaire contained an instructional-manipulation check (IMC) that asked participants to select a specified option. The final Google Form was available for four weeks (15 January–12 February 2025). Before proceeding, each participant viewed an information sheet explaining the study purpose, voluntary nature of participation, anonymity, and data-protection measures. A total of 312 submissions were received. After an initial review, 11 forms were removed for having more than 10% missing data. Lastly, 7 responses were eliminated due to failure on the attention-check question. This resulted in a final usable dataset of 294 responses. The final sample size exceeds the minimum threshold required for partial least squares structural equation modelling (PLS-SEM), with 8 latent constructs and sufficient statistical power (Hair et al., 2021).

Data were analysed using SmartPLS 3, a widely used software for variance-based structural equation modelling (PLS-SEM). It was selected due to its robust handling of complex models with formative and reflective constructs, small-to-medium sample sizes, and non-normal data distributions, making it particularly suitable for exploratory and predictive research designs like the current study.

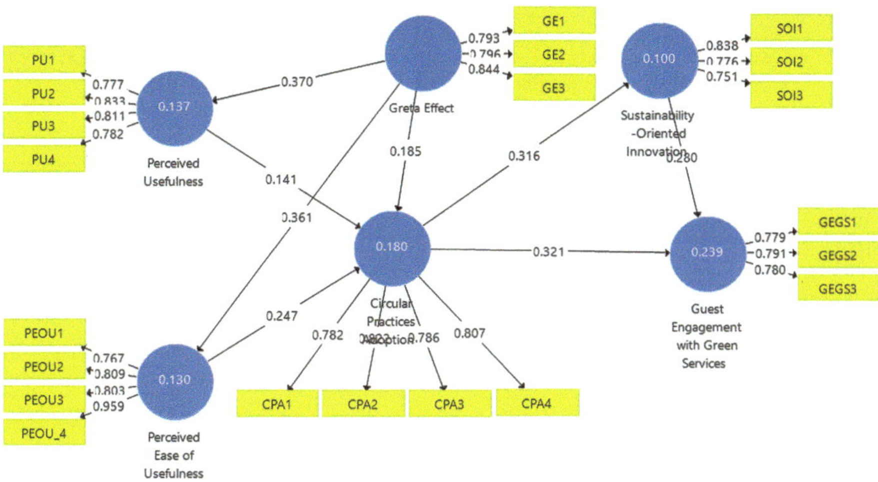

Fig. 2 Measurement model assessment using SmartPLS3. *Source* Author's own work

4 Data Analysis

4.1 Measurement Model Analysis

Factor loadings. All indicators loaded strongly on their intended latent variables (0.751–0.959; see Fig. 2 and Table. 1), comfortably exceeding the 0.708 threshold (Hair et al., 2021). The lowest loading was SOI3 = 0.751, still well above the cut-off, while the highest was PEOU_4 = 0.959, confirming indicator reliability.

Internal consistency and convergent validity. Cronbach's α ranged from 0.687 (GEGS) to 0.858 (PEOU), composite reliability (CR) from 0.827 to 0.903, and average variance extracted (AVE) from 0.614 to 0.702 (Table 2). All values exceed the recommended threshold, demonstrating adequate internal consistency and convergent validity across constructs.

Discriminant validity. Heterotrait–Monotrait (HTMT) ratios ranged from 0.103 to 0.544, all well below the conservative 0.85 criterion, supporting discriminant validity. The highest HTMT (SOI–GEGS = 0.544) is still comfortably within bounds (Table 3), indicating that each construct is empirically distinct.

4.1.1 Structural Model Analysis

Predictive power. The model explains a modest to moderate proportion of variance in the endogenous constructs (Table 4):

Table 1 Factor loading

	Circular Practices adoption	Greta effect	Guest engagement with green services	Perceived ease of usefulness	Perceived usefulness	Sustainability-Oriented Innovation
CPA1	0.782					
CPA2	0.822					
CPA3	0.786					
CPA4	0.807					
GE1		0.793				
GE2		0.796				
GE3		0.844				
GEGS1			0.779			
GEGS2			0.791			
GEGS3			0.78			
PEOU1				0.767		
PEOU2				0.809		
PEOU3				0.803		
PEOU_4				0.959		
PU1					0.777	
PU2					0.833	
PU3					0.811	
PU4					0.782	
SOI1						0.838
SOI2						0.776
SOI3						0.751

Source Computed from data

Table 2 Internal consistency and convergent validity

	Cronbach's alpha	Composite reliability	Average variance extracted (AVE)
Circular practices adoption	0.813	0.876	0.639
Greta effect	0.742	0.853	0.659
Guest engagement with green services	0.687	0.827	0.614
Perceived ease of usefulness	0.858	0.903	0.702
Perceived usefulness	0.814	0.877	0.641
Sustainability-Oriented innovation	0.699	0.832	0.623

Source Computed from data

Table 3 HTMT ratio

	Circular practices adoption	Greta effect	Guest engagement with green services	Perceived ease of usefulness	Perceived usefulness
Circular practices adoption					
Greta effect	0.412				
Guest engagement with green services	0.54	0.297			
Perceived ease of usefulness	0.386	0.39	0.185		
Perceived usefulness	0.312	0.469	0.181	0.202	
Sustainability-Oriented innovation	0.406	0.164	0.544	0.196	0.103

Source Computed from data

Table 4 R^2 value

	R Square	R Square Adjusted
Circular practices adoption	0.18	0.172
Guest engagement with green services	0.239	0.234
Perceived ease of usefulness	0.13	0.127
Perceived usefulness	0.137	0.134
Sustainability-Oriented innovation	0.1	0.097

Source Computed from data

These values meet the "substantial" (>0.26) to "weak" (>0.02) guidelines for exploratory research and confirm that the antecedents retain explanatory relevance. Path significance A 5000-subsample bootstrapping procedure in SmartPLS 3 produced the standard-error estimates for hypothesis testing (Fig. 3). Eight of the nine direct hypotheses (H1a–H5) are significant at $p < 0.01$ (Table 5). The indirect path (H6) is significant as the indirect effect CPA → SOI → GEGS was positive, and its 95% bootstrap confidence interval did not straddle zero, indicating partial mediation.

Effect sizes and collinearity. f^2 values ranged from 0.04 (small) to 0.12 (medium); VIFs were all <3.0, indicating no multicollinearity concerns.

The measurement model meets all reliability, convergent, and discriminant-validity requirements. The structural results confirm that The Greta Effect enhances both perceived usefulness and ease of use of AI, which in turn foster circular practice adoption. Circular practices directly improve SOI and guest engagement with green services, while also exerting an indirect influence on GEGS through SOI, demonstrating partial mediation. Together, the findings lend robust empirical support to the theorised framework and highlight the pivotal role of circular practices and innovation in engaging guests with hotels' green service offerings.

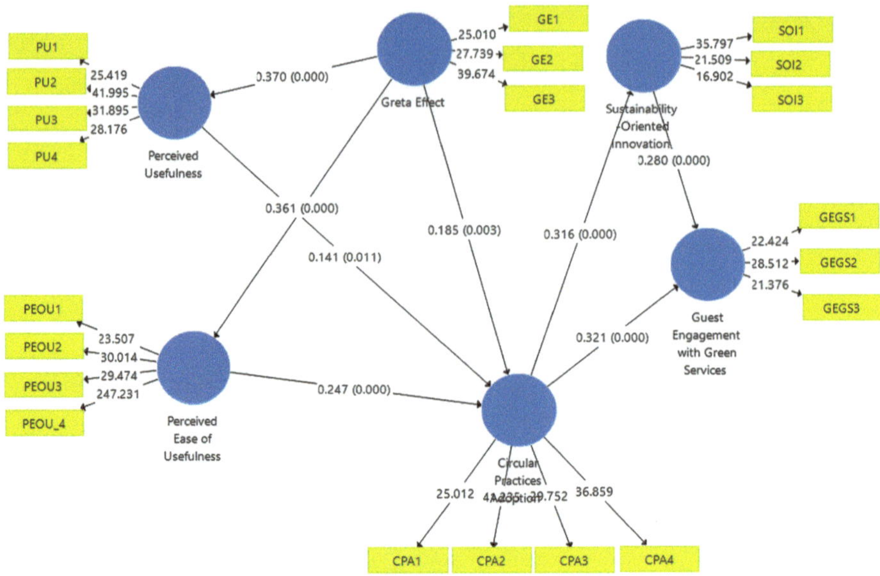

Fig. 3 Result of path analysis. *Source* Author's own work

Table 5 Hypothesis testing (bootstrapped, $n = 294$)

Hypothesis	Path	β (O)	*t*	*p*	Result
H1a	GE → PU	0.37	7.543	<0.001	Supported
H1b	GE → PEOU	0.361	7.473	<0.001	Supported
H1c	GE → CPA	0.185	3.103	0.002	Supported
H2a	PU → CPA	0.141	2.582	0.01	Supported
H2b	PEOU → CPA	0.247	4.122	<0.001	Supported
H3	CPA → SOI	0.316	5.821	<0.001	Supported
H4	SOI → GEGS	0.28	4.939	<0.001	Supported
H5	CPA → GEGS	0.321	6.303	< 0.001	Supported
H6	CPA → SOI → GEGS	0.088*	3.418	0.001	Supported

Source Computed from data
* Indirect effect β calculated from SmartPLS 3 bootstrapping; CI 95% = 0.038–0.150 (does not include 0)

5 Discussion

The present study aimed to examine the influence of the Greta Effect (GE) on circular practices adoption (CPA) in the hospitality sector through the mediating influence of technology-related perceptions and innovation orientation. The findings from the measurement model confirmed that all constructs demonstrated satisfactory levels of reliability and validity. Factor loadings exceeded the threshold of 0.70, and both convergent and discriminant validity were well established, indicating that the constructs were accurately and distinctly measured. The R^2 values suggest that moderate variance in key dependent constructs, CPA (18%), GEGS (23.9%), and SOI (10%), is explained by their respective predictors, highlighting the model's explanatory power.

From the structural model, it is evident that the Greta Effect significantly influences both perceived usefulness (PU) and perceived ease of usefulness (PEOU), which in turn positively affect circular practices adoption. This indicates that guests who are environmentally conscious are more likely to perceive AI technologies as useful and easy to use, thereby encouraging sustainable practices. Furthermore, CPA significantly contributes to both sustainability-oriented innovation (SOI) and guest engagement with green services (GEGS). The mediation role of SOI between CPA and GEGS was also supported, suggesting that the adoption of circular practices fosters innovation, which subsequently enhances guest participation in green initiatives. The robust T-statistics and significant p-values from the bootstrapping results reinforce the strength of these relationships.

6 Theoretical Implications

This work presents some interesting theoretical additions to the body of knowledge on sustainable hospitality and technology acceptance. First, it expands the knowledge of the Greta Effect by proving its impact on cognitive constructions like perceived usefulness (PU) and perceived ease of usefulness (PEOU) of AI technologies as well as on behavioural consequences like circular practices adoption. This puts environmental awareness as a major factor influencing models of technology acceptability in hotel environments. Second, the study creates a fresh paradigm that catches the change from internal operational sustainability efforts to external guest engagement by including circular practices adoption (CPA), sustainability-oriented innovation (SOI), and guest engagement with green services (GEGS). At last, the research empirically validates the mediating function of SOI, therefore supporting innovation theory by demonstrating that sustainable practices have to be matched with creativity to increase visitor participation in environmentally friendly services.

7 Managerial Implications

For hotel managers, especially those running 3- to 5-star green-certified hotels, the results provide insightful analysis. First of all, the Greta Effect emphasises the need of focussing on and involving ecologically concerned visitors. Hotels could showcase their green credentials and use marketing techniques appealing to the environmental ideals of their visitors. Second, the close ties among PU, PEOU, and CPA imply that investments in user-friendly AI solutions such smart thermostats, eco-feedback systems, and automated check-in kiosks may greatly improve the acceptance of sustainable operating procedures. Moreover, because CPA promotes both innovation and guest involvement, managers must create a culture that supports circular economy concepts, therefore guaranteeing that innovation is not just technical but also matched with sustainability aims. Interactive touchpoints that raise awareness of green practices and inspire involvement in them will help visitors to further be informed about this.

8 Limitations and Future Research Directions

This research is not without restrictions even with the insightful analysis. The sample included only of visitors of 3 to 5-star hotels, therefore limiting the generalisability of the results to other kinds of lodging or areas. Furthermore, the data was gathered using a self-reported Google Form, which despite the inclusion of attention-check questions may contain typical method bias. To verify the suggested model even further, future studies could consider a larger and more varied sample including visitors from low-cost hotels and other geographical areas. Furthermore, there might be longitudinal studies to evaluate changes in attitudes and behaviour over time. Investigating various mediators or moderators including environmental awareness, cultural orientation, or digital literacy might provide closer understanding of the interactions between sustainability, technology acceptance, and visitor behaviour.

References

Adams, R., Jeanrenaud, S., Bessant, J., Denyer, D., & Overy, P. (2016). Sustainability-oriented innovation: A systematic review. *International Journal of Management Reviews., 18*(2), 180–205.

Ainger, F., & Fanetti, S. (2024). Youth climate activism Political and regulatory outcomes. In V. E. Albanese, S. Fanetti, & R. Minazzi (Eds.), *Social mobilisation for climate change* (1st ed., pp. 111–128). Routledge.

Ajzen, I. (1991). The theory of planned behaviour. *Organisational Behaviour and Human Decision Processes., 50*(2), 179–211.

Al Halbusi, H., Al-Sulaiti, K., Abbas, J., Al-Sulaiti, I. (2022). Assessing factors influencing technology adoption for online purchasing amid COVID-19 in Qatar: Moderating role of word of mouth. *Frontiers in Environmental Science, 10.*

AlGerafi, M. A. M., Zhou, Y., Oubibi, M., & Wijaya, T. T. (2023). Unlocking the potential: A comprehensive evaluation of augmented reality and virtual reality in education. *Electronics, 12*(18), 3953.

Alhwaiti, M. (2023). Acceptance of artificial intelligence application in the Post-Covid Era and its impact on faculty members' occupational well-being and teaching self efficacy: A path analysis using the UTAUT 2 model. Appl Artif Intell 37(1).

Al-kfairy, M., Alomari, A., Al-Bashayreh, M., Alfandi, O., & Tubishat, M. (2024). Unveiling the Metaverse: A survey of user perceptions and the impact of usability, social influence and interoperability. *Heliyon., 10*(10), Article e31413.

Avula, M., Reddy, C. K. K., & Sithole, T. (2024). Artificial intelligence's potential to improve operational efficiency and customer experience in the hospitality and tourism sectors. In M. B. Talukder, S. Kumar, & P. K. Tyagi (Eds.), *Impact of AI and tech-driven solutions in hospitality and tourism* (pp. 351–382). IGI Global.

Ayodele, A. A., Panama, A. E., & Akemu, E. (2017). Green awareness and consumer purchase intention of environmentally-friendly electrical products in Anambra, Nigeria. *Journal of Economics and Sustainable Development., 8*(22), 98–112.

Bakshi, N., & Singh, A. K. (2024). AI-Enabled smart buildings reducing the carbon footprint and enhancing guest experiences in eco-friendly hotels. In M. B. Talukder, S. Kumar, & P. K. Tyag (Eds.), *Hotel and travel management in the AI Era* (pp. 79–98). IGI Global.

Bashynska, I. (2023). AI-Driven personalization in advertising: Transforming consumer engagement through sustainability and circular economy. *Scientific Journal of Bielsko-Biala School of Finance and Law., 27*(4), 106–112.

Beldad, A. D., & Hegner, S. M. (2018). Expanding the technology acceptance model with the inclusion of trust, social influence, and health valuation to determine the predictors of German users' willingness to continue using a fitness app: A structural equation modeling approach. *International Journal of Human-Computer Interaction., 34*(9), 882–893.

Bocken, N. M. P., de Pauw, I., Bakker, C., & van der Grinten, B. (2016). Product design and business model strategies for a circular economy. *Journal of Industrial and Production Engineering., 33*(5), 308–320.

Brown, P., Bocken, N., & Balkenende, R. (2019). Why do companies pursue collaborative circular oriented innovation? *Sustainability, 11*(3), 635.

Chathoth, P. K., Ungson, G. R., Altinay, L., Chan, E. S. W., Harrington, R., & Okumus, F. (2014). Barriers affecting organisational adoption of higher order customer engagement in tourism service interactions. *Tourism Management., 42*, 181–193.

Chen, M. F., & Tung, P. J. (2014). Developing an extended theory of planned behavior model to predict consumers' intention to visit green hotels. *International Journal of Hospitality Management., 36*, 221–230.

Cheng, M. (2016). Current sharing economy media discourse in tourism. *Annals of Tourism Research., 60*, 111–114.

Cialdini, R. B., Reno, R. R., & Kallgren, C. A. (1990). A focus theory of normative conduct: Recycling the concept of norms to reduce littering in public places. *Journal of Personality and Social Psychology., 58*(6), 1015–1026.

Coles, T., Zschiegner, A. K., & Dinan, C. (2014). A cluster analysis of climate change mitigation behaviours among SMTEs. *Tourism Geographies., 16*(3), 382–399.

Davis, F. D. (1989). Perceived usefulness, perceived ease of use, and user acceptance of information technology. *MIS Quarterly, 13*(3), 319–340.

Evans, S., Vladimirova, D., Holgado, M., Van Fossen, K., Yang, M., Silva, E. A., et al. (2017). Business model innovation for sustainability: Towards a unified perspective for creation of sustainable business models. *Business Strategy and the Environment., 26*(5), 597–608.

Florido, C., Jacob, M., & Payeras, M. (2019). How to carry out the transition towards a more circular tourist activity in the hotel sector. *The Role of Innovation. Administrative Sciences., 9*(2), 47.

Geissdoerfer, M., Savaget, P., Bocken, N. M. P., & Hultink, E. J. (2017). The circular economy—A new sustainability paradigm? *Journal of Cleaner Production., 143*, 757–768.

Gopalakrishna Pillai, S,. Arasli, F., "Kai," Haldorai, K., & Rahman, I. (2025). Unlocking sustainable performance through circular economy principles. *Journal of Hospitality and Tourism Insights, 8*(5), 1970–1991.

Hair, J. F., Astrachan, C. B., Moisescu, O. I., Radomir, L., Sarstedt, M., Vaithilingam, S., et al. (2021). Executing and interpreting applications of PLS-SEM: Updates for family business researchers. *Journal of Family Business Strategy., 12*(3), Article 100392.

Han, H., & Hsu, L. T., (Jane), & Lee, J. S. (2009). Empirical investigation of the roles of attitudes toward green behaviors, overall image, gender, and age in hotel customers' eco-friendly decision-making process. *International Journal of Hospitality Management, 28*(4), 519–28.
Han, H., & Hyun, S. S. (2018). What influences water conservation and towel reuse practices of hotel guests? *Tourism Management, 64*, 87–97.
Hu, P. J., Chau, P. Y. K., Sheng, O. R. L., & Tam, K. Y. (1999). Examining the technology acceptance model using physician acceptance of telemedicine technology. *Journal of Management Information Systems., 16*(2), 91–112.
Ivanov, S., & Webster, C. (2019). Conceptual framework of the use of robots, artificial intelligence and service automation in travel, tourism, and hospitality companies. In *Robots, artificial intelligence, and service automation in travel, tourism and Hospitality*. Emerald Publishing Limited, pp. 7–37.
Jana, A., Shandilya, G., & Srivastava, P. (2025). Eco-conscious luxury: Decoding the preferences of affluent travellers in sustainable hospitality. *Journal of Hospitality and Tourism Insights., 8*(5), 1717–1739.
Jones, P., & Wynn, M. G. (2019). The circular economy, natural capital and resilience in tourism and hospitality. *International Journal of Contemporary Hospitality Management., 31*(6), 2544–2563.
Juvan, E., Hajibaba, H., & Dolnicar, S. (2018). Environmental sustainability. In S. Dolnicar (Ed.), *Peer-to-Peer accommodation networks: Pushing the boundaries* (pp. 267–278). Goodfellow Publishers Limited.
Kirchherr, J., Reike, D., & Hekkert, M. (2017). Conceptualizing the circular economy: An analysis of 114 definitions. *Resources, Conservation and Recycling., 127*, 221–232.
Klewitz, J., & Hansen, E. G. (2014). Sustainability-oriented innovation of SMEs: A systematic review. *Journal of Cleaner Production, 65*, 57–75.
Krauter, J. (2024). Bridging the uncanny valley: Improving AI chatbots for effective leadership mentoring. *Open Journal of Leadership., 13*(03), 342–384.
Kristensen, H. S., Mosgaard, M. A., & Remmen, A. (2021). Circular public procurement practices in Danish municipalities. *Journal of Cleaner Production., 281*, Article 124962.
Kulkov, I., Kulkova, J., Rohrbeck, R., Menvielle, L., Kaartemo, V., & Makkonen, H. (2024). Artificial intelligence—Driven sustainable development: Examining organizational, technical, and processing approaches to achieving global goals. *Sustainable Development, 32*(3), 2253–2267.
Lüdeke-Freund, F., Gold, S., & Bocken, N. M. P. (2019). A review and typology of circular economy business model patterns. *Journal of Industrial Ecology., 23*(1), 36–61.
Mansoor, M., Jam, F. A., & Khan, T. I. (2025). Fostering eco-friendly behaviors in hospitality: Engaging customers through green practices, social influence, and personal dynamics. *International Journal of Contemporary Hospitality Management., 37*(5), 1804–1826.
Melnyk, V., Carrillat, F. A., & Melnyk, V. (2022). The influence of social norms on consumer behavior: A meta-analysis. *Journal of Marketing., 86*(3), 98–120.
Merli, R., Preziosi, M., Acampora, A., & Ali, F. (2019). Why should hotels go green? Insights from guests experience in green hotels. *International Journal of Hospitality Management., 81*, 169–179.
Mochla, V., & Tsourvakas, G. (2024). Factors that affect ethical consumption and eWOM of Millennials and Z generations. *Journal of Contemporary Marketing Science., 7*(3), 233–251.
Morosan, C., & DeFranco, A. (2016). Modeling guests' intentions to use mobile apps in hotels: The roles of personalization, privacy, and involvement. *International Journal of Contemporary Hospitality Management., 28*(9), 1968–1991.
Najafi, V. (2024). Exploring front desk employees' perceptions of technology in hotel information systems: An integration of technology acceptance and self-efficacy models. *Jamk University of Applied Sciences.*
Nelson, R. R., & Sampat, B. N. (2001). Making sense of institutions as a factor shaping economic performance. *Journal of Economic Behavior & Organization., 44*(1), 31–54.

Rajendran, R. K., Mohana, P. T., Immanuel, V. A., Blessing, N. R. W., Chitrarasu, K., Vetrivel, S., et al. (2024). The Intersection of Technology and Social Sustainability. In C. R. G. Popescu (Ed.), *Building inclusive global knowledge societies for sustainable development* (pp. 457–470). IGI Global.

Rane, N., Choudhary, S. P., & Rane, J. (2024). Acceptance of artificial intelligence: Key factors, challenges, and implementation strategies. *Journal of Applied Artificial Intelligence., 5*(2), 50–70.

Rattalino, F. (2018). Circular advantage anyone? Sustainability-driven innovation and circularity at Patagonia, Inc. *Thunderbird International Business Review, 60*(5), 747–755.

Rout, K., Sahoo, P. R., Behera, P., & Bhuyan, A. (2024). Cognitive, affective, and demographic factors and their impact on human behavior to adopt LLM-based generative artificial intelligence. In K. Kanti Ray, Bhuwandeep, (Eds.), *Future of management: Embracing sustainability, diversity, and inclusivity*. 1st ed. Routledge, pp. 71–82.

Seyfi, S., & Hall, C. M. (2023). Beyond Saint Greta: Generation Z and sustainable tourism behaviours and practices. In *Handbook on tourism and behaviour change*. Edward Elgar Publishing, pp. 98–109.

Seyfi, S., Elhoushy, S., Kuhzady, S., Vo-Thanh, T., & Zaman, M. (2025). Bridging the green marketing communication gap: Assessing image coherence in green hotels. *International Journal of Tourism Research, 27*(2).

Shaik, A. S., Alshibani, S. M., Jain, G., Gupta, B., & Mehrotra, A. (2024). Artificial intelligence (AI)-driven strategic business model innovations in small- and medium-sized enterprises. Insights on technological and strategic enablers for carbon neutral businesses. *Business Strategy and the Environment, 33*(4), 2731–2751.

Sørensen, F., & Bærenholdt, J. O. (2020). Tourist practices in the circular economy. *Annals of Tourism Research., 85*, Article 103027.

Srivastava, P., Mishra, N., Singh, N., & Ramkissoon, H. (2024). Beyond carbon footprints: The 'Greta Thunberg Effect' and tourist hotel preferences. *Journal of Travel and Tourism Marketing., 41*(4), 578–595.

Tavares de Carvalho, P., Raimundo, R. J., & Dias Lopes, J. (2024). Improving guest and owner satisfaction through a circular economy: An agritourism case study. *Tourism and Hospitality, 5*(4), 887–905.

Warren, C. (2018). Smart advice and resource use feedback makes a smarter guest and a smarter host. In *2nd united national world tourism organisation Oviedo*. Asturias, Spain.

Warren, C., & Coghlan, A. (2016). Using character strength-based activities to design pro-environmental behaviours into the tourist experience. *Anatolia., 27*(4), 480–492.

Yu, J., Baah, N. G., Kim, S. (Sam), Moon, H., Chua, B. L., & Han, H. (2024). Effects of green authenticity on well-being, customer engagement and approach behavior toward green hospitality brands. *International Journal of Contemporary Hospitality Management, 36*(9), 3129–3150.

Zahidi, F., Kaluvilla, B. B., & Mulla, T. (2024). Embracing the new era: Artificial intelligence and its multifaceted impact on the hospitality industry. *Journal of Open Innovation: Technology, Market, and Complexity., 10*(4), Article 100390.

Zaki, K., & Farrag, M. (2024) The impact of circular economy on environmental performance in the tourism and hospitality industry: The role of low-carbon behavior and eco-friendly behavior. *The International Journal of Tourism and Hospitality Studie, 7*(2), 285–304.

Part IV
AI, Finance, and the Digital Economy

Chapter 17
Automated Decisions and Market Fairness in Decentralised Finance (DEFI): Exploring the Ethical Dimensions of AI-Powered Trading

Animesh Singh

Abstract The intersection of decentralised finance (DeFi) and artificial intelligence (AI) is a major change in the financial sector. This quick development of decentralised finance (DeFi) has revolutionised conventional financial systems by introducing computer-based, automatic, and algorithm-based trading systems. AI-based trading platforms have become the key actors of this new ecosystem, providing greater efficiency and liquidity and, at the same time, casting significant ethical concerns. In this paper, the ethical aspects of the issue of automated decisions in DeFi are discussed, including the fairness of the market, the biases of the algorithms, and the responsibility in AI-based trading. Based on a thorough overview of the modern-day literature, we examine the complexity within the distributed systems that are not controlled by a central authority. By so doing, we point out that automated trading decisions may unwittingly spread systemic biases, alter market forces, and put into question the very notions of fairness that financial markets are based on. The results highlight the need to have more open algorithmic operations and better checks and balances, and promote a reasonable balance between technological innovation and the integrity of the market.

Keywords Decentralised finance (DeFi) · Automated trading · Artificial intelligence (AI · Algorithmic bias · Market fairness · Ethics · Accountability

1 Introduction

The financial industry is undergoing a period of massive change with decentralised finance (DeFi) and artificial intelligence (AI) disrupting the very structure and working methods of the market. DeFi is a novel blockchain-based and distributed

A. Singh (✉)
USB-MBA, Chandigarh University, Mohali, Punjab, India
e-mail: animesh.researcher02@gmail.com

K. Kumar et al. (eds.), *AI-Driven Circular Economy for Carbon Neutrality in Emerging Economies*, Environmental Footprints and Eco-design of Products and Processes,
https://doi.org/10.1007/978-3-032-30298-4_17

ledger system ecosystem that creates financial products and services without any middlemen. Decentralised finance (DeFi) is a new field of study in the confluence of blockchain technology, digital assets, and financial services. The objective behind the design of DeFi protocols is the disintermediation through a blend of conventional and innovative service models. At the time when the industry was expanding fast, the total value of all digital assets locked in DeFi protocols increased over 2019 less than $1 billion to over 15 billion by the end of 2020 and even more in May 2021 (Defiant, 2020). However, DeFi is yet to reach its infancy. Even in skills alone, it is exciting to imagine what can be done with all of these opportunities when smart contracts, along with their ability to automate the process and remove so much red tape, can become the way to democratise access and efficiency. The consequences after these systems engage with the data-infused dynamic of the AI are alluring but terrifying. As the use of AI algorithms to trade strategies without human intervention becomes a possibility, it is an enabling agent with convoluted rules of decision-making that are quantitative in nature. But it has a place of both advantage and danger. On the one hand, it has liberalised the market, generated more liquidity, reduced the cost of transactions, and expanded the context of access to financial services. Conversely, it raises systemic and moral issues of fairness and transparency, any bias of systems, and the responsibility of autonomous systems (Chopra, 2024). In this paper, the author centres on the discussion of ethics regarding the implementation of AI in the process of making decisions within decentralised finance systems. We examine the bias generation by algorithmic trading processes to introduce market integrity issues and accountability challenges and to entrench them. The literature review incorporated a variety of existing materials, such as scholarly articles, industry reports, and cases, to analyse the wide range of research on this topic and find innovative technologies used in ethical management. The paper explains ways of taking advantage of this disruptive technological development in governance in the DeFi industry, in addition to identifying challenges that jeopardise equitable financial ecosystems. The study is intended to recap its research goals by reviewing literature findings, which includes to critically analyse the role of AI in automating trading decisions within DeFi ecosystems; to assess the impact of automated decisions on market fairness and equity; to explore the ethical challenges associated with the deployment of AI in DeFi trading; to propose ethical and governance frameworks that can assist in implementation of AI in DeFi.

2 Background and Literature Review

DeFi is a revolutionary financial industry. DeFi was initially a result of the developments in blockchain technology, and it has developed into its own ecosystem where financial services such as lending, borrowing, and trading are carried out through smart contracting. Contrary to the conventional methods of financial institutions, where intermediaries have to be used, DeFi works based on distributed networks, which makes settlement less time-consuming and diminishes transaction fees. Researchers have reported the potential of DeFi mechanisms, including liquidity pools through to decentralised exchanges, to reinvent trust in finance (Gan et al., 2023). Despite the benefits of decentralisation of the system, it does not have a single entity to control its functioning, which presents any biases and shortcomings it already possesses. This is a grave concern to researchers and other regulatory bodies that focus on the moral consequences of the decentralised systems (Mishra, 2024). Decentralised exchanges such as Uniswap and Balancer that do not depend on centralised regulation allow cryptocurrency users to trade tokens (Lin et al., 2020). Decentralised exchanges are operated by automated market maker (AMM) protocols that create pricing mechanisms via liquidity pools that create trust less trading solutions (Angeris et al., 2021). The decentralised financial platforms Aave, Compound, and MakerDAO allow users to borrow money using their assets or lend their bitcoin holdings to earn interest (Angeris et al., 2021). The platforms are automated and follow the use of smart contracts that carry out the liquidation of collateral and the calculation of interest based on the requirements in the market. Yield farming and liquidity mining are DeFi protocols that allow users to put money into assets in return for token rewards or dividend payouts on transaction fees (Ozbayoglu et al., 2020). Yield farming can enable users to passively earn money as they contribute to ensuring the market of Bitcoin DeFi is liquid and stable (Sutton & Holt, 2018). Yield farming is a common practice on Yearn Finance and Curve Finance, and Bancor is another example of a famous platform. Some of the cryptocurrencies, such as DAI, USDC, and USDT, are designed to have a stable value in comparison to an asset of reference (Mita et al., 2019). Stablecoins are an important part of DeFi as they allow developing tokens that monitor the value of real-world assets, such as gold or stocks, on the blockchain (Bianchi & Babiak, 2022). This is because artificial intelligence and machine learning have revolutionised the financial industry. The ability of artificial intelligence to operate on large volumes of data results in improved accuracy of forecasting and an increase in efficiency of operations. Real-time market analysis performed by the AI systems is used to generate instantaneous decisions that remove all human errors and judgemental biases by the traders. The study proves that adequate AI application brings more liquid and stable markets. However, it does indicate that these kinds of implementations might contribute to the increase in inequality in the market. (Horváth, 2023). Artificial intelligence has already entered the general financial services sector, transforming various fields in the current financial sector. Financial institutions now have the capacity to automate their operations, increase their decision-making processes, and improve their customer services. These solutions are implemented with the help of machine learning and deep learning, along with natural language processing (Bahrammirzaee, 2010). The debates on the ethical issues regarding automated systems are mostly limited to the question of accountability with regard to unfairness, as well as the detection of bias. Application of trading algorithms reveals the effects of bias that are based on the historical imbalances or a dataset that induces systemic inequities. When the automated systems are subjected to little bias in the development process, this seemingly insignificant fact will result in a whole lot of impact on the market once the system is operationalised

(Kakarala & Rongali, 2025). The identification and rectification of bias in decentralised financial systems is a challenging issue due to the fact that these financial systems rely on various algorithms that run in the background. Decentralised systems have a high level of uncertainty in terms of accountability. The lack of central authorities in decentralised finance poses a problem of identifying the party that should take responsibility for the trading errors and market manipulation incidents. The authors focus on market systems development that combines human oversight and technical oversight to guarantee reasonable operations (Park & Chang, 2023). The rise of AI in DeFi is a significant academic draw due to the fact that it ensures improved security and automation of financial systems constructed based on blockchain technology. The use of AI analytics in DeFi will enable risk managers to spot the vulnerabilities in the protocol and indicate abnormal behaviour and minimise the newest security risks (Zaheer et al., 2024). Machine learning models to provide liquidity assist the decentralised exchanges in reforming their liquidity pools according to modifications in market conditions (Wang et al., 2022). AI can facilitate the problem of decentralised autonomous organisations (DAOs) to enhance their governance by adopting the system of automated decision-making and dynamic behaviour trends analysis (Chen & Patel, 2023). AI predictive models can be implemented in DeFi to improve the process of capital borrowing and lending because of the evaluation of risks incurred by the borrowers and the prediction of the interest rates (Kim et al., 2022). AI technology adoption in DeFi systems has significant challenges that are related to the probability of data privacy infringement and algorithm bias, in addition to the risk of providing control to artificial intelligence. The proposed study aims to discover opportunities of the intersections between decentralised finance (DeFi) and artificial intelligence (AI) by conducting a literature review and analytical synthesis of the latest academic publications. In our research gathering, we considered and incorporated reports as posted by financial institutions, white papers by blockchain companies, and regulatory documents. This is not an ideal technique for formulating theoretical ideas using literature reviews because it has certain limitations to its usefulness. The developments of DeFi and AI have two impacts: research can become outdated and empirical because of the real-world examples, and data gathering is more complex since DeFi works in a decentralised setting. The broad scope of the study helps us to analyse various ethical issues that require our attention.

3 The Convergence of AI and DeFi in Automated Trading

The DeFi sites employed artificial intelligence that contributes to the increased speed of their trading in the same manner that it is executed in the financial sector. The AI trading solutions have capabilities that are superior to that of human beings with respect to analysing market data and executing trades at lower costs. All these factors lead to the improved liquidity of the market and the increased speed of the processing speed of transactions. The results of the research indicate that AI-based systems are faster to respond to market signals that contribute to the enhancement of high-frequency trading activities in any standard market structure (Li et al., 2021) The liquidity in the market has its pace and volume due to the influence of decentralised networks that involve liquidity pooling and algorithmic order matching. The current positive trends have the potential of resulting in a future whereby technological benefits will result in an unequal competitive advantage (Nartey, 2024). The scenario casts basic questions on the fairness of the market as well as the fair access to financial services.

The introduction of rapid trade execution has transformed market dynamics to establish a new era of predictable high-frequency trading, which affects the traditional markets (Kumar et al., 2025). Stable markets emerge from liquid conditions and efficient transactions, but automated decision-making speed presents a series of potential dangers (Singh et al., 2025). Errors within decentralised systems that lack centralised decision-makers can spread across the entire chain when algorithms or data show defects. Traditional market history, including the flash crash, proves that high-speed trading systems can produce major market instabilities (Demirer et al., 2019). The risks involved in DeFi systems become more serious since minimal oversight authorities exist, and independent traders perform quick, consecutive trades through autonomous actions. The unpredictable outcomes of unmonitored AI algorithm interactions generate new behaviours which elude system designers' predictions. The unexpected effects from these interactions could lead to market destabilisation and short-term liquidity challenges. The benefits of AI systems must receive equal attention to the risks of system-wide failures that could harm the public's trust in the technology.

4 Ethical Dimensions of Automated Trading

The process of training artificial intelligence systems with biased historical data leads to algorithmic bias, which is the main issue (Singh et al., 2024). Automated trading systems exhibit different manifestations of this issue. An AI system which learns from historical market data showing racial and socio-economic market disparities will continue to promote specific market segments while devaluing others. The issue of market fairness and price formation emerges as a key concern. Decentralised trading methods face challenges because proprietary algorithms use information that creates obstacles to reducing bias (Gu et al., 2022). The lack of information about data origins and modelling processes, together with data source justification, makes it impossible to find and fix biased outcomes. Unchecked, slight instances of bias throughout time will accumulate to impact the entire market rather than just individual traders. Developing autonomous systems with fairness as the top priority requires that all systems should detect bias as their primary function. Fairness within any financial system must be upheld across all marketplaces. Fairness in a market requires both the prevention of fraud and the equal treatment of all participants (Prakash & Tiwari, 2025).

The technological systems and their organisational structure are the two aspects of the introduction of the AI-driven DeFi systems that are problematic. The programmes of making decisions algorithmically lack transparency, hence adding to the asymmetry of information in the market (Turillazzi et al., 2023). A number of existing systems have implemented operation configurations that determine unequal opportunities of accessing proprietary algorithms. The decentralised nature of all DeFi protocols implies that no one can be accountable to devise fair operational standards of protocol implementation. The fairness procedures are implemented and spread to a number of dispersed nodes. This absence of a central force to provide the normal or equal treatment of those involved in a market when it comes to performing algorithms creates uncertainties regarding ethics (Kim et al., 2022). The ethical issues can be resolved only by the establishment of governance systems, which might reach the level of acceptable outcomes without forgetting technical methods. Any financial system should observe equity in every market place. Fraud as well as fair treatment of all parties in a market must be just. The technological systems and their organisational structure are the two aspects of the introduction of the AI-driven DeFi systems that are problematic. The programmes of making decisions algorithmically lack transparency, hence adding to the asymmetry of information in the market. Other systems which are operating already have operational partitions that produce disparities to obtain the benefit of proprietary access to algorithms (Fletcher & Le, 2021). The decentralised nature of all DeFi protocols implies that no one can be accountable to devise fair operational standards of protocol implementation. The fairness procedures are implemented and spread to a number of dispersed nodes. Gigantic loads of central powers to secure the standard or equal treatment of market participants in the algorithmic performance outcomes cause ambiguous questions of morality. The ethical issues can be resolved only by the establishment of governance systems, which might reach the level of acceptable outcomes without forgetting technical methods.

5 Regulatory and Governance Challenges

The traditional financial institutions are the primary interest to the regulatory bodies as they are the central bodies that are subjected to a specific legal framework. The conventional financial regulators have the responsibility of regulating the decentralised systems that cross borders not to mention the development of the new technological platform. Artificial intelligence trading technologies in DeFi markets introduce rapid dynamism in how the financial transactions are carried out. The computerised system providing negative feedback to the customer poses challenges in determining the liability, and the regulators are not able to monitor automated systems which contravene the very idea of market integrity (Kozimov, 2024). A number of suggestions have been present during the recent years, concerning the necessity to establish global regulatory methods of decentralised networks. Regulatory levels that should safeguard the consumer and fair markets in the technological development are still in its nascent stage. A balance must be reached by the regulatory bodies so as to promote innovation and encompass ethical financial practices simultaneously, as it will be imperative to have the highest amount of knowledge regarding technological features and the economic environment of the DeFi systems. The existence of machine learning applications in the DeFi lending platform has been instrumental in providing substantial contribution to the credit assessment and risk management processes as well as the loan management processes. The DeFi systems are powered by AI as the technology that evaluates the reliability of the borrowers based on real-time information regarding the wallets and blockchain transactions and smart contract interactions that are opposed to the services offered by the traditional bank that used the data regarding the credit history (Smith & Johnson, 2023). DeFi lending platforms implement AI, which essentially responds to market fluctuations by increasing or decreasing collateral to reduce the risk of liquidating the borrower with the main objective of maintaining consistency in the lending pool (Doe et al., 2022). Interest rates are easily determined through sophisticated computer algorithms which are actively engaged in the management of the supply and demand of liquidity by actively monitoring real-time data (Tripathi et al., 2022). The artificial intelligence technologies come in handy in terms of lowering the interest rates on the loans, as well as helping the environmental protection programmes (Lee, 2021). The presence of difficulties with the sphere of data privacy and the presence of algorithmic biases and regulatory requirements that cannot allow adopting coprophagic AI-enhanced lending systems in the service of the poor and the unbanked characterise the sphere of the DeFi. The development of DeFi platforms gives momentum to research on other types of governance that enhance accountability. The model of the decentralised autonomous organisations is an example of alternative systems of governance. The mechanism, however, is an effective self-regulation mechanism that presents other concerns. The decision-making process of the DAO through sharing of coins with few of their holders may alter the process of decision-making by the process that disturbs the operating framework of decentralisation and introduces bias (Kozimov, 2024). In some DAOs, moral outcomes are not the consequences of the utilisation of democratic principles. The judgement on these specific platforms is made on financial gains and not the moral concerns. The supporters believe that the disparity between the financial advantages of the algorithmic transparency votes and the customary law will be eradicated over time (Boonpeam et al., 2021). The dual approach in the system provides a way whereby the responsibility is distributed between the human factors and the technology factors. The necessity of the change in the governance structure of the distributed systems is also crucial as the new technologies and the market conditions are different.

6 Mitigation Strategies and Ethical Recommendations

The first action to reduce the ethical concerns focuses on the establishment of transparency in the process of designing and implementing algorithms. The disclosure of the proprietary algorithms by organisations does not reduce the competitive advantage since the description is adequate. The publication of the training data specifications and the model parameters and risk management data allow the developers to build trusted bonds with the consumers and allow the third-party audit. By using outside algorithm testing, the developers will be able to detect and rectify the possible bias to ensure that such problems do not propagate into the system. It is not only an effective way to minimise distortions in the market but also gives the users confidence that the system is operating based on fair and objective parameters (Eswaran et al., 2024; Kaith & Sachdeva, 2024). Besides transparency, the independent auditing structure should be established. The auditing process must perform a periodical evaluation to assess the algorithmic fairness and data reliability and environmentally test the system performance at varying conditions in the market. One such practice is the use of third-party organisations that issue food safety certifications to assess trading algorithm in relation to generally accepted ethical standards (Srivastava et al., 2024). These external audits should be incorporated by the legal organisations in their detection procedures as they determine the linkage between the technical activities and legal duties (Srivastava et al., 2018). Audits plus regulatory standards are an optimistic view of ensuring the upholding of ethical conducts in the trade systems powered by AI (Bhambhwani & Huang, 2024). Among the most overwhelming tasks in this respect is to find a proper balance between encouraging creativity and applying the required ethical safeguards. Due to a lot of its traditional resistance to change often attributed to the financial industry, it is ever evolving rapidly in technology. To address both needs, there will be a need to be cooperative. Effective communication forums, public–private partnerships and transdisciplinary committees would assist the regulators, technologists, and market actors to work together to establish ethical principles and regulatory concepts. Such cooperative systems would also have the benefit of a quick response to changes in technology that the ethical behaviour also keeps changing in tandem with the market innovations. Ethical stewardship of decentralised finance systems posed by artificial intelligence requires ethical leadership not only by the developers and regulators. The masses are turned into an evolutional asset of promoting development through their education. The description of algorithmic trading trade-offs and ethical issues allow the consumer to be held accountable in their decision-making. Grassroots supervision will be able to grow democracy in the decentralised governance of financial markets through blockchain-based vote-casting systems and dispute resolution mechanisms (Avgouleas & Kiayias, 2020). The decentralised financial markets require all participants to establish an ethical responsibility culture among traders to multinational companies since this will dictate the survival of the system.

7 DeFi and Risk

The fact that decentralised finance (DeFi) has emerged as one of the trends in modern financial systems is a relatively new phenomenon, via the use of publicly accessible blockchains and smart contracts. Ethereum is the main foundation for DeFi apps, which copy everyday financial services like lending, borrowing, trading, and managing assets (Aramonte et al., 2021). These apps prioritise automation, clear visibility, and open access for anyone without needing permission. The advocates of DeFi often claim that there is a decline in systemic risk due to the decentralisation process caused by the elimination of custodial agents and the substitution of institutional trust with cryptographic validation and algorithmic execution (Carter & Jeng, 2021). On the other hand, a few experts objected to this assumption, arguing that decentralisation does not eliminate risk, but rather merely transfers the risk to more technical, economic, and governance scales only dimly perceived by the participants. As they begin to grow in size, various types of concentration begin to creep into practice, such as depending upon large blockchains, a small few centralised issuers of stablecoins, trending oracle services, liquidity centres, and governance run by large token holders. These growths have been accompanied by what has been described as a decentralisation illusion, where systems that are being marketed as decentralised are systemically dependent and asymmetry of power (Deshmukh et al., 2021). Because of this, the incentive of greater resilience offered by DeFi needs to be considered.

Technically and functionally, DeFi poses another environment of risks, which is anchored with the application of smart contracts (Oliver et al., 2022). Smart contracts enable computerised transactions, which are programmable, yet they do not use human intermediaries, and directly encode financial logic in the form of software, which is immutable or can be modified partially after exposure. This implies that a malfunction in the code or unintended effects between contracts or bad incentive designs may be irreversible, e.g. loss or embezzlement of money forever. Empirical evidence exists, showing that coding weaknesses, re-entrancy attacks, and intentionally engineered exploits have cost billions of dollars to DeFi protocols, the material effects of software-based finance (Liu et al., 2022). Even though immutability can be considered as a protection against arbitrary interference, it also constrains the reactions to the discovery of vulnerability, necessitating the developers and users to plan intricate migrations to novel contracts. Conversely, upgradeable contracts cannot be claimed to be immutable, since the privileged actors may change the contract logic, sometimes under pressure of external legal or political authority, once more bringing discretionary control (Liu et al., 2022). Besides the smart contracts, the DeFi systems use the features of the blockchain level, such as the ordering of transactions and the off-chain data provision. The profit attached to a reordering, front-running, or sandwich attack in extracting maximum extractable value may be made by validators or other advanced market participants at the expense of ordinary users who do not receive technical protection (Salami, 2021a, 2021b). Oracles are also among the parts of dependency since they furnish outside data, such as asset prices, but they also possess the disadvantage of manipulation and the lack of coordination and concentration of power. Such technical and operational procedures mean that DeFi does not make trust disappear; it is simply passed on to the providers of the infrastructure, governance participants, and technically skilled participants whose authority may be transparent and difficult to debate (Salami, 2021a, 2021b).

The risks of DeFi are strictly correlated with the financial behaviours, incentives, and market dynamics which become achievable through permissionless and highly composable protocols (Chen & Bellavitis, 2020). The DeFi markets are characterised by volatility of prices, engagement in speculation, and rapidly moving capital, which amplify the risks of the traditional market and add other forms of instability (Wieandt & Heppding, 2022). DeFi modularity helps make protocols compatible to operate within interrelated networks to manage innovation, but leads to the formation of thick webs of dependency, liquidity, collateral assets, and stablecoin repositories. The stress created by a single protocol could therefore propagate very quickly across the ecosystem, creating cascading failures, which are difficult to contain. Many DeFi projects have relied on systems with high-yield generating mechanisms that promote high yields, which in most situations are underpinned by an ever-present inflow of new funding and not by economic activities (Chen & Bellavitis, 2020). Such designs can cause people to suddenly leave when trust is lost or markets change, as seen when algorithmic stablecoins collapsed, wiping out large amounts of money and disrupting other connected systems.

Stablecoins have played an important role in the DeFi environment as units of account, payment instruments, and a store of value, although the risk characteristics of the instruments vary considerably depending on the design. Fiat-pegged stablecoins are controlled by central companies, meaning users must trust the issuer to manage reserves honestly (Klages-Mundt et al., 2020). Because these issuers can freeze funds or follow regulations, they introduce counterparty and regulatory risks into systems that are meant to be decentralised. Crypto-backed and algorithmic stablecoins are not as dependent on custodians, but prone to collateral volatility, liquidity mismatches, and confidence-based runs (Klages-Mundt et al., 2020). Such liquidity provision schemes introduce further economic risks involving subjecting their participants to the temporality of loss and flash loans, permitting the immediate use of the pricing inefficiency and protocol weaknesses. All these dynamics strengthen the experimental nature of DeFi markets and underline the fact that innovation has been outpacing the development of effective risk-reducing measures.

Legal and emergent risks complicate the DeFi landscape as the systems of decentralisation come into conflict with regulatory systems that were designed to facilitate intermediated finance. The traditional financial regulation is rooted in identifiable actors that can be licensed, regulated, and held accountable in comparison with DeFi. These uncertainties have been more convenient in escaping the action. A lot more are caught in the crossfire of unexpected enforcement action as the regulators are probing the governance frameworks, interface providers, and the bearers of the tokens (Zetzsche et al., 2020). Criminal cases on legal dimensions of decentralised autonomous organisations demonstrate that the regulators may treat DAOs as unincorporated associations, and decentralised systems are challenging the alleged limited liability assumption (Buterin et al., 2018).

At the same time, the separation between fixed, non-modifiable protocols, and adjustable front-end interfaces creates points of control that weaken claims of regulatory neutrality, because interface providers can limit access, comply with sanctions, or change how users interact with the system when required by regulators. The newness, complexity, and rapid speed of development of DeFi infrastructures are the sources of the emergent risks that demand users to be highly technically aware and offer high operational security, but without much specific protection against irreversible error (Deshmukh et al., 2021). The transparency of blockchains, which is also said to be a protective feature, is often too complex due to smart contract interactions and inter-protocol dependencies, and much of it is inaccessible to non-expert users. Transparency does not, therefore, only imply any substantial meaning or informed consent. Combined, the practical, operational, economic, legal and emergent risks of DeFi mean that the prospect of a safer, more autonomous financial system is conditional upon the prospect of identifying ways to resolve the governance risks, being more aware of risks, and responding to regulatory interaction rather than merely a by-product of decentralised design.

8 Discussion

The financial industry undergoes an absolute revolution due to artificial intelligence (AI) and decentralised finance (DeFi), which possesses both beneficial and detrimental features. The use of AI technology in automated trading systems has provided fast trade execution, which has achieved efficiency in the market. A technology that runs out of control poses a threat of increasing the existing biases and, at the same time, generating unexpected economic effects. The training of machine learning models is a significant issue due to the fact that the historical data is fundamentally biased and thus initiates algorithmic bias (Gai et al., 2021). Systemic inequality in financial markets is caused by the fact that the biases that existed in the past are not disclosed. Market bias is harder to detect and correct with a decentralised system because it has less transparency. Trade equity inefficiency arises when the biases on the financial side are not identified, and this creates systemic detriment to various economic sectors. The financial system requires prompt policy interventions in order to ensure that it remains strong and stable. The AI systems will be at a lower risk since the algorithms are designed transparently and audited by external bodies working in different capacities to ensure that the systems operate fairly and fight discrimination. Application of both the traditional regulating bodies as well as decentralised autonomous organisations (DAOs) creates a system of governance that provides efficiencies in the operation, as well as upholding ethical practices. The formulation of the financial regulations requires the contribution of the technologists along with ethicists and policymakers in developing adaptive rules that remain abreast of the market changes. The decentralised systems that will employ AI to provide the fairness and accountability functions will create an all-inclusive benefit to all its users. Simulation models can help researchers to follow the development of algorithmic biases in various market regions. Market upheavals and biases are not predestined and, on the contrary, depend on the loopholes in checks and balances and the poor design of the system. Nevertheless, with consideration of careful development, thorough testing, and pre-emptive regulation, the challenges may be turned into opportunities to create a more equitable and transparent financial ecosystem.

9 Conclusion

Finance is rapidly changing because of decentralised finance (DeFi) and artificial intelligence (AI), which improve efficiency and encourage innovation, but also create complex ethical challenges. The case demands immediate intervention into the issues of fairness and bias as well as the issues of accountability since the AI-based trading systems have become the primary actors in the market. Academic work and the field applications have been found to support this debate. The rate and efficiency of AI in terms of financial transactions results in the liquidity improvement but can create unexpected effects on financial systems (Zohar et al., 2020). Due to the decentralised nature of the activities of DeFi platforms and the fact that the decision-making process occurs based on algorithms, it is difficult to detect and eliminate biased decision-making patterns. Lack of central control in DeFi systems poses challenges on holding responsible parties' responsibility that translates into ineffective response to process failures and market disruptions. AI-based trading operations are susceptible to risk reduction based on transparency systems and independent audit and governance systems that maintain balance. Regulators of financial markets should form international partnerships to address the issues of decentralised market by adopting dynamic policies to manage the technological changes. Educating the community and actively involving them are essential for developing and maintaining ethical rules. Both financial systems will be able to evolve in a responsible way when the administrators keep a right balance between innovation, stability, and equity. Decentralised finance has created new financial opportunities by followers of DeFi but these developments bring up some basic problems in morality that must be resolved with definite answers. Decentralised finance offers new ways to use financial services without banks, using blockchain and smart contracts. However, it involves serious risks such as technical errors, market crashes, governance problems, and legal uncertainty. DeFi is innovative but still unstable and needs stronger safeguards to protect users. The critical issue is to strike the appropriate balance between security and liquidity that will also enhance justice and transparency. The future financial system requires responsible leadership that develops innovative solutions while ensuring fairness and sustainability through strong integrity and accountability.

References

Angeris, G., Kao, H.-T., Chiang, R., Noyes, C., & Chitra, T. (2021). An analysis of Uniswap markets. *Cryptoeconomic Systems, 1*(1).

Aramonte, S., Huang, W., & Schrimpf, A. (2021). *DeFi risks and the decentralisation illusion.* Bank for International Settlements. https://www.bis.org/publ/qtrpdf/r_qt2112b.htm

Avgouleas, E., & Kiayias, A. (2020). *The architecture of decentralised finance platforms: A new open finance paradigm* (Edinburgh School of Law Research Paper No. 2020/16). https://doi.org/10.2139/ssrn.3666029

Bahrammirzaee, A. (2010). A comparative survey of artificial intelligence applications in finance: Artificial neural networks, expert systems and hybrid intelligent systems. *Neural Computing and Applications, 19*(8), 1165–1195.

Bhambhwani, S. M., & Huang, A. H. (2024). Auditing decentralized finance. *The British Accounting Review, 56*(2), 101270. https://doi.org/10.1016/j.bar.2023.101270

Bianchi, D., & Babiak, M. (2022). On the performance of cryptocurrency funds. *Finance Research Letters, 44*, 102391. https://doi.org/10.1016/j.frl.2021.102391

Boonpeam, N., Werapun, W., & Karode, T. (2021). The arbitrage system on decentralized exchanges. In *Proceedings of the 2021 18th International Conference on Electrical Engineering/Electronics, Computer, Telecommunications and Information Technology (ECTI-CON)* (pp. 768–771). IEEE.

Buterin, V., Hitzig, Z., & Weyl, E. G. (2018). *Liberal radicalism: A flexible design for philanthropic matching funds* (SSRN Scholarly Paper No. 3243656). Social Science Research Network. https://doi.org/10.2139/ssrn.3243656

Carter, N., & Jeng, L. (2021). *DeFi protocol risks: The paradox of DeFi* (SSRN Working Paper No. 3866699). https://doi.org/10.2139/ssrn.3866699

Chen, X., & Patel, S. (2023). AI-driven risk assessment in DeFi: Enhancing fraud detection and security auditing. *Blockchain & AI Review, 8*(3), 112–134.

Chen, Y., & Bellavitis, C. (2020). Blockchain disruption and decentralized finance: The rise of decentralized business models. *Journal of Business Venturing Insights, 13*, e00151. https://doi.org/10.1016/j.jbvi.2019.e00151

Chopra, P. (2024). Ethical implications of AI in financial services: Bias, transparency, and accountability. *International Journal of Scientific Research in Computer Science, Engineering and Information Technology, 10*(5), 306–314. https://doi.org/10.32628/cseit241051017

DappRadar. (2020, December 28). *Exclusive: DeFi year in review.* The Defiant. https://thedefiant.substack.com/p/exclusive-defi-year-in-review-by-1f2

Demirer, R., Leggio, K. B., & Lien, D. (2019). Herding and flash events: Evidence from the 2010 flash crash. *Finance Research Letters, 31*, 260–265.

Doe, J. K., Van de Wetering, R., Honyenuga, B., & Versendaal, J. (2022). Extended contextual validation of stakeholder approach to firm technology adoption: Moderating and mediating relationships in an innovation ecosystem. *Society and Business Review, 17*(4), 506–540.

Deshmukh, S., Resas, D., Gogel, D., Geest, A., & Sillaber, C. (2021). *Decentralized finance (DeFi) policy-maker toolkit.* World Economic Forum. https://www.weforum.org/whitepapers/decentralized-finance-defi-policy-maker-toolkit

Eswaran, U., Eswaran, V., Eswaran, V., & Murali, K. (2024). Security, risk management, and ethical AI in the future of DeFi. In *AI-driven decentralized finance and the future of finance* (pp. 48–88). IGI Global. https://doi.org/10.4018/979-8-3693-6321-8.ch003

Fletcher, G. G. S., & Le, M. M. (2021). The future of AI accountability in the financial markets. *Vanderbilt Journal of Entertainment & Technology Law, 24*, 289–332.

Gai, K., Qiu, M., & Zhao, S. (2021). Personalized financial services using machine learning: The future of AI in DeFi. *AI & Finance Journal, 7*(1), 52–64. https://doi.org/10.1145/3444878

Gan, R., Lin, X., & Wang, L. (2023). Why trick me: The honeypot traps on decentralized exchanges. *Proceedings of the ACM Conference*, 17–23. https://doi.org/10.1145/3605768.3623546

Gu, W., Li, Y., & Liu, S. (2022). AI meets decentralized finance (DeFi): A survey. *Artificial Intelligence Review, 55*(4), 2901–2936. https://doi.org/10.1007/s10462-022-10180-x

Horváth, D. (2023). Money in the digital age: Exploring the potential of central bank digital currency. *Sustainable Futures, 6*, 100136. https://doi.org/10.1016/j.sftr.2023.100136

Kaith, A., & Sachdeva, G. (2024). Bibliometric insights into green human resource management: A connection with business ethics and values. *PURUSHARTHA-A journal of Management, Ethics and Spirituality*, 17(1), 51–63. https://doi.org/10.21844/16202117104

Kakarala, M., & Rongali, S. (2025). Existing challenges in ethical AI. *World Journal of Advanced Research and Reviews, 25*(3), 549–554. https://doi.org/10.30574/wjarr.2025.25.3.0554

Kim, J., Lee, D., & Thompson, P. (2022). AI-based anomaly detection in blockchain transactions. *Journal of Financial Analytics, 12*(2), 56–74.

Klages-Mundt, A., Harz, D., Gudgeon, L., Liu, J.-Y., & Minca, A. (2020). Stablecoins 2.0. *Proceedings of the ACM Conference on Advances in Financial Technologies*, 59–79. https://doi.org/10.1145/3419614.3423261

Kozimov, N. (2024). Shareholder proposals on climate change and environment. *SSRN Electronic Journal*. https://doi.org/10.2139/ssrn.5058412

Kumar, K., Singh, R., Choudhary, M., & Sharma, A. (2025). Climate change mitigation strategies in environmentally sensitive industries. In *Climate change and social responsibility* (Vol. 25, pp. 185–204). Emerald Publishing Limited.

Lee, J. Y. (2021). A study on metaverse hype for sustainable growth. *International Journal of Advanced Smart Convergence, 10*(3), 72–80.

Li, L., Ouyang, Z., Yue, J., Wang, Z., & Zheng, Z. (2021). A survey on AI for DeFi. *Frontiers in Artificial Intelligence, 4*, 780164. https://doi.org/10.3389/frai.2021.780164

Lin, J., Xu, J., & Lan, L. (2020). *A dynamic approach for evaluating the liquidity of decentralized exchanges* (arXiv:2011.11248). http://arxiv.org/abs/2011.11248

Liu, L., Chen, C., & Wang, B. (2022). Predicting financial crises with machine learning methods. *Journal of Forecasting, 41*(5), 871–910. https://doi.org/10.1002/for.2840

Mishra, P. (2024). The ethical implications of AI in education. *Journal of Informatics Education and Research, 4*(2). https://doi.org/10.52783/jier.v4i2.1827

Mita, M., Ito, K., Ohsawa, S., & Tanaka, H. (2019). What is stablecoin? *Proceedings of the IIAI-AAI Conference*, 60–66. https://doi.org/10.1109/IIAI-AAI.2019.00023

Nartey, J. N. (2024). The convergence of decentralized finance (DeFi) and artificial intelligence (AI). *International Journal of Blockchain and Trust Applications, 2*(2), 78–89. https://doi.org/10.18178/IJBTA.2024.2.2.78

Oliver, J., Chipolina, S., & Szalay, E. (2022, June 3). Luna crash sends a chill through decentralised finance market. *Financial Times*.

Ozbayoglu, A. M., Gudelek, M. U., & Sezer, O. B. (2020). Deep learning for financial applications: A survey. *Applied Soft Computing, 93*, 106384. https://doi.org/10.1016/j.asoc.2020.106384

Park, E., & Chang, Y. (2023). The future of decentralized finance. *Journal of Emerging Technologies, 19*(4), 77–95.

Prakash, A., & Tiwari, A. K. (2025). The role of institutional factors in achieving SDG 16: A thematic review of drivers of corporate anti-corruption action. *Environment, Sustainability, and Governance Insights, 1*(1), 34–52.

Salami, I. (2021a). Challenges and approaches to regulating decentralized finance. *American Journal of International Law, 115*, 425–429. https://doi.org/10.1017/aju.2021.66

Salami, I. (2021b). Decentralized finance: The case for a holistic approach to regulating the crypto industry. *SSRN Journal*. https://doi.org/10.2139/ssrn.3711777

Singh, R., Kumar, K., & Khan, S. (2024). A comprehensive view of artificial intelligence (AI)-based technologies for sustainable development goals (SDGs). In *Artificial intelligence enabled management: An emerging economy perspective* (pp. 183–200). De Gruyter.

Singh, R., Tiwary, K. S., Kumar, K., & Kumar, A. (2025). Empowering a sustainable future: Blockchain's role in economy, organizations, and finance towards smart circular economy. In *Blockchain technologies for smart circular economy and organisational sustainability* (pp. 271–283). Springer Nature Switzerland.

Smith, S., & Johnson, G. (2023). A systematic review of the barriers, enablers and strategies to embedding translational research within the public hospital system focusing on nursing and allied health professions. *PLoS ONE, 18*(2), e0281819.

Srivastava, S. K., Anshul, A., Pathak, P., & Bansal, J. (2018). Responsible consumption for curbing food wastage: An Exploratory enquiry. *PURUSHARTHA-A Journal of Management, Ethics and Spirituality, 11*(2), 104–116.

Srivastava, S. K., Sinha, A. K., & Kumar, S. (2024). Workplace spirituality empowered by spiritual leadership to promote organizational effectiveness: A bibliometric analysis. *PURUSHARTHA-A journal of Management*, Ethics and Spirituality, 17(1), 1–22. https://doi.org/10.21844/16202117101

Sutton, S. G., & Holt, M. (2018). The changing face of accounting in a digital world. *International Journal of Accounting Information Systems, 28*, 1–9. https://doi.org/10.1016/j.accinf.2017.10.004

Tripathi, V., Singh, K., Parihar, S., & Srivastava, S. K. (2022). A bibliometric analysis on supply chain collaboration. *PURUSHARTHA-A journal of Management, Ethics and Spirituality*, 15(2), 66–103. https://doi.org/10.21844/16202115205

Turillazzi, A., Tsamados, A., Genç, E., Taddeo, M., & Floridi, L. (2023). *Decentralised finance (DeFi): A critical review of related risks and regulation* (SSRN Working Paper No. 4593242). Social Science Research Network.

Wang, K. H., Zhao, Y. X., Jiang, C. F., & Li, Z. Z. (2022). Does green finance inspire sustainable development? Evidence from a global perspective. *Economic Analysis and Policy, 75*, 412–426.

Wieandt, A., & Heppding, L. (2022). Centralized and decentralized finance: Coexistence or convergence? *SSRN Electronic Journal*. https://doi.org/10.2139/ssrn.4046173

Zaheer, A., Singh, A., Mishra, Reepu, K.K., & Vardari, L. (2024). The future of alchemy: Emerging trends and technologies metaverse in insurance–A virtual customer experience. In S. Taneja, P. Kumar, Reepu, M. Kukreti, & E. Özen (Eds.), *Data alchemy in the insurance industry: The transformative power of big data analytics*. Springer.

Zetzsche, D. A., Arner, D. W., & Buckley, R. P. (2020). Decentralized finance. *Journal of Financial Regulation, 6*(2), 172–203. https://doi.org/10.1093/jfr/fjaa010

Zohar, A., Lichtenstein, B., & Sim, D. (2020). Scalability challenges of integrating AI with blockchain technology. *IEEE Blockchain Conference Proceedings, 21*(4), 212–226. https://doi.org/10.1109/BCF.2020.0456

Chapter 18
The Ethical Landscape of DeFi: Challenges and Opportunities in a Tokenised Economy for Emerging Economies

Justine Mbudaya, David Chikwere, Kundayi Chinyongo, and Munyaradzi Chibaro

Abstract The rapid expansion of decentralised finance (DeFi) within the tokenised economies of emerging markets has generated complex ethical and economic questions, particularly regarding financial inclusion, regulatory adequacy and systemic resilience. Although DeFi promises to democratise access to financial services through blockchain-enabled transparency and disintermediation, existing research provides limited integrated analysis of its ethical implications alongside its developmental potential. Addressing this gap, the chapter aims to examine how DeFi's emerging architectures influence ethical financial conduct, regulatory capacity and inclusive growth in developing economies. The study adopted a desktop research approach supported by qualitative content analysis of contemporary academic literature, policy reports and practical case studies on DeFi deployment. Findings reveal that DeFi presents a dual trajectory. It holds promise for empowering underserved populations, decreasing dependency on traditional financial systems and stimulating economic innovation. Concurrently, it poses heightened risks related to fraud, weak governance, technological vulnerabilities and the exacerbation of pre-existing financial disparities. The chapter recommends a balanced framework that integrates

J. Mbudaya (✉)
BA ISAGO University, Gaborone, Botswana
e-mail: justine.mbudaya@baisago.ac.bw

D. Chikwere
Leeds Business School, Leeds Beckett University, Leeds, UK
e-mail: D.chikwere@leedsbeckett.ac.uk

K. Chinyongo
Midlands State University, Gweru, Zimbabwe
e-mail: chinyongok@staff.msu.ac.zw

M. Chibaro
University of Botswana, Gaborone, Botswana
e-mail: chibarom@ub.ac.bw

K. Kumar et al. (eds.), *AI-Driven Circular Economy for Carbon Neutrality in Emerging Economies*, Environmental Footprints and Eco-design of Products and Processes,
https://doi.org/10.1007/978-3-032-30298-4_18

ethical safeguards with innovation incentives, supported by targeted policy guidelines, capacity building initiatives and technological risk mitigation strategies to ensure responsible implementation. The study argues that for DeFi to realise its transformative potential in emerging markets, ethical governance, financial security and inclusive access must be established concurrently. Only then can decentralised technologies meaningfully contribute to sustainable and equitable economic development..

Keywords Blockchain ethics · Decentralised finance · Emerging economies · Financial inclusion · Tokenised economy

1 Introduction

The rise of decentralised finance (DeFi) is indicative of this changing form and function of global finance, notably in emerging economies that are characterised by underdeveloped and even inaccessible legacy banking systems for the majority of people (Ozili, 2022; Zetzsche et al., 2020). Based on blockchain, DeFi promises to democratise financial services by bypassing traditional finance players and providing access across borders to decentralised lending, borrowing and investing platforms (Chen & Bellavitis, 2020; Chohan, 2021). Although such developments are expected to facilitate greater participation and efficiency, they will also pose new intricate ethical issues that should be addressed critically.

These are promising but fragile environment in which to roll out DeFi due to rapidly growing financial markets, middle-income populations and a large number of unbanked people in developing economies (Demirgüç-Kunt et al., 2018). Several countries in Africa, Asia and Latin America exhibit shared structural impediments such as low levels of financial inclusion, exchange rate volatility and fragmented capacities for regulation (Beck et al., 2007; Demirgüç-Kunt et al., 2022). It is in this context that DeFi holds great promise for economic enfranchisement, while also posing the potentials of deepening disparities or even leading to new types of digital marginalisation (Harvey et al., 2021).

Although DeFi has been growing at an alarming pace, the ethical, regulatory and socio-economic considerations that determine DeFi's influence in the Global South are underexplored in scholarly and policy discourse including areas of conventional banking like tokenised financial systems. Existing research often focuses on technological potential or economic opportunities and neglects distributive justice, procedural fairness and inclusion of technology as well as system sustainability (Schär, 2021). This gap highlights the importance of a comprehensive analysis of how DeFi's transformative potential comes up against ethical weaknesses in these markets.

To fill this gap, this chapter critically reviews the ethical dynamics that presents itself as constructive and risks in DeFi utilisation in emerging economies. It studies best practices, cases and legislative stories looking for principles that may assist in the sustainable development of tokenised economic systems. To achieve that goal, the chapter utilises insights identified among users, financial institutions and regulators as well as broader society to develop an extensive and context-specific understanding of ethical considerations in shaping DeFi's contribution in the Global South.

2 Literature Review

2.1 Theories Underlying DeFi Ethics

From a theoretical point of view, several ethical frameworks including Utilitarianism, Deontology and Virtue Ethics, can be employed to analyse the ethics of decentralised finance. Utilitarians would focus on maximising the sum of everyone's welfare, and their assessment of DeFi would be in terms of its net welfare benefit to society (Philippon, 2019). DeFi's potential to increase inclusion in the financial system and decrease transaction costs should be weighed against concerns about systemic stability and fraud (Baur et al., 2018).

Deontological theories maintain evidence of duties and rights, by asking whether DeFi systems adhere to basic rules, such as transparency, consent and fair play (Aramonte et al., 2021). Conversely, virtue ethics are interested in character traits and dispositions which are definitionally in the purview of a DeFi system to encourage or discourage (Narayanan & Clark, 2017).

The above ethical frameworks should be applied especially to blockchain-based financial systems (Cong & He, 2019). The decentralised architecture of DeFi presents ethical issues that could be different from what traditional financial ethics had imagined (Zetzsche et al., 2020). The lack of central authorities leads to questions about accountability and responsibility when systems fail or cause harm (Hacker & Thomale, 2018). Likewise, pseudonymity of the blockchain transactions raises issues between privacy rights and regulatory functions (Auer, 2019).

2.2 Financial Inclusion and Digital Divide

Despite increasing attention in the literature about financial inclusion in emerging countries and the promise of digital financial services to reach unbanked populations (Schueffel, 2016; United Nations, 2018), DeFi carries forward this promise by creating more open access and removing gatekeepers (Zetzsche et al., 2020). Nevertheless, there are also significant digital divides that could constrain DeFi's potential inclusiveness (Demirgüç-Kunt et al., 2018).

Smartphone penetration, adequate access to the internet and digital literacy remain major challenges in many low- and middle-income countries (Demirgüç-Kunt et al., 2022). Multiple studies in different developing markets show that the financial inclusion benefits of mobile money have been achieved unequally (Arner et al., 2017).

Urban, educated and higher-income populations have tended to benefit more than rural, less educated and lower-income groups (Beck et al., 2007). Questions arise as to whether the adoption of DeFi may replicate or even exacerbate these inequalities (Cong & He, 2019).

2.3 Regulatory and Governance Challenges

Data along these lines is less scarce for emerging economies, where DeFi regulation is mostly uncharted and presents both opportunities and risks (OECD, 2020; Ozili, 2022). In some cases, regulators have adopted positive attitudes, seeing DeFi as a way to bypass standard financial infrastructure (Chen & Bellavitis, 2020). Others have chosen a preservative approach out of concerns for financial stability and consumer protection (Financial Stability Board, 2022).

DeFi protocols are also controlled in novel ways that existing regulatory systems fail to encompass (Harvey et al., 2021). Decentralised Autonomous Organisations (DAOs) and tokenised governance models offer a completely new form of decision-making that does not align well with traditional legal norms (Chohan, 2024). While the world is connected, DeFi protocols also face jurisdictional issues as transactions and control decisions are made among multiple jurisdictions (Zetzsche et al., 2020).

2.4 Environmental and Sustainability Considerations

The environmental costs of blockchain technology and proof-of-work consensus in particular have been increasingly portrayed as a moral issue (Schär, 2021). The new blockchains are applying less energy-consuming methods of consensus, but the carbon print of DeFi is still not marginal (Iqbal et al., 2022). An example of environmental impact of the development of DeFi adoption in EU could be a very important ethical issue for those countries as they are thematically supported by their ground sustainability belief (European Commission, 2020).

3 Methodology

The study used a qualitative desktop research and systematic content analysis method to trace the ethical terrain of DeFi in the developing world. The method was intended to cover current practices or developments, challenges and future opportunities for DeFi in order to answer the questions of what ethical considerations need to be considered when developing and adopting future DeFi.

3.1 Data Collection

The review used multiple sources of literature to provide a comprehensive view into the subject. Academic references were taken from peer-reviewed contributions on blockchain technology, financial ethics and development economics. Reports from industry leaders, consultancies, regulatory bodies and other international organisations (OECD, 2020; World Economic Forum, 2021) offered readable insight on what was DeFi adoption or regulation in the developing countries. In addition, case studies in the areas of Sub-Saharan Africa, Southeastern Asia, Latin America and Eastern Europe were considered to reflect diversity of DeFi on a regional level (Zetzsche et al., 2020). White papers and policy documents from government and regulatory authorities highlighted the urgent recognition that authorities need to regulate and monitor DeFi (IOSCO, 2022).

3.2 Content Analysis Framework

The authors used qualitative content analysis to generate patterns, themes and trends from the literature. Thematic coding was employed to identify recurring ethical issues, orientations and recommendations (Remus & Schuppli, 2021). The framework of analysis was based on four main axes: inclusivity and access, security and trust, governance and accountability and sustainability and systemic impact (Schär, 2021). Each dimension was then broken down into distinct ethical elements which were inspired by the literature review.

3.3 Analytical Approach

The methods used in the analysis included descriptive assessment and critical examination of ethical arguments and empirical studies. Descriptive analysis was applied to narrate visually the terrain of DeFi adoption in developing nations and uncover observable trends, challenges and prospects influencing the space. The ethical logic and empirical claims around DeFi were scrutinised. The study also used comparative approaches investigating how advanced DeFi adoption and regulation were in various developing countries. This comparative approach assisted in delineating the situational aspects that affect ethical outcomes and thus can help inform best practices.

4 The Dual Nature of DeFi: Opportunities and Challenges

DeFi also holds the promise of interoperability across blockchains that could help tear down financial sector silos, greatly promoting innovation and building vibrant financial ecosystems (Carter & Jeng, 2021).

4.1 *Opportunities of DeFi in Emerging Economies*

4.1.1 Democratising Financial Access

Democratising financial access entails making financial services and opportunities accessible and inclusive for people of all income levels, backgrounds and locations. The most attractive promise of DeFi to developing economies is in the democratisation of financial services (Harvey et al., 2021; Ozili, 2022). Many developing countries have very high barriers to entry into banking networks, high transaction costs and limited geographical reach in conventional banking networks (Demirgüç-Kunt & Klapper, 2018). In ICT terms, DeFi offers the potential to democratise lending/borrowing, trading, investing to any individual connected to the Internet, without the need for intermediaries of the traditional MFS (Chen & Bellavitis, 2020).

The dematerialisation and democratisation of finance provided by DeFi is relevant especially for people from sections of society that have previously not been served by the formal financial sector (Zetzsche et al., 2020). Small shop owners, rural farmers and those who lack historical credit can possibly gain access to capital from DeFi lending protocols (Schär, 2021). Similarly, in nations with unsafe currencies, citizens can save in stable cryptocurrencies and invest internationally in ways that were not accessible before (Adrian & Mancini-Griffoli, 2021).

Nevertheless, the potential of democratisation of DeFi is constrained by a number of problems (Morgan, 2022). Many of the candidate users are hampered by the technical challenge (Narayanan et al., 2016). The concepts of a private key, smart contract and a liquidity pool are perceived through the prism of complexity of bitcoin or other cryptocurrency that might be incomprehensible to a great part of even educated individuals in the less developed economy (OECD, 2020). Not only privacy benefits come with the pseudonymous nature of the blockchain transactions, but they also present the problem of building trust and resolving conflicts (Van Wegberg et al., 2018).

4.1.2 Economic Empowerment Through Tokenisation

The issuances of tokens, by tokenising assets and activities, are creating the economic empowerment of today's emerging economies (OECD, 2020; World Economic Forum, 2021). Real estate, commodities, intellectual property or any other asset could be structured as a token and traded on platforms, such as DeFi platforms, which could help unlock value believed to be illiquid until now (Hornuf et al., 2021). This democratisation of asset ownership and trade might allow more people to partake in economic growth (Werner & Schär, 2022).

Tokenisation allows new economic organisation and value creation (Cong & He, 2019). Community currency tokens may also be used to represent claim rights on local projects or local enterprises within the community, advancing a rationale of collective ownership and collective decision-making to a greater extent (Morrar & Al-Ansi, 2020). The establishment of tokens offers new options for earning a living based on skills, knowledge or social capital without finding a traditional job or intermediary (Berg et al., 2019).

The promise of tokenisation carries over to governance and social organisation (Chohan, 2024). Tokenised voting systems can be used to create more participative decision-making in communities, organisations and even official services. This would be especially important in less developed nations where traditional governance mechanisms may be limited or uninterested in citizens (UNDP, 2019).

4.1.3 Reducing Traditional Banking Constraints

DeFi protocols are agnostic from traditional banking infrastructure, allowing this model to potentially circumvent most of the factors that constrain financial inclusion in emerging markets (Schär, 2021). There is a potential for cross-border transactions to be done without correspondent banks, leading to lower transaction costs (Auer & Boar, 2021). Lending decisions can be made automatically, depending only on on-chain collateral, not arbitrary credit rating, which can provide broader access to credit (Harvey et al., 2021).

This is particularly important for small and medium enterprises (SMEs) in developing countries, as traditional banking limitations are by and large decreased for them (International Finance Corporation, 2019). There is still a scarcity of working-capital finance, trade finance and investment capital among the SMEs from the traditional banking sources (World Bank Group, 2017). DeFi platforms can offer other forms of financing with a greater accessibility and flexibility than traditional bank loans (Ozili, 2022).

Yet the circumvention of conventional banking systems introduces new risks and challenges as well (Financial Stability Board, 2022). Conventional banks are important for deposit insurance, regulatory supervision and consumer protection reasons (Zetzsche et al., 2020). DeFi may offer less protection, which raises the possibility of more significant loss or fraud (Van Wegberg et al., 2018).

4.2 Ethical Challenges in DeFi Adoption

4.2.1 Security Vulnerabilities and Fraud Risks

The decentralised and frequently anonymous operating environment of DeFi also creates substantial security challenges that raise ethical issues, especially for the most vulnerable segments of the population of emerging economies (Schär, 2021; Werner & Schär, 2022). The vulnerabilities in smart contracts have caused billions of dollars in losses, and there is little room for the users to get proper remedies (Harvey et al., 2021). The complexity of DeFi protocols challenges the ability of ordinary users to accurately evaluate their risks and enables information asymmetries that can be exploited by malicious actors (Werner & Schär, 2022).

The risk of fraud in DeFi is particularly salient in emerging markets where regulation is scarce and financial literacy is low (Ozili, 2022; U.S. Department of the Treasury, 2023). DeFi has been replete with the failures of Ponzi schemes, fraudulent tokens and exit scams which have preyed on the greedy or those ignorant to the blockchain world (Financial Stability Board, 2022). Fraud victims have no recourse as the liability of a fraud is not appreciable on a blockchain once a transaction is authenticated (Narayanan et al., 2016).

The security vulnerabilities of DeFi come with some interesting questions about responsibility and liability (Zetzsche et al., 2020). Who is responsible for losses if smart contracts malfunction or are hacked? How can end users be kept safe while not compromising the decentralised nature of these systems? Their importance is even higher in developing countries where an internet user may not have the financial resources to bear such non-reversible losses and where the number of legal remedies available is limiting (Zetzsche et al., 2020).

4.2.2 Regulation and Consumer Protection

When there are no clear-cut regulatory constructs for DeFi in the collective emerging markets in general, this raises some serious ethics concerns (OECD, 2020; Zetzsche et al., 2020). Proper oversight in these environments can present unsuspecting customers with risks they are ill-equipped to comprehend or even worse, when the bad actors get a free ride (Financial Stability Board, 2022). With an internationally decentralised system such as DeFi, it is difficult for national regulators to regulate, and gaps in regulation can be exploited (IOSCO, 2022).

Regulatory consumer protection policies embedded in traditional financial systems, such as deposit insurance, dispute resolution procedures and fiduciary duties, are often non-existent or not enforced in DeFi (Harvey et al., 2021). For the more disadvantaged strata who may be less capable of assessing risks and absorbing losses, it is simply despicable (Demirgüç-Kunt & Klapper, 2018). Yet the absence of Know your Customer (KYC) and Anti-Money Laundering (AML) requirements in the majority of DeFi protocols has cleared the path for criminal behaviour as well (Van Wegberg et al., 2018).

Regulatory uncertainty in DeFi: Users and provider's accessibility is strongly affected by a lack of regulatory clarity in DeFi (Hornuf et al., 2021). Individuals and businesses having to navigate DeFi might be uncomfortable with the level of their legality, valid service providers may not have space to operate due to regulation and legal cloud (Ozili, 2022). This uncertain regulatory status could serve to restrict the potential good of DeFi while giving free reign to bad actors (Auer & Rua, 2019).

4.2.3 Technological Gaps and Digital Divide

The digital prerequisites to DeFi involvement introduce a new dimension of digital divide which can further drive inequities among people in the developing world (United Nations, 2020; World Bank, 2021). Reliable internet access, compatible devices and even technical know-how is limited in many parts of the world (Demirguc-Kunt & Klapper, 2018). These technology barriers could exclude the very people who could most benefit from DeFi and who could afford it the most (Morgan, 2022). Usability of most DeFi applications is still complicated and intimidating for non-technical users (Werner & Schär, 2022; Zetzsche et al., 2020). Handling private keys, estimating transaction fees and interacting with complicated user interfaces, all require a high degree of technical knowledge (Narayanan et al., 2016). This level of complexity may erect sufficiently high walls that prevent DeFi from being used by all but the educated and technologically literate, which could serve to aggravate rather than lessen gaps in financial inclusion (Schär, 2021).

The digital and financial skills needed by DeFi users to engage in DeFi safely are high (Werner & Schär, 2022). That is, users must know not only how to handle the tech, but how to evaluate risks, how to control their engagement and how to guard themselves against fraud (Ozili, 2022). In case of emerging countries, lack of proper education and training systems means a lot of users can engage with DeFi without comprehensive assessment of the risk they are considering (U.S. Department of the Treasury, 2023).

4.2.4 Market Volatility and Financial Stability

High volatility in cryptocurrency markets exposes DeFi users to high risks, especially users in developing economies who might not be financially endowed enough to withstand losses (Baur et al., 2018; Financial Stability Board, 2022). The high degree of speculation for many DeFi assets and the widespread borrowing in DeFi systems could exacerbate market dynamics and potentially introduce systemic risks (Cong & He, 2019).

The interconnectedness of DeFi protocols makes that failures in one protocol can propagate throughout the entire space (Werner & Schär, 2022). This interconnectedness, along with high leverage and insufficient liquidity in some markets, produces systemic risks that have implications for the whole financial system (Harvey et al., 2021). These systemic effects can be particularly damaging for developing countries with less developed financial stability arrangements (World Bank, 2021).

The use of algorithmic stablecoins and other experimental financial products in DeFi also increases stability risks (Schär, 2021). Algorithmic stablecoins' breakdowns recently indicate that these instruments may fail to deliver the stability they overpromise, threatening their end users with losing their money at a large scale (Auer & Tercero-Lucas, 2022). The complexity of such instruments makes it complex for the regulators and users to evaluate the risks correctly (Schär, 2021).

5 Ethical DeFi Development Critical Success Factors

5.1 *Inclusive Design and Accessibility*

More inclusive DeFi protocols are an opportunity that can be used to overcome some of the ethical issues mentioned above (Arner et al., 2020; United Nations, 2020). The concepts of inclusive design can be applied to the development of user interfaces that are easy to use and accessible to users with any degree of technical expertise and culturally varied audiences (Arner et al., 2016). The utilisation of DeFi solutions by communities with various backgrounds can be made easily with the help of multilingual interfaces, easy-to-use services and culturally aware design features (Lapointe & Fishbane, 2019).

Mobile-first DeFi app development is especially beneficial for developing countries where mobile phones are more prevalent than personal computers (Werner & Schär, 2022; World Bank, 2021). Mobile apps can offer a simplified interface which abstracts away technical details but still exposes DeFi services (Ozili, 2022). If implemented and integrated with mobile money services already in operation in some of these countries, it could further lower barriers to entry and utilise the existing familiarity of users with digital financial services (Ozili, 2022).

The use of community-driven development approaches can further provide guarantees that DeFi protocols work in such a way that best meets the needs of users in these emerging economies, as opposed to providing a solution that was developed for different contexts (World Economic Forum, 2022). Local communities are one of the most effective resources for identifying user need, understanding the cultural and preferences surrounding the market and learning of the principal challenges of the context that are invisible to an outsider (Chen & Bellavitis, 2020).

5.2 Education and Financial Literacy

Improving general education and financial literacy are critical channels to foster the proper utilisation of DeFi (Demirgüç-Kunt & Klapper, 2018; U.S. Department of the Treasury, 2023). These classes would have to cover not only the technological side of DeFi protocol use but financial themes such as risk management, portfolio diversification and fraud prevention (Narayanan et al., 2016). Interventions in schooling approaches need to be culturally affirming and must use modes of communication that reach specific sets of population (United Nations, 2020).

Peer-delivered education interventions could be especially advantageous in emerging markets contexts where formal education may not be readily available (Khera et al., 2021). Community leaders and entrepreneurs, as well as those who first adopt DeFi in communities they serve can provide education and training (Arner et al., 2020). This could potentially create a trusting relationship and test whether education is operationally doable and culturally appropriate (World Bank, 2021).

Learning about DeFi can also be gamified by employing interactive teaching methods to create a more engaging and active model (Alammary et al., 2019). Simulated game play, applications and an active teaching programme might assist the learner in their understanding of complex and challenging concepts without the exposure of one's own money (Werner & Schär, 2022). These strategies are potentially attractive for future generations, considering that young people are using more digital tools (World Bank, 2021).

5.3 Governance of Innovation and Stakeholders' Engagement

Another way to promote ethics is to have more fair governance mechanisms that represent the stakeholders and users of the DeFi protocols more (Chohan, 2021; Harvey et al., 2021). Traditional token-based governance will result in power accumulation to the rich token users' allocations, which will harm small users and minorities (Werner & Schär, 2022). Other governance structures that prioritise utilisation, expertise or community contribution might lead to a more inclusive decision-making process (Chohan, 2024).

It is also possible to have multi-stakeholder governance structures that consider the interest of different user communities in the development of protocols and governance decision-making (Matsuo & Ushida, 2020). The approach might involve representation of users, developers, regulators and civil society organisations (OECD, 2020). Trust and legitimacy are elicited by frequent consultation of stakeholders and decision-making transparency (Cong & He, 2019).

The coupling of legacy governance solutions and blockchains is an additional innovative lever (Tinn et al., 2020). Hybrid systems, which comprise on-chain voting with off-chain deliberation and traditional democratic procedures, may be better capable of benefiting from the virtues of both approaches yet overcoming limitations of only technical governance systems (Werner & Schär, 2022).

5.4 Collaboration and Standards Development Between Regulatory Agencies

Collaborative agendas on regulation design are fertile grounds for promoting an ethical adoption of DeFi as well as to address valid regulatory concerns (Adrian et al., 2022; BIS, 2022; Ozili, 2022). Such sandboxes, when properly regulated, can create safe test environments for experimentation and train regulators on emerging technologies and their potential to disrupt (Auer & Rua, 2019). A setting for multi-stakeholder working groups with regulators, business, NGOs and academic researchers could be envisaged to develop input into appropriate regulation (Wharton FinTech Policy-Maker Toolkit, 2021).

Writing international standards and best practice guidance for DeFi might also contribute to the resolution of these jurisdictional problems arising from the global character of these systems (OECD, 2020). International organisations, such as the Financial Stability Board, the Bank for International Settlements, regional development banks, can facilitate this coordination (IOSCO, 2023).

Meanwhile self-regulation efforts by industry members may fill the gap in formal regulatory structures and promote social norms and standards for ethical conduct (Arner et al., 2019). Responsible practice can be promoted without stifling innovation and flexibility through mechanisms such as industry associations, certification programmes and voluntary codes of conduct (GDF, 2020). The Table 1 shows DeFi regulatory approaches across emerging economies

6 Conclusions

Based on the findings, the chapter concludes that Decentralised Finance (DeFi) has the capability to revolutionise the developing world by increasing access to finance, reducing the cost of transactions and promoting cross-border economic activity in markets that are not well served by traditional banks. The ethical hazards and systemic vulnerabilities and for which the risks and weaknesses, include fraud, non-accountability, market manipulation and digital exclusion, are an enormous hurdle for DeFi even-handed integration into emerging regions. This absence of common rules makes the landscape volatile, whereby an unintended inequality can strengthen, or users, particularly those with weak digital competences, can be exposed to exploitation. Tokenisation can democratise ownership and investment, but without safeguards it can also lead to more volatility, scams and wealth disparities. The cultural, infrastructure and institutional divide in these markets should inform how DeFi systems

Table 1 DeFi regulatory approaches across emerging economies

Region	Country	Approach	Key initiative	Impact assessment	References
Africa	Kenya	Experimental	Blockchain regulatory sandbox, M-Pesa integration pilots	Greater mobile-DeFi linkage, increased access to digital finance benefits. Even though a concrete 15% growth is hard to attribute solely to such pilots from a single source, the general trend of digital finance adoption in Kenya is clear and technology-led based on mobile money	(Osabutey & Jackson, 2024; Ozili, 2023; World Bank, 2022)
	South Africa	Comprehensive	Crypto Asset Regulatory Framework (2022)	Targeted more towards regulatory transparency and investor protection, with risk management given top priority like illegal activity. While sophisticated impact metrics in the form of illegal activity reduction are being attained, the framework lays a base for safer and more compliant crypto activity	(FSCA, 2022; PwC, 2023)
	Nigeria	Restrictive-Progressive	Initial ban on crypto transactions followed by a regulated framework (e.g. CBN circular 2021; SEC guidelines 2022)	Mixed results: The first ban resulted in an abrupt fall in exchange-based P2P volumes but simultaneously caused the expansion of OTC and informal channels. Later initiatives towards a regulated framework are poised to legalise crypto activities, expecting enhanced formal adoption as well as improved regulation	(CBN, 2021; Chainalysis, 2021; IMF, 2023; SEC Nigeria, 2022)

(continued)

Table 1 (continued)

Region	Country	Approach	Key initiative	Impact assessment	References
Southeast Asia	Singapore	Innovation-friendly	MAS Payment Services Act, regulatory sandbox	The MAS Payment Services Act (PSA) came into force in 2020 and offers a wide-ranging regulatory framework for payment services, including digital payment tokens and stablecoins. Regulatory sandbox efforts in Singapore have promoted a pro-innovation culture. This strategy has also led to Fintech FDI growth, with indications of robust levels of investment growth and solidified its position as a regional financial tech and innovation centre	(Monetary Authority of Singapore, 2019; PwC, 2023; Deloitte, 2023)
	Thailand	Balanced	Regulatory sandbox, CBDC pilot programmes	Increased institutional adoption, controlled innovation	(IMF, 2021)
	Philippines	Emerging	BSP cryptocurrency regulations (2023)	Early-stage implementation, limited data	(Chen & Bellavitis, 2020)
Latin America	El Salvador	Progressive	Bitcoin Legal Tender Law (2021)	Increased adoption rates, reduced remittance costs (30%), economic volatility	(Chohan, 2021; IMF, 2022)
	Brazil	Regulatory	Marco Legal das Criptomoedas (2022)	Increased institutional participation, improved transparency	(Brazil, 2022)
	Argentina	Reactive	Response to hyperinflation, partial regulations	Increased stablecoin adoption, currency hedging	(IMF, 2023; BIS, 2020)

Source: Literature analysis

are architected and administered. Generic models are also likely to perform poorly (both ethically and in terms of accuracy).

7 Recommendations

Governments and regional entities should establish innovation-friendly environments that allow DeFi projects to be safely tested without stifling creativity. Local education programmes should be strengthened to cover blockchain basics, personal data protection and token economics,empowering communities to effectively participate in the DeFi ecosystem. Ethical concerns, such as auditability, transparency and user protections, need to be a part of the design process and not just get slotted in at the end. Inclusive mechanisms should be developed to direct financial innovation towards broader development goals, fostering active collaboration among tech innovators, civil society and regulatory organisations to ensure DeFi serves the public interest. Inspiring local DeFi solutions and making sure policies should encourage the projects targeting region-based issues such as remittance cost, microfinance, supply chain transparency and promoting community stake. Adoption of fair tokenomics and anti-exploitation safeguards including quadratic funding mechanisms, crypto distribution transparent models and regular smart contract audits to build a fairer ecosystem.

References

Adrian, T., & Mancini-Griffoli, T. (2021). The rise of digital money. *Annual Review of Financial Economics, 13*(1), 57–77. https://doi.org/10.1146/annurev-financial-110220-072521

Adrian, T., Erceg, C., Giesecke, S., & Patel, N. S. (2022). Regulating crypto: The right rules for digital assets. *IMF Finance & Development, 59*(2), 54–57.

Alammary, A., Alhazmi, S., Almasri, M., & Gillani, S. (2019). Blockchain-based applications in education: A systematic review. *Applied Sciences, 9*(12), 2400. https://doi.org/10.3390/app9122400

Aramonte, S., Huang, W., & Schrimpf, A. (2021). DeFi risks and the paradox of decentralisation (BIS Bulletin No. 43). Bank for International Settlements. https://www.bis.org/publ/bisbull43.htm

Arner, D. W., Barberis, J., & Buckley, R. P. (2016). The evolution of fintech: A new post-crisis paradigm. *Georgetown Journal of International Law, 47*(4), 1271–1319.

Arner, D. W., Barberis, J., & Buckley, R. P. (2017). FinTech, RegTech and the reconceptualization of financial regulation. *Northwestern Journal of International Law & Business, 37*(3), 371–413.

Arner, D. W., Barberis, J., & Buckley, R. P. (2019). The evolution of fintech: A new post-crisis paradigm? *Georgetown Journal of International Law, 47*(4), 1271–1319.

Arner, D. W., Buckley, R. P., & Zetzsche, D. A. (2020). Sustainability, FinTech and financial inclusion. *European Business Organization Law Review, 21*(1), 7–35. https://doi.org/10.1007/s40804-020-00183-y

Auer, R. (2019). Beyond the do-it-yourself blockchain: The anatomy of stablecoins (BIS Working Papers No. 794). Bank for International Settlements.

Auer, R., & Boar, C. (2021). Central bank digital currencies: Motives, economic implications and the research frontier (BIS Working Papers No. 976). Bank for International Settlements.
Auer, R., & Rua, D. B. (2019). The global reach of central bank digital currencies (BIS Working Papers No. 865). Bank for International Settlements.
Auer, R., & Tercero-Lucas, D. (2022). Distrust or speculation? The socioeconomic drivers of U.S. cryptocurrency investments (BIS Working Papers No. 951). Bank for International Settlements.
Bank for International Settlements. (2020). Annual economic report 2020. BIS.
Bank for International Settlements. (2022). Annual economic report 2022. BIS.
Baur, D. G., Hong, K., & Lee, A. D. (2018). Bitcoin: Medium of exchange or speculative assets? *Journal of International Financial Markets, Institutions and Money, 54*, 177–189. https://doi.org/10.1016/j.intfin.2017.12.004
Beck, T., Demirgüç-Kunt, A., & Levine, R. (2007). Finance, inequality and the poor. *Journal of Economic Geography, 7*(3), 325–355. https://doi.org/10.1093/jeg/lbm008
Berg, C., Davidson, S., & Potts, J. (2019). *Understanding the blockchain economy: An introduction to institutional cryptoeconomics*. Edward Elgar Publishing.
Brazil. (2022). Lei nº 14.478, de 21 de Dezembro de 2022. Diário Oficial da União.
Carter, N., & Jeng, L. (2021). DeFi protocol risks: The paradox of DeFi. In B. Coen & D. Maurice (Eds.), *Regtech, suptech and beyond: Innovation and technology in financial services*. Risk Books. https://doi.org/10.2139/ssrn.3866699
Central Bank of Nigeria. (2021). Circular to all deposit money banks (DMBs), non-bank financial institutions (NBFIs) and other financial institutions (OFIs) on dealing in cryptocurrencies and facilitating payments for cryptocurrency exchange (Circular No. FPR/DIR/GEN/CIR/06/001). CBN.
Chainalysis. (2021). The 2021 geography of cryptocurrency report. Chainalysis.
Chen, Y., & Bellavitis, C. (2020). Blockchain disruption and decentralized finance: The rise of decentralized business models. *Journal of Business Venturing Insights, 13*, Article e00151. https://doi.org/10.1016/j.jbvi.2019.e00151
Chohan, U. W. (2021). Decentralized finance (DeFi): An emergent alternative financial architecture (Critical Blockchain Research Initiative Working Paper). SSRN.
Chohan, U. W. (2024). The governance of decentralized autonomous organizations (DAOs) (Critical Blockchain Research Initiative Working Paper). SSRN.
Cong, L. W., & He, Z. (2019). Blockchain disruption and smart contracts. *The Review of Financial Studies, 32*(5), 1754–1797. https://doi.org/10.1093/rfs/hhz007
Deloitte. (2023). The next wave: Emerging digital life in South and Southeast Asia. Deloitte China.
Demirgüç-Kunt, A., & Klapper, L. (2018). Financial inclusion in developing countries: An overview (Policy Research Working Paper No. 8624). World Bank.
Demirgüç-Kunt, A., Klapper, L., Singer, D., & Ansar, S. (2018). The global findex database 2017: Measuring financial inclusion and the fintech revolution. *World Bank*. https://doi.org/10.1596/978-1-4648-1259-0
Demirgüç-Kunt, A., Klapper, L., Singer, D., & Ansar, S. (2022). The global findex database 2021: Financial inclusion, digital payments and resilience in the age of COVID-19. *World Bank*. https://doi.org/10.1596/978-1-4648-1897-4
European Commission. (2020). A new circular economy action plan: For a cleaner and more competitive Europe. European Commission.
Financial Sector Conduct Authority. (2022). Declaration of crypto assets as a financial product (FAIS Notice 90 of 2022). FSCA.
Financial Stability Board. (2022). Assessment of risks to financial stability from crypto-assets. FSB.
Global Digital Finance. (2020). Code of conduct for digital assets. GDF.
Hacker, P., & Thomale, C. (2018). Crypto-securities regulation: ICOs, token sales and cryptocurrencies under EU financial law. *European Company and Financial Law Review, 15*(4), 645–696. https://doi.org/10.1515/ecfr-2018-0021
Harvey, C. R., Ramachandran, A., & Santoro, J. (2021). DeFi and the future of finance (SSRN Scholarly Paper ID 3711777). *SSRN*. https://doi.org/10.2139/ssrn.3711777

Hornuf, L., Schmitt, M., & Schürenkrämer, M. (2021). The tokenization of assets: Promises and challenges. *Journal of Corporate Finance, 70*, Article 102089. https://doi.org/10.1016/j.jcorpfin.2021.102089

International Finance Corporation. (2019). MSME finance gap: Assessment of the shortfalls and opportunities in financing micro, small and medium enterprises in emerging markets. IFC.

International Monetary Fund. (2021). Thailand: Financial sector assessment program Technical note on fintech and digitalization. IMF.

International Monetary Fund. (2022). El Salvador: 2022 Article IV consultation. IMF.

International Monetary Fund. (2023). Nigeria: 2023 Article IV consultation. IMF.

International Organization of Securities Commissions. (2022). Decentralized finance report. IOSCO.

International Organization of Securities Commissions. (2023). Policy recommendations for crypto and digital asset markets. IOSCO.

Iqbal, N., Singh, D., & Kumar, S. (2022). Environmental sustainability and blockchain technology: A comprehensive review. *Environmental Science and Pollution Research, 29*(19), 27463–27501. https://doi.org/10.1007/s11356-022-18966-8

Khera, P., Ng, M. S., Ogawa, M. S., & Sahay, M. R. (2021). Is digital financial inclusion unlocking growth? IMF Working Paper No. 2021/167. International Monetary Fund. https://doi.org/10.5089/9781513583082.001

Lapointe, C., & Fishbane, L. (2019). The blockchain ethical design framework. *Innovations: Technology, Governance, Globalization, 12*(3–4), 50–71. https://doi.org/10.1162/inov_a_00275

Matsuo, S., & Ushida, H. (2020). Decentralized autonomous organizations: Governance structures and economic implications. *Frontiers in Blockchain, 3*, 25. https://doi.org/10.3389/fbloc.2020.00025

Monetary Authority of Singapore. (2019). Payment services act 2019. Singapore Statutes Online.

Morgan, P. J. (2022). Fintech and financial inclusion in Southeast Asia and India. *Asian Economic Policy Review, 17*(2), 183–208. https://doi.org/10.1111/aepr.12379

Morrar, R., & Al-Ansi, A. (2020). Community currencies and local economic development: A systematic review. *Local Economy, 35*(4), 332–351. https://doi.org/10.1177/0269094220937659

Narayanan, A., & Clark, J. (2017). Bitcoin's academic pedigree. *Communications of the ACM, 60*(12), 36–45. https://doi.org/10.1145/3132259

Narayanan, A., Bonneau, J., Felten, E., Miller, A., & Goldfeder, S. (2016). Bitcoin and cryptocurrency technologies: A comprehensive introduction. Princeton University Press.

Organisation for Economic Co-operation and Development. (2020). The tokenisation of assets and potential implications for financial markets. *OECD Publishing*. https://doi.org/10.1787/d57c4c2f-en

Osabutey, E. L. C., & Jackson, T. (2024). Mobile money and financial inclusion in Africa: Emerging themes, challenges and policy implications. *Technological Forecasting and Social Change, 202*, Article 123339. https://doi.org/10.1016/j.techfore.2024.123339

Ozili, P. K. (2022). Financial inclusion research around the world: A review. *Forum for Social Economics, 51*(4), 429–457. https://doi.org/10.1080/07360932.2020.1715238

Ozili, P. K. (2023). *Financial inclusion and fintech in Africa.* Routledge.

Philippon, T. (2019). The great reversal: How America gave up on free markets. Harvard University Press.

PwC. (2023). Global crypto regulation report 2023. PwC.

Remus, U., & Schuppli, C. (2021). Qualitative content analysis in information systems research: Guidelines and illustrations. *Communications of the Association for Information Systems, 49*, 171–196. https://doi.org/10.17705/1CAIS.04909

Schueffel, P. (2016). Taming the beast: A scientific definition of fintech. *Journal of Innovation Management, 4*(4), 32–54. https://doi.org/10.24840/2183-0606_004.004_0004

Schär, F. (2021). Decentralized finance: On blockchain- and smart contract-based financial markets. Federal Reserve Bank of St. *Louis Review, 103*(2), 153–174. https://doi.org/10.20955/r.103.153-74

Securities and Exchange Commission Nigeria. (2022). New rules on issuance, offering platforms and custody of digital assets. SEC Nigeria.
Tinn, K., Reay, J., & Van Wegberg, P. (2020). Blockchain governance: What we can learn from the economics of corporate governance. *Journal of British Blockchain Association, 3*(1), 1–8. https://doi.org/10.31585/jbba-3-1-(2)2020
U.S. Department of the Treasury. (2023). 2023 DeFi illicit finance risk assessment. U.S. Department of the Treasury.
United Nations Development Programme. (2019). Digital transformation and governance. UNDP.
United Nations. (2018). World economic situation and prospects 2018. United Nations.
United Nations. (2020). Sustainable development goals report 2020. United Nations.
Van Wegberg, R., Oerlemans, P., & Van Deventer, J. (2018). Bitcoin money laundering: Mixed results? An explorative study on money laundering of cybercrime proceeds using bitcoin. *Journal of Financial Crime, 25*(2), 419–435. https://doi.org/10.1108/JFC-11-2016-0067
Werner, M., & Schär, F. (2022). Principles of decentralized finance: A structured literature review. *Journal of Financial Market Infrastructures, 10*(3–4), 47–72.
Wharton FinTech. (2021). *Policy-maker toolkit: Building a robust ecosystem for fintech innovation.* Wharton School of Business.
World Bank Group. (2017). MSME finance gap: Assessment of the shortfalls and opportunities in financing micro, small and medium enterprises in emerging markets. World Bank Group.
World Bank. (2021). World development report 2021: Data for better lives. *World Bank.* https://doi.org/10.1596/978-1-4648-1600-0
World Bank. (2022). Kenya economic update: Navigating the pandemic. World Bank.
World Economic Forum. (2021). Decentralized finance (DeFi) policy-maker toolkit. WEF.
World Economic Forum. (2022). Why decentralized finance is good news for the unbanked. World Economic Forum.
Zetzsche, D. A., Arner, D. W., & Buckley, R. P. (2020). Decentralized finance. Journal of Financial. *Regulation, 6*(2), 172–203. https://doi.org/10.1093/jfr/fjaa010

Chapter 19
Time-Varying Interactions Between Artificial Intelligence and Sustainable Finance in the Post-pandemic Era

Mohit Verma, Mukhtaruddin Mukhtaruddin, and Bikram Paul Singh Lehri

Abstract Given technological innovation's significant influence on investment, it becomes important to investigate the effect of artificial intelligence (AI) indices within the broader framework of sustainable finance. As the global investment environment increasingly becomes sustainability-oriented, it is important to understand the interaction between technology-driven assets and environmentally oriented financial instruments. This chapter is a conceptual and narrative review, focusing on the interactions between AI-related equity indices and key elements of sustainable finance, including clean energy markets, green bonds, and environmental pricing indices. It takes the form of a time-varying lens that assesses how the COVID-19 pandemic has reset priorities in sustainable investing. The chapters reveal major themes, theoretical understanding and the prospects. This chapter offers a framework for studying the connection between AI and sustainable finance, and for understanding technology-enabled, environmentally responsible financial ecosystems. The chapter offers valuable evidence to inform policymaking, strategic investment planning, and institutional risk management of sustainable development.

Keywords Artificial intelligence · Sustainable finance · Technology adoption · TVP VAR · Indian stock markets · Emerging markets

M. Verma (✉)
JGLS, O.P. Jindal Global University, Sonipat, Haryana, India
e-mail: phdmohitv@gmail.com

M. Mukhtaruddin
Faculty of Economics, Universitas Sriwijaya, Palembang, Indonesia
e-mail: mukhtaruddin67@unsri.ac.id

B. P. S. Lehri
University School of Business, Chandigarh University, Mohali, India
e-mail: lehribikram@gmail.com

K. Kumar et al. (eds.), *AI-Driven Circular Economy for Carbon Neutrality in Emerging Economies*, Environmental Footprints and Eco-design of Products and Processes,
https://doi.org/10.1007/978-3-032-30298-4_19

1 Introduction

The upsurge in the development of artificial intelligence (AI) methods has transformed the financial business sector concerning the distribution of money, the prevention of financial risks, and the development of investment strategies (Eliwa et al., 2021; Ghosh et al., 2021; Krueger et al., 2020; Ozili, 2023; Zhang et al., 2020). Simultaneously, the need to transition to sustainability due to climate change, societal disparities, and the issue of responsible management has made sustainable finance the model of reference for long-term economic sustainability. Since these two terms overlap, the input peculiarities of AI-related financial products, in particular, AI equity indices, i.e., the sustainability-related asset (clean energy stocks, green bonds, and environmental pricing mechanisms) should be researched (Liu, 2023; Liu et al., 2023). The listing companies in the form of AI indices include those that are specialized in either the development or deployment of AI technologies (Su et al., 2021).

These indices are not merely market measures, but the perception of the investor in the transformations in technology and the shifting values of the waves of change. However, AI intersects with sustainable finance, raising significant questions: Could AI help develop more sustainable investments? What does it imply to AI portfolios and their effect on ESG performance? and Is AI increasing or decreasing systemic financial and environmental risks? These questions have been increased by the coronavirus pandemic. It has, on the one hand, shaken the global economies, revealed frailty in the supply chain and tightened social rifts, which are central to sustainable and responsible investing strategies (Tripathi et al., 2022). Nonetheless, the crisis simultaneously accelerated the use of AI across the most important industries of finance, health care, and logistics (Baker et al., 2022). Sustainability and technological innovation became the highlight of the post-pandemic recovery when investors had already changed their strategies in such an uncertain era.

The chapter is trying to explore the evolving correlation between AI indices and sustainable finance. The relationships are not fixed, and they are dynamically responsive to technological changes, legislation, and macroeconomic shocks (Saci et al., 2024; Xu et al., 2022). Our approach is a narrative review to look at (a) what was pre-pandemic experimenting and the consequences of these relationships, and (b) how these relationships changed during three defining periods: pre-pandemic experimenting, pandemic-induced reconfiguring, and post-pandemic calibrating. The three significant themes presented in this literature, identified in the chapter, are: Integration of AI and ESG—The chapter considers the application of AI solutions to enhance the quality, transparency, and monitoring of ESG data performance (Agrawal, 2025; Joselin et al., 2025; Li et al., 2021; Singh et al., 2024). Technology-Driven Green Finance, which investigates the possibility of using AI to price assets cost-effectively, optimize a portfolio and assess the impact of green financial products and services in the green bond sector and renewable energy (Zhang, 2024; Zhang et al., 2020). And last, Risk, Regulation, Ethics—Responses to the challenges of using AI in sustainable finance, such as algorithmic bias, non-transparent decision-making, and the significant environmental impact of energy-intensive, high-demand computing models (Kaith & Sachdeva, 2024; Tripathy et al., 2025; Verma & Mukhtaruddin, 2023).

The union of AI and sustainable finance is both supportive and knotty. The application of AI technologies must align with the principles of fairness, transparency, and environmental responsibility, while AI can make a sustainable investment strategy more precise, flexible, and responsive (Eliwa et al., 2021; Kumar et al., 2025; Prakash & Tiwari, 2025; Singh et al., 2025a, 2025b; Zumente & Bistrova, 2021). To sum up, the chapter conceptualized the relationship between AI indexes and Sustainable Finance dimensions in a dynamic, transforming environment. It offers policymakers, investors, and scholars a framework for the next frontier of technology-powered, sustainability-oriented finance, grounded in timely developments and theoretical perspectives.

2 Theoretical Frameworks and Key Concepts

Grasping the area where AI and Sustainable Finance intersect, and especially post-pandemic, requires some grounding in a few foundational elements. So, the theoretical and definitional foundations to which the discussion adheres are introduced in this section.

2.1 Artificial Intelligence in Finance

Artificial intelligence is the ability of machines or systems to perform tasks that usually require human intelligence, for example, learning from data, reasoning, problem solving, and language processing (Russell & Norvig, 2021). In the domain of financial research, AI has gradually advanced from rule-based approaches to advanced learning algorithms, which are able to handle large-scale, high-dimensional and sometimes unstructured data. This technological change has profound ramifications in the context of decision-making under uncertainty, real-time analytics, and dynamic optimization. There are several branches of AI to consider for finance:

Machine Learning (ML): Makes predictive analytics possible by recognizing patterns in historical and real-time data. Credit scoring, fraud detection, and portfolio optimization are some tasks that ML is used (Krauss et al., 2017). In sustainable finance, ML can evaluate ESG (Environmental, Social, Governance) risk factors, detect greenwashing, and predict sustainability trends.

Natural Language Processing (NLP): Enables systems to comprehend and interpret human language. NLP is extensively used in parsing company reports, news feeds, social media, and sustainability disclosure to extract ESG signals (Li et al., 2021).

Deep Learning and Neural Networks: Such designs handle identifying complex non-linear relationships and are applied to studying climate risk, carbon footprints, and environmental monitoring from satellite imagery and sensor information (Rolnick et al., 2019).

Given AI's capacity to handle the "three Vs" in big data—volume, velocity, and variety. The use of AI in financial markets has accelerated in recent years. Since sustainable finance depends on dynamic and varied information sources, AI becomes a strategic catalyst to interpret data-led sustainability perspectives.

2.2 Sustainable Finance

Sustainable finance is a practice by which Environmental, Social, and Governance (ESG) factors are embedded into decision-making processes and capital allocation (OECD, 2020). Unlike traditional finance, which focuses on risk-return trade-offs, sustainable finance introduces a third dimension, i.e., impact. This is consistent with the objective for long-term value creation, systemic risk reduction, and socio-environmental responsibility. The environmental interests discuss climate change mitigation, the control of pollution, the conservation of biodiversity, and the sustainable management of natural resources. Social comprises human rights, labor practices, community relations, and product responsibility. Governance includes ethical business practice, board diversity, anti-corruption project, and shareholder right (Ghosh et al., 2021). Sustainable finance is not one-dimensional. It spans multiple strategies, including: ESG Integration, Impact Investing, Thematic Investing, and Socially Responsible Investing (SRI). The development of sustainable finance has been triggered by complacency of a range of actors, among others, growing stakeholders' demand, policy (e.g., EU Sustainable Finance Taxonomy), and recognition of ESG issues as financially material (Krueger et al., 2020; Singh et al., 2025b). As part of fiduciary duty and long-term risk management, institutional investors today are increasingly considering ESG risks.

2.3 Time-Varying Relationship

AI and sustainable finance themselves are naturally dynamic. Technologies change, legal frameworks move, social norms are different, and external shocks that create something like pandemics or climate crises happen and reshape that financial landscape anyway. This dynamism calls for a time-dependent analysis, which reflects changes in priorities, instruments, and outcomes over time. Monitor and report AI progress and the implementation of sustainable finance (Lo, 2017; Singh & Mittal, 2019). For example, applications submitted prior to 2020 were primarily on ESG data scraping and categorization, while those post-pandemic are on impact analysis and system robustness. Determine shifts in sustainability paradigms—from climate-centric theories to comprehensive, justice-based theories, which include social equity and participatory governance.

Assess policy and market responses to disasters. The COVID-19 crisis, for instance, magnified digital adoption and surfaced systemic weaknesses, thus amplifying attention to sustainability (Naidoo, 2020). To the extent that the designers view the new financial system as capable of adjusting or being managed, such an approach fits into the framework of adaptive finance, whereby systems must have the ability to adapt to both exogenous pressures and endogenous learning (Lo, 2017). A further reason is that the model links with dynamic capabilities theory (Teece, 2018) that highlights the role of organizational agility and learning in harnessing AI for sustainable change. Theoretically, this time-varying analysis gets conceptualized within the framework of complex theory, or more precisely, complex adaptive systems theory, of which AI is both a means of managing complexity and a cause of shifting system behavior. The sustainability decisions and the reversible information give each other a feedback mechanism to build a dynamic ecosystem of automated finance.

3 Interactions, Evolution, and Post-Pandemic Narratives

The integration of AI and SF is slowly changing the financial world landscape of the world. Thus, although the interaction of AI with sustainable finance is evidently catalytic, it is also reciprocal the AI, by allowing real-time and scalable realization of high information value through large volumes of data, and through forms of novel analytical power and visualization, sustainable finance goals and principles, in their turn, influence the ethical parameters, the extent, and application of AI systems. The COVID-19 pandemic has also sped up this shift, as it has boiled funders' interest in resiliency, substance, and accountability when it comes to financial processes, and tech solutions (Tripathy et al., 2025; Verma & Mukhtaruddin, 2023). We narrate an endogenous story that connects in Chapter Time-Varying AI and Sustainable Flood: The Central Point 5 an endogenous way how AI improves sustainable finance, how sustainability principles similarly affect AI development, how these connections changed during and also post-pandemic, and how particular sectors update these novelties.

3.1 AI as a Catalyst for Sustainable Finance

AI in driving ESG into investment. Artificial intelligence is a game-changer for pushing ESG into investment, in reducing friction in admitting that it is one. The key contribution of AI technology to sustainable finance is the improvement of the quality, availability, and timeliness of ESG data, which in the past have been confronted with quality inconsistency, heterogeneity, and subjectivity (Dorfleitner et al., 2017).

AI technologies, in particular machine learning (ML) and natural language processing (NLP), can now read through and interpret huge volumes of unstructured data emanating from corporate sustainability reports, news articles, regulatory disclosures, and even social media. For instance, NLP models can infer sentiment and ESG-related controversies hence providing real-time ESG scoring with dynamic investment strategies (Arjaliès & Mundy, 2013). For example, products such as Truvalue Labs use AI to digest ESG signals culled from more than 100,000 sources in nearly real time to create a forward-looking ESG lens (Truvalue Labs, 2020).

Besides, AI can enhance active surveillance and adaptive intervention. AI models will forecast the risks of social responsibility in case they are trained with past and present data of the performance that is optimized by ESG, i.e., risk of liability to pay fines to the authorities and weather conditions. It allows the portfolio managers and the risk analysts to be proactive and model the possible scenarios (Ghosh et al., 2021). Regarding green investments, AI is the basis of innovation and creation of new climate-conscious indices, carbon-footprint-based asset screening, and customized advisory solutions that align things with their preferences in terms of sustainability outcomes (Baker et al., 2022).

Impact measurement, an important but inaccessible element of sustainable financial practice, is also improved by AI. Unlike traditional accounting-based measurement, AI could also measure non-financial performance, such as biodiversity impact, compliance with labor standards, or supply chain emissions, using geospatial information and remote sensing information systems (Friede et al., 2015).

3.2 AI for Sustainability

Even though AI can be deployed to meet ESGs, the agenda setting of sustainability is also a crucial aspect that defines the creation and application of AI. It is this symbiotic relationship that is vital to avoid setting the same sustainability goal that AI tools are aimed at contributing to, to be achieved. Banks are emerging to understand that machine-generated decision-making has the potential to be either history-repeating or the development of new, invidious categories (Barocas et al., 2019). In credit scoring, such as with AI models, they can reinforce the socioeconomic bias in the record they are later provided with, unless equitable limitations are imposed. This shows why the concept of ethical design is critical in the development of AI to guarantee that AI technologies will remain sustainable both in terms of governance and socially. Along the same lines, explainability is of increasing significance. As AI tools become more and more part of financial systems, stakeholders, including investors and regulators, want to be transparent on how ESG scores are computed or investment decisions made by AI models. Some of the tools that are on the increase in ESG, in a bid to interpret predictive analytics of ML, include SHAP (Shapley Additive Explanations) and LIME (Local Interpretable Model-agnostic Explanations) (Doshi-Velez & Kim, 2017).

On the environmental dimension, the rise of Green AI is an appeal to an increasing realization of the carbon price of training giant AI models at scale. It was proved that the emissions of a large single transformer model are greater than the emissions of five cars throughout their lifetime (Strubell et al., 2019). The developers are thereby seeking energy-efficient models and cloud providers that are supposed to utilize renewable energy. Moreover, institutions are evaluating the AI implementation based on a sustainability lens to determine the benefit of the AI system to society as a means of evaluating its environmental cost.

3.3 Post-Pandemic Evolution

The AI-sustainability intersection has not been the same. Its development can be considered in three phases, namely pre-pandemic experimentation, acceleration of the pandemic, and post-pandemic incorporation. Prior to the pandemic, AI in sustainable finance was to a great extent immature and poorly organized. The first users of AI used it to sift alternatives sources of data, such as social media or satellite imagery, to obtain ESG information. However, the ESG metrics were never standardized and it was not clear what regulation would demand of them and the amount of cross-disciplinary groups that would do so in order to close gaps. It was the pandemic, which acted as a significant disrupter and catalyst, COVID-19. Conversely, it also exposed the weaknesses in the system, and the end of the need to have resiliency of financial systems. On the positive side, it gave a push to the digitization process, and financial institutions started to use AI-based tools to provide remote work, conduct the onboarding of clients digitally, and monitor risks in real time. Most importantly, the flight to ESG-related products was higher during the crisis, and this kind of asset performed better than non-ESG types (Rolnick et al., 2019). AI played a significant role in steering investors through the tumbling markets, notifying them of the ESG scandals as they happened and optimizing the portfolios.

AI and Sustainability: Rethinking post-pandemic intersections AI and sustainability are backsliding on a structural reformation post-pandemic. It is a departure of pilot use cases to enterprise-wide implementation. In the case of the ESG research and reporting as well as risk management platforms, AI is there on both sides, but a hybrid indeed. Today, however, most of the biggest asset managers in the world are also trying to use AI to manage billions of dollars of ESG funds, combining both structured financial data and unstructured data to analyze sustainability disclosures (Kumar & Prakash, 2019; Kumar et al., 2021). Moreover, rules on ESG reporting and responsible AI are being actively developed in Europe, Asia, and North America regulators, which strengthens this two-fold change (Eliwa et al., 2021; OECD, 2020).

3.4 Finance Use Cases

The implications of AI in sustainable finance vary within financial industries, which possess varied priorities, data setting and regulatory setting. AI technology is also being utilized by banks to estimate its exposure to ESG risk in loans. This involves a combination of normal credit measures with non-financial measures, e.g., water consumption, labor policies, and carbon. The AI models have the ability to sift through the disclosures of borrowers and triggering red flags on the ESG abuses and the probability of defaulting of the scheme due to climate or reputational risks (Baker et al., 2022). A sustainability-linked loan (SLL) is a novel product in which the paid margin is tied to the borrower's ESG performance. In real time, AI systems can monitor a borrower's sustainability KPI performance and, based on the KPIs, adjust prices or issue compliance notices.

Asset and Wealth Management: AI in asset management is changing the process of ESG investing by using smart indexing, thematic investing, and automatic rebalancing. An example is that AI models make a selective choice of portfolios depending on the events of shareholder activism, the news sentiment, and corporate sustainability. In partnership with advisors and fintechs that incorporate ESG algorithms, geolocating and lifestyle-matching sustainable funds or other ESG assets so that the mass affluents can get access to sustainable investment options according to the preferences (Khan et al., 2024; Kulmala et al., 2023).

ESG Impact Measurement: AI-powered analytics solutions are addressing the ESG impact measurement which is a perpetual headache to asset managers. These tools consume various feeds of data, including environmental impact analysis, to stakeholder analysis, and offer multi-dimensional perspectives of the implication of investment decisions.

Insurance: The insurance industry is more commonly using AI for modeling climate risk, as severe weather events become more frequent. Pairing AI with geospatial and climate projection data allows insurers to more effectively assess property and catastrophe risks, price coverage accordingly and develop new products (e.g., climate-resilient microinsurance) (OECD, 2020). Insurers also consider the ESG risk of corporate clients, applying AI to analyze whether underwriting a project or a company is consistent with the insurer's sustainability goals. Some have embraced "green underwriting" models that reward clients with sustainable business practices. Table 1 summarizes the application of AI in sustainable finance.

4 Challenges and Ethical Dilemmas

If the incorporation of artificial intelligence (AI) into sustainable finance has the potential to be transformative, it also raises significant challenges and ethical considerations that need to be explored. The issues pertain to biases in both data

Table 1 Applications of AI in sustainable finance

AI technique	Function in finance	Sustainable finance applications	Example tools/Use cases
Machine learning (ML)	Pattern recognition, prediction modeling	ESG risk assessment, green portfolio optimization	ESG score prediction models, Truvalue Labs, MSCI ESG tools
Natural language processing (NLP)	Text extraction and sentiment analysis	Screening sustainability reports, social media for greenwashing detection	Parsing news, company filings; sentiment analysis on ESG
Deep learning	Nonlinear, large data pattern extraction	Climate risk modeling, environmental impact estimation	Satellite imagery for deforestation tracking, biodiversity AI
Explainable AI (XAI)	Model transparency and interpretability	Explainable ESG ratings, compliance with ethical AI in finance	SHAP, LIME for ESG model interpretability
Green AI	Energy-efficient AI computing	Minimizing carbon footprint in AI systems	Carbon-aware model training, sustainable data centers

Source Author's own creation

and algorithms, a lack of standardized ESG frameworks, and changing regulatory dynamics.

4.1 Data and Algorithmic Bias

AI is only as good as the data it is trained on. And it's a particularly big issue in sustainable finance, where ESG data is at a disadvantage due to the inconsistencies and lacunae present in it. The majority of ESG datasets are self-reported, unaudited, and differ by sector and by region, introducing machine learning challenges of model training as well as prediction accuracy. Furthermore, the records of historical finance are often tainted with entrenched socioeconomic prejudices. And when AI is trained on such data, it is at risk of encoding those biases, hindering social sustainability goals (Doshi-Velez & Kim, 2017). For example, if historical data is an expression of discriminatory credit issuing practices, credit scoring models will also tend to discriminate against minority groups.

4.2 Lack of Standardization and Interoperability

Standardization and Interoperability in Experience Data. As a part of today's W-LBPM, the standardization of W-LBPM in the form of OpenMobile Interactive

Messaging Interface and WMMS lacks standardization for experience data in W-LBPMs and other Experience-oriented Systems. Another important obstacle is the lack of widely-accepted standards for ESG reporting and AI ethics in finance. At present ESG taxonomies differ significantly and materiality assessments vary by region and can lead to inconsistency in assessing sustainability performance (Eliwa et al., 2021). Pre-production ESG Scoring: In IP, there's no standardization, and differing AI models created for ESG scoring are usually "black boxes" with no transparency on their functions. The absence of this transparency erodes stakeholder trust and hinders regulatory compliance (Doshi-Velez & Kim, 2017; Ghosh et al., 2021; Krueger et al., 2020). This issue is most acute in the field of automated investment determination, in which fiduciary duty requires an apparent demonstration of integrating sustainability into portfolio management.

4.3 Regulatory and Governance Stipulations

In both AI and sustainable finance, much regulatory ground remains to be covered. Even though efforts such as the EU Sustainable Finance Disclosure Regulation (SFDR) and the AI Act can help steer us (Zumente & Bistrova, 2021), numerous jurisdictions still lack broader governance of AI (Zumente & Bistrova, 2021). Ethical issues such as greenwashing, AI responsibility, and data privacy are also poorly integrated throughout most of the guidelines. Such an uncontrolled situation results in the investment not advancing the ethical development of AI with a perspective on sustainable finance, and risks that financial institutions will descend into illegality and infamy. Concisely, unless a keen watch is maintained on the area of data governance, algorithmic fairness, and regulatory coordination, using AI in sustainable finance may not deliver on its commitment and on the contrary, may even end up operating counter-cyclically to the very ESG values it is supposed to advance.

5 Implications and Future Research Directions

As AI is more closely interlinked with the sustainable finance landscape, many research and policy gaps appear requiring systematic attention. Understanding these gaps is important not only for advancing academic research, but also to inform regulation and institutional practices that facilitate ethical, effective, and futureproof AI use in sustainable finance.

5.1 Interdisciplinary and Systems-Based Models

Among the most urgent of these is the creation of interdisciplinary models that bring together environmental, finance, data science, and ethical reasoning. Many contemporary AI-based financial models are also divorced from the context within which sustainability goals are housed, viewing ESG metrics as fixed inputs rather than as dynamic, co-dependent variables within an environment (Escrig-Olmedo et al., 2019; Khan et al., 2024). Future research needs to take systems thinking as a basis to model complex feedback loops, such as the interrelation between environmental degradation and credit risk, or between social inequality and returns on investment, upon which to further develop a more coherent AI design (Kulmala et al., 2023).

5.2 AI for Sustainable Policy and Governance

AI, in turn, may enable innovative policy approaches and simulations in real-time, such as for instance with respect to carbon pricing schemes, climate adaptation measures, food waste, or allocation decisions on green subsidies (Krauss et al., 2017; Srivastava et al., 2018, 2021). Policy officials could use these applications to test scenarios and evaluate the efficacy of policy before rolling it out. But such uses need to be accompanied by transparency in algorithm design and stakeholder involvement to prevent AI from reinforcing current power imbalances in climate governance (Barocas et al., 2019).

5.3 Standardization and Regulatory Harmonization

The world needs a form of harmonization between the global regulations in order to spur standardization in ESG reporting, AI accountability, and green taxonomy. Irrational frameworks of ESG disclosure and disparate AI ethics principles may create disorientation among investors and regulators (Rolnick et al., 2019). Research may serve to encourage ordinary harmonization (e.g., by linking the EU AI Act to the EU Taxonomy on Sustainable Activities) but can also be used to promote harmonization across jurisdictions.

5.4 Capacity Building and Education

Lastly, raising the awareness of financial professionals on AI and ESG will be of paramount importance. Educational opportunities in AI and sustainability today may be offered in distinct streams (at best), but the rise of their intersection is being overlooked. The change in the curricula will be needed to support cross-disciplinary competences, such as ethical AI, sustainable finance indicators, and data governance (Srivastava et al., 2024; Zhang et al., 2020). The institutional investors as well as the asset managers must have the technical fluency and the sustainability mindset to work on this terrain. The research and policy imperatives discussed in this paper will give stakeholders the instrument to bend this curve, which will result in a future where AI is no longer a piece of technology, but a component of the building blocks of sustainable, inclusive financial systems.

6 Conclusion

The cross-point between AI and sustainable finance may be viewed as one of the most promising fields of our post-pandemic world, as well as the most threatening ones. Not only does AI enhance the performance of ESG investing by providing more efficient data processing and impact metrics, but it also compels the financial system to redefine what value creation means. And in a world where environmental risks, social justice and the examination of governance are more important than ever, AI offers a platform to smarter, more adaptable and more inclusive fiscal structures. However, this promise cannot be fulfilled unless we make active actions to address the ethical, regulatory, and societal issues in order to ensure that AI actually enhances, and does not harm, the sustainability objectives it is expected to hold.

References

Agrawal, D. (2025). A comprehensive review of investing and financing for sustainable tourism projects. *Environment, Sustainability, and Governance Insights, 1*(1), 53–70.

Arjaliès, D. L., & Mundy, J. (2013). The use of management control systems to manage CSR strategy: A levers of control perspective. *Management Accounting Research, 24*(4), 284–300. https://doi.org/10.1016/j.mar.2013.06.003

Baker, M., Bergstresser, D., Serafeim, G., & Wurgler, J. (2022). Financing the response to climate change: The pricing and ownership of U.S. green bonds. *Brookings Papers on Economic Activity*, Spring, 91–141.

Barocas, S., Hardt, M., & Narayanan, A. (2019). *Fairness and machine learning*. https://fairmlbook.org/

Dorfleitner, G., Utz, S., & Wimmer, M. (2017). Sustainable, responsible and impact investing and investment funds: A literature review. *Journal of Business Ethics, 138*(2), 325–343. https://doi.org/10.1007/s10551-015-2610-6

Doshi-Velez, F., & Kim, B. (2017). Towards a rigorous science of interpretable machine learning. *arXiv*. https://arxiv.org/abs/1702.08608

Eliwa, Y., Aboud, A., & Saleh, A. (2021). ESG practices and the cost of debt: Evidence from EU countries. *Critical Perspectives on Accounting, 79*, Article 102097. https://doi.org/10.1016/j.cpa.2019.102097

Escrig-Olmedo, E., Izquierdo, M., Ferrero-Ferrero, I., Rivera-Lirio, J., & Torres, M. (2019). Rating the raters: Evaluating how ESG rating agencies integrate sustainability principles. *Sustainability, 11*(3), 915. https://doi.org/10.3390/su11030915

Friede, G., Busch, T., & Bassen, A. (2015). ESG and financial performance: Aggregated evidence from more than 2000 empirical studies. *Journal of Sustainable Finance & Investment, 5*(4), 210–233. https://doi.org/10.1080/20430795.2015.1118917

Ghosh, A., Ghosh, S., & Mitra, S. (2021). Artificial intelligence in risk management: Applications in ESG and climate finance. *Journal of Risk and Financial Management, 14*(8), 364. https://doi.org/10.3390/jrfm14080364

Joselin, D., Singh, A., & Tiwary, K. S. (2025). The nexus between CSR practices and financial performance: Empirical evidence from the manufacturing sector in India. *Environment, Sustainability, and Governance Insights, 1*(1), 1–17.

Kaith, A., & Sachdeva, G. (2024). Bibliometric insights into green human resource management: A connection with business ethics and values. *PURUSHARTHA-A Journal of Management, Ethics and Spirituality, 17*(1), 51–63. https://doi.org/10.21844/16202117104

Khan, M., Huo, C., Zahid, R., & Maqsood, U. (2024). The automated sustainability auditor: Does artificial intelligence curtail greenwashing behavior in Chinese firms? *Business Strategy and the Environment, 33*(8), 9015–9039. https://doi.org/10.1002/bse.3961

Krauss, C., Do, X. A., & Huck, N. (2017). Deep neural networks, gradient-boosted trees, random forests: Statistical arbitrage on the S&P 500. *European Journal of Operational Research, 259*(2), 689–702. https://doi.org/10.1016/j.ejor.2016.10.031

Krueger, P., Sautner, Z., & Starks, L. T. (2020). The importance of climate risks for institutional investors. *Review of Financial Studies, 33*(3), 1067–1111. https://doi.org/10.1093/rfs/hhz137

Kulmala, M., Lintunen, A., Lappalainen, H., Virtanen, A., Yan, C., Ezhova, E., et al. (2023). Opinion: The strength of long-term comprehensive observations to meet multiple grand challenges. *Atmospheric Chemistry and Physics, 23*(23), 14949–14971. https://doi.org/10.5194/acp-23-14949-2023

Kumar, K., & Prakash, A. (2019). Developing a framework for assessing sustainable banking performance of the Indian banking sector. *Social Responsibility Journal, 15*(5), 689–709.

Kumar, K., Prakash, A., & Singh, K. (2021). How National Education Policy 2020 can transform future generations in India. *Journal of Public Affairs, 21*(3), Article e2500.

Kumar, K., Singh, R., Choudhary, M., & Sharma, A. (2025). Climate change mitigation strategies in environmentally sensitive industries. *In Climate Change and Social Responsibility* (Vol. 25, pp. 185–204). Emerald Publishing Limited.

Li, F., Wang, H., & Zhang, Q. (2021). Textual analysis of ESG reports using NLP. *Journal of Sustainable Finance & Investment, 11*(2), 123–140.

Liu, L. (2023). Research on the relationship between ESG and firm value. *Advances in Economics, Management and Political Sciences, 65*(1), 245–254. https://doi.org/10.54254/2754-1169/65/20231651

Liu, Z., Zhang, X., Wang, J., Shen, L., & Tang, E. (2023). Coupling coordination between digital economy and green finance. *PLoS ONE, 18*(10), Article e0291936. https://doi.org/10.1371/journal.pone.0291936

Lo, A. W. (2017). *Adaptive markets: Financial evolution at the speed of thought.* Princeton University Press.

Naidoo, R. (2020). The COVID-19 pandemic and environmental sustainability. *Nature Sustainability, 3*(6), 409–410. https://doi.org/10.1038/s41893-020-0560-6

OECD. (2020). OECD business and finance outlook 2020: Sustainable and resilient finance. *OECD Publishing.* https://doi.org/10.1787/eb61fd29-en

Ozili, P. (2023). Bank loan loss provisioning for sustainable development. *Journal of Sustainable Finance & Investment*, 1–13. https://doi.org/10.1080/20430795.2022.2163847

Prakash, A., & Tiwari, A. K. (2025). The role of institutional factors in achieving SDG 16: A thematic review of drivers of corporate anti-corruption action. *Environment, Sustainability, and Governance Insights, 1*(1), 34–52.

Rolnick, D., Donti, P. L., & Kaack, L. H., et al. (2019). Tackling climate change with machine learning. *arXiv*. https://arxiv.org/abs/1906.05433

Russell, S., & Norvig, P. (2021). *Artificial intelligence: A modern approach* (4th ed.). Pearson.

Saci, F., Jasimuddin, S., & Zhang, Z. (2024). ESG performance and systemic risk sensitivity. *Management of Environmental Quality, 35*(6), 1274–1294. https://doi.org/10.1108/MEQ-02-2023-0060

Singh, R., Kumar, K., & Khan, S. (2024). A comprehensive view of artificial intelligence (AI)–based technologies for sustainable development goals (SDGs). *Artificial Intelligence Enabled Management: An Emerging Economy Perspective*, 183, 9783111172408-012.

Singh, R., Sharma, A., & Kumar, K. (2025a). Navigating carbon neutrality: Strategies for sustainable future and environmental challenges. In *Climate neutrality through smart eco-innovation and environmental sustainability* (pp. 69–80). Springer Nature Switzerland.

Singh, R., Tiwary, K. S., Kumar, K., & Kumar, A. (2025b). Empowering a sustainable future: Blockchain's role in economy, organizations, and finance towards smart circular economy. In *Blockchain technologies for smart circular economy and organisational sustainability* (pp. 271–283). Springer Nature Switzerland.

Singh, S., & Mittal, S. (2019). Drivers of CSR practices among family firms in India. *International Journal of Organizational Analysis, 27*(4), 947–971.

Srivastava, S. K., Anshul, A., Pathak, P., & Bansal, J. (2018). Responsible consumption for curbing food wastage: An exploratory enquiry. *PURUSHARTHA-A Journal of Management, Ethics and Spirituality, 11*(2), 104–116.

Srivastava, S. K., Singh, S., & Srivastava, M. (2021). Food waste at household and social gatherings: Drivers and possible remedies. *PURUSHARTHA-A Journal of Management, Ethics and Spirituality, 14*(2), 117–133.

Srivastava, S. K., Sinha, A. K., & Kumar, S. (2024). Workplace spirituality empowered by spiritual leadership to promote organizational effectiveness: A bibliometric analysis. *PURUSHARTHA-A Journal of Management, Ethics and Spirituality, 17*(1), 1–22. https://doi.org/10.21844/16202117101

Strubell, E., Ganesh, A., & McCallum, A. (2019). Energy and policy considerations for deep learning in NLP. *arXiv*. https://arxiv.org/abs/1906.02243

Su, X., Liu, X., Motlagh, N., et al. (2021). Intelligent and scalable air quality monitoring with 5G edge. *IEEE Internet Computing, 25*(2), 35–44. https://doi.org/10.1109/MIC.2021.3059189

Teece, D. J. (2018). Business models and dynamic capabilities. *Long Range Planning, 51*(1), 40–49. https://doi.org/10.1016/j.lrp.2017.06.007

Tripathi, V., Singh, K., Parihar, S., & Srivastava, S. K. (2022). A bibliometric analysis on supply chain collaboration. *PURUSHARTHA-A Journal of Management, Ethics and Spirituality, 15*(2), 66–103. https://doi.org/10.21844/16202115205

Tripathy, U., Verma, M., & Misra, A. K. (2025). *ESG performance and corporate metrics: Mapping research trends*. Advance online publication.

Truvalue Labs. (2020). *How AI and alternative data are transforming ESG investing*. https://www.truvaluelabs.com

Verma, M., & Mukhtaruddin, M. (2023). Environmental responsibilities and firm financial performance. *Indonesian Journal of Sustainability Accounting and Management, 7*(1), 27–40.

Xu, X., Chen, Y., Zhao, J., & Wang, J. (2022). Effect of ESG news on the Chinese stock market. *Journal of Global Information Management, 30*(1), 1–21. https://doi.org/10.4018/JGIM.313916

Zhang, Q. (2024). Recent developments on ESG and corporate value. *Highlights in Business, Economics and Management, 24*, 2214–2218. https://doi.org/10.54097/e8t1m912

Zhang, Z., Yang, T., & Liu, Y. (2020). SABlockFL: Blockchain-based smart agent system. *International Journal of Crowd Science, 4*(2), 133–147. https://doi.org/10.1108/IJCS-12-2019-0037

Zumente, I., & Bistrova, J. (2021). ESG importance for long-term shareholder value. *Journal of Open Innovation: Technology, Market, and Complexity, 7*(2), 127. https://doi.org/10.3390/joitmc7020127

Chapter 20
Generative AI a Boon or a Threat: Rising Impact of Deepfake on Circular Economy in Southeast Asia

Sonal Tikku and Ritesh Dwivedi

Abstract Circular economy is changing the global landscape. From the traditional concept of produce and discard rampantly used in linear economy, the world is moving towards circular economy with focus on reuse, regenerate and recycle. This fundamental change in core flow of resources has created a positive impact in global scenario. The proposal explores the rise and spreads of Deepfakes in Asian subcontinent. The chapter will first start with the history of generative AI and its impact on financial sector. Later on, the chapter will create an understanding of AI tool's positive impact on banking and financial sector and various other AI tools used in today's modern banking. From positive impact of AI, we will understand various cyber frauds coming hand in hand on increase of AI in today's financial world. Deepfake with its manipulative techniques has created mistrust in sustainability claims of various companies. With Greenwashing, Deepfake has eroded the trust and caused reputational damages. The chapter will also touch base upon the societal impact of Deepfake on the South Asian economy and how it is impacting trust of common people. The chapter also sheds light on how misinformation using Deepfake is impacting circular economy. In this chapter, we have mainly focussed on Southeast Asian countries such as China, Taiwan, Vietnam, India and Philippines. The chapter will also study the future path of Deepfake and how the same can be used in positive way in promoting circular economy.

Keywords Deepfakes · Southeast Asia · Generative AI · Artificial intelligence · Financial sector · Circular economy

S. Tikku
Amity School of Insurance, Banking and Actuarial Science, Amity University, Noida, India
e-mail: stikku@amity.edu

R. Dwivedi (✉)
Symbiosis Institute of Business Management, Noida, India
e-mail: ritesh.dwivedi@sibmnoida.siu.edu.in

K. Kumar et al. (eds.), *AI-Driven Circular Economy for Carbon Neutrality in Emerging Economies*, Environmental Footprints and Eco-design of Products and Processes,
https://doi.org/10.1007/978-3-032-30298-4_20

1 Introduction

Circular economy is changing the global landscape. In recent years there has been a significant shift worldwide from linear economy to circular economy, with the focus changing to recycle and regenerate. AI has spread its roots towards circular economy with its use of digital information system. Artificial intelligence is engulfing the financial world, and no part of subcontinent is untouched. AI has disrupted the world with the speed of processing high data volumes and performing tasks like humans. Even though technology has its advantages, lately it is being used for frauds, which has created fear in the minds of the people. Generative AI is based Generative AI being a subset of Artificial intelligence, is rampantly used for fraudulent activities. It uses AI for creation of content that can be in the forms of videos, images, texts, and audio in request to the user prompt. As per reports from Mc Kinsey, one-third organisations are using generative AI in one of their departmental functions. By 2026, more than 80% of organisations would have deployed generative AI in their workflows (Mc Kinsey).

Deepfake frauds are a combination of deep learning and fakes, and it involves hypersensitive videos which involve real persons. Frauds done by Deepfakes utilise neutral networks that analyse datasets to replicate the person's facial expression, behaviour, voice and speech patterns. The real videos and authentic voices are manipulated; hence, detection of Deepfake frauds is very difficult. Deepfake has spread its roots in Asia–Pacific region with 1530% increase from 2022 to 2023. Vietnam stands at the tops the Deepfake frauds at 25.35%, followed by Japan at 23.4%. Deepfake criminals are using biometric identification that helps criminals to steal the personal and financial information of the victims (Surasit, 2024). Deepfakes have increased significantly across the globe by a figure of 10X from 2022 to 2023. In North America, the increase in Deepfake is 1740%, 1530% in APAC, 780% in Europe, which includes the United Kingdom, 450% increase in Middle east and Africa, and 410% in Latin Across. Across the globe, the country which is highly affected by Deepfakes is Spain and the highest number of Deepfake frauds occur in Latin America (Sum sub, 2022).

2 Literature Review

Big Data and machine learning are revolutionising the banking sector. The use of these technologies has helped banking industry manifolds by increasing processing speed, efficiency and taking customer experience to the next level. From customer service, data processing, loan disbursement and trading to regulatory compliance, all the departments of banks are impacted by the AI. Chatbots, Virtual Assistants and Robotic automation processes have changed the face of financial industry by minimising cost, improving operational efficiency and delivering within timelines (Achary, 2021). Generative AI has led to creation of Deepfakes. Through the use of algorithms, Deepfakes produce fake audio visuals, which resemble the real-life person and give an impression of authentic videos. The level of similarity is so high that there is very little line of distinction between original and fake videos. Such deceptive videos when spread through social media and internet impact not only the individual but also the society as a whole. These Deepfake contents do not require any specialised skill for production as they are created using free and open-source software which even a newbie can create (Kosarkar & Sakarkar, 2024). In Southeast Asia, Deepfakes have created far bearing impacts. In this region, for the incidence of Deepfake misuse, the laws related to privacy and harassment fail to address the crime created by usage of Generative AI. This region having a culture where victims avoid complaining due to peer pressure and fear of reputational loss. The current laws fail to understand the beneficial and useful impact of Deepfakes manipulations. There is a need for having common legislation in this region for countering Deepfake problems (Adam & Chen, 2025). Deepfakes are highly used in social engineering scams. With manipulation of audio, video or both the scammers impersonate real-life person and illicit transfer of funds or sharing of confidential information. Financial institutions can overcome this problem by enhancing system security through multi-level authentication, especially for high-net-worth clients. The spread of malicious information using Deepfake technology is nibbling the public trust. Recently, there have been cases where this misinformation spread by Deepfake has deeply impacted investor confidence and led to destabilisation of the stock market (Monetary Authority of Singapore, 2025).

A study of Asian countries shows that previous exposure to Deepfake has influence on user's perception of news to be taken as correct or not. Those who are previously exposed to Deepfakes are more likely to consider the Deepfake news as correct instead of those users who did not have any Deepfake exposure before. Across the Asian countries, the factors influencing belief in Deepfakes are age, political interest, and traditional media use. Excessive use of social media for checking latest news or information also influences user's belief on Deepfake news. The policymakers can combat Deepfake through psychological strategies to reduce the effect of Deepfake in the minds of public; psychology influences Deepfake acceptance. In the growing world of technology, manipulating videos and audios using Deepfake technology creates distrust among voters for a particular political party. Trust of public institutions is completely affected by spread of misinformation by political rivals, hackers, criminals and foreign nations (Purwadi et al., 2022). Women have always been a vulnerable section of Society. Pornographic Deepfake, a Deepfake targeting females in explicit digital content without consent is rampant nowadays. There is a need for Governments and Regulators across the continents to view this offence on a serious note and take stringent action against the Deepfake developer and perpetuator (Miriam et al., 2025).

Terror organisations worldwide are using Deepfakes and artificial intelligence (AI) as new tools for improving their activities of terror and spreading propaganda. AI has helped in countering terrorism, but it has also helped terrorists in achieving their aim of misinformation spread in more innovative ways. The regulatory framework to combat this issue is still at a nascent stage as AI has positive as well as negative implications. Hence, regulators have to balance the act before penning down regulations so as to understand their implications on both the effects of AI. Southeast Asian region is very diverse due to its multiethnic groups, multilingual and multireligious population (Dymples, 2022; Wan Rosli, 2022).

3 Objectives

The main objectives of study are enlisted below:

- Firstly, the study aims to raise awareness of rise in the use of Deepfakes for crimes in Southeast Asia.
- Secondly, the study focusses on the need for the countries in Southeast Asia to come together to fight Deepfake issue together with common laws across the nations.
- Thirdly, the study also analyses the present status of countermeasures being taken against Deepfake crimes.
- Finally, the study aims to create better understanding of various implications due to the rise in Deepfakes in various sectors such as Healthcare, regulatory, legal, ethical, education, to name a few.

4 Methodology

In the captioned study, a systematic literature review was conducted. In this review, an appropriate database was selected based on keywords. Further in-depth review was done of the identified database. The papers in the database for this study were taken from the period 2015 to 2025. These papers were scrutinised to remove duplicate papers. Key words were used to identify relevant research papers related to the topic. The research papers considered in this study were taken from secondary sources such as Scopus indexed journals and Google Scholar. Also, data was collected from various articles which were available online on official websites of United Nations, KPMG and other reputed organisations. This study is qualitative research which is done to understand Deepfake in context of Southeast Asian countries.

5 Deepfake Technology and Mechanism

Deepfake terminology is an amalgamation of two words "Deep learning and Fake" that refers to the hyper-realistic videos created, which are similar to real-life people. Deepfakes are also known as synthetic media, as in this technique, the face of a real person is superimposed on some other person's face. Creation of Deepfakes involves neural networks that analyse vast data sets to reproduce a person's facial expressions, behaviour, voice, and speech patterns. Deepfakes engineers' real footage and sounding are not easily detected (Surasit, 2024). Generative adversarial network (GAN) and autoencoders (AE) are used by Deepfakes. GAN trains two networks, a generator and a discriminator, together, where the generator is engaged to generate images which appear as real images. Deepfakes are mainly of four categories and are enlisted below:-

- **Deepfake Pornography**: In this category, the face of an individual is transferred onto porn actor's body. The major problem is the influx of sexual privacy, humiliation and abuse.
- **Deepfakes in Political Campaigns**: Deepfake usage in political campaign has a negative impact of eroding the trust of people in political institution.
- **Commercial use of Deepfakes**: The usage of Deepfakes for commercial purposes is beneficial to society, but subject to the permission of the rights holder.
- **Creative Deepfakes**: The use of Deepfakes is fostering creativity by students or artists by encouraging creativity in the form of debates, interaction and creative educational material. There is also a negative side to it which is the possibility of bullying of children at school (Meskys et al., 2020).

5.1 Diverse Forms of Deepfakes

Deepfakes have penetrated the society in various ways. Below are enlisted different forms of Deepfakes:

- **Text-based Deepfakes**: AI is used to generate various texts such as emails, social media posts, articles and blogs which emulate writing style of a particular individual personality lead undeniably paramounting to misleading content creation.
- **Face Swapping Deepfakes**: Advanced AI techniques help in creating seem to be realistic videos using the advanced video techniques by swapping the face of a person in the video with another person's face.
- **Voice Synthesis Deepfakes**: AI is used to generate synthetic voice recordings that mimic the speech patterns and voice intonations to create fake audio messages and that seems to reach high levels of accuracy.
- **Object Manipulation Deepfakes**: AI techniques have helped in manipulating videos of objects by making changes in their appearances or behaviours. This leads to fallacious event presentations causing spread of misinformation.
- **Hybrid Deepfakes**: These Deepfakes involve intermingling various techniques such as swapping of face with altered voice or movements of body to result in more convincing outcomes.
- **Real-Time Deepfakes**: With further advancement in AI technology, Deepfakes can now be created in real time to recreate content from altering voice to video in real time. Deepfake live allows real-time impersonation during live elections

videos, live streams and video conferences (Deepfake Forensics, 2023; Dixit, 2023)

5.2 *Deepfakes Creation Techniques*

Automated learning techniques such as generative antagonist network (GAN) are used to synthesise images by superimposing images and audios on an existing image or audio. Through advanced AI tools, synchronisation of body movements and facial expression takes place. Sophisticated AI algorithms are used in Deepfake creations, which develop impact as real as life (Marino & Monaca, 2023). With easy availability of AI tools, Deepfakes have evolved to being very sophisticated posing difficulty to identify them easily. This rise in Deepfake has led to problems like intimidation of people, threat, misinformation and adverse effect on state affairs (Ramluckan, 2024). Factors to be taken into consideration while classifying content as Deepfake:

(i) Type of data carrier
(ii) In what manner AI is used for creation of fake material
(iii) Extent of manipulation
(iv) Check whether Deepfake involves existing or non-existent person.
(v) Extent to which consumers think the content posted by Deepfake is true (Sloot & Yvette, 2022).

AI has become easily accessible to all. This has led to increase in Deepfake crimes. GANs create media which cannot be detected by cyber security experts. These experts have 3000 functional models for Deepfake detection, but the main challenge is that these models are able to identify only simple Deepfakes. Nowadays AI had reached to another level of sophistication which can alter maps, imagery and x-rays and create almost realistic contents which are difficult to detect (Mathews, 2023).

6 Evolution of Deepfakes

Deepfake technology evolved in 2017 with the posting of creativity on Reddit online platform by a software developer.

Chronological evolution and development of Deepfakes (Chataut & Upadhyay, 2024):

I. **Video Manipulation Techniques**: The 1990s was an era when video manipulation techniques saw tremendous development. Adone premiere (in 1991) and After Effects (in 1993) surfaced as tools for advanced video editing. These tools acted as forebearers for future video manipulation techniques such as Motion tracking and Chroma keying, which were rudimentary attempts to future face swapping techniques.

II. **Machine Learning Advancements**: In the early 2000s, machine learning was developing at a fast pace in various fields such as neural networks and statistical learning methods.

III. **Facial Recognition Technology**: Modern Deepfakes have a very important component which is facial recognition technology. In early 2000s, there was immense development in video manipulation and 3D face modelling. In Bregler et al. (1997)evolved video rewrite system which made existing video appear as if the person was speaking, even though he had not spoken those in contexts. In 2003, Blanz and Vetter introduced methods to create 3D face models which were developed from a single face. These models were later used for swapping faces in Deepfake videos.

IV. **Emergence of Face Swapping Techniques Mid 2010**: Thies et al. (2018), introduced face-to-face method for real-time face capture and enactment. This acted as a stepping stone towards real-time face-swapping technique used in Deepfakes.

V. **Emergence of Deepfake**: The year 2010 saw the development of Deepfakes, with the word Deepfake coined in year 2017 by a Reddit user. He also shared the face-swapped videos. Recently, three key crucial technologies which are important for developing Deepfake videos are generative adversarial networks (GANs), deep neural networks (DNNs) and large language models (LLMs). These tools together help in creating pragmatic Deepfakes. In 2000s to 2010, graphic techniques were introduced which developed further in 2013 with invention of machine learning, which focussed on facial features. Generative adversarial networks came into existence in 2016 which led to invention of Deepfakes in 2017. In 2018, fake videos were made using Deepfakes and this led to formation of legislations against Deepfakes in 2019. From the year 2019 to 2021, Deepfakes developed with integration of voice and video cloning and swapping of faces. In 2022, regulatory authorities came up with further laws to reduce the impact of Deepfakes on society as a whole (Chataut & Upadhyay, 2024).

7 Deepfake Applications and Uses

In today's fast-moving world, Deepfakes act as tools for creativity and destruction. Applications of Deepfakes are enlisted below:

7.1 Positive Applications of Deepfakes

- Deepfakes have impacted circular economy as it has helped in bringing down the requirement of travel needs to create awareness about sustainable products. Now with use of synthetic media training, marketing, and stakeholder engagement can be achieved easily.

- Use as Educational Tool: Deepfakes by use of AI-generated content aid in creating educational contents which facilitate learning. Instructional videos used in education use Deepfake technologies to create contents in different languages.
- Entertainment Industry: Deepfake helps directors of movies by creating post visual effects. This helps in reduction of cost of production and better production time management.
- Therapeutic applications: Deepfakes are used in mental well-being of individuals.
- Sustenance of long-distance relationships: Deepfakes help to reinforce synchronous presence, which is useful for family members who are located remotely/long distance from the individual.
- Digital Story Telling: Deepfakes help in digital story telling by creating interactive videos where the contemporary figure is used.

7.2 Negative Application of Deepfakes

- The increased occurrence of Deepfake has created trust issues for digital media. With the use of Deepfake, fake certifications claiming sustainability have deeply impacted the economy. This has led to mistrust in the eyes of the people.
- Proliferation of Misinformation: In political scenarios, Deepfakes act as the medium for spread of misinformation. Manipulation of video recording of politicians and showing them as saying something which the person would not have said, affects public sentiments and can in turn change the electoral results in a negative manner.
- Companies engaged in Greenwashing use Deepfake contents in giving misinformation about their products, thus influencing consumer demand for genuine products.
- Deepfake using synthetic content has impacted companies genuinely engaged in using circular practices. The public is now losing faith in their environmentally friendly products
- Private equity companies, due to misinformation spread through Deepfakes, investors are reluctant to invest in these ventures.
- Non-consensual explicit: Deepfakes have been used to target women by superimposing their faces onto explicit material, affecting women emotionally and ultimately leading to reputational loss.
- Fraudulent Financial Transactions: Deepfakes are being used by cybercriminals for impersonation as senior company executives during manipulated audio or video calls. These criminals then convince the company employees to part with sensitive information or carry out fraudulent transactions of high-value amounts, causing financial or reputational loss to the enterprise.
- Enhanced Social Engineering attack: Deepfakes are also acting as facilitators for negative elements to plant malware or persuade employees to grant access to core technology systems of organisations (Caporusso, 2021; KPMG Website; Thukrul et al., 2024; Yessimova & Shevyakova, 2024).

8 Impact and Implication

Deepfake technology with its free distribution on various platforms such as social media and internet has impacted society, individuals, governments at large. Sky-high misuse of Deepfakes on social media has become a grave concern. Below mentioned are various impacts of Deepfakes and their implications:

- **Legal and Regulatory Implications**

Nowadays to tackle Deepfake pitfalls, legal frameworks are looking into enforcing legal penalties for the misuse of visual data. The legislative bodies are being proactive by encouraging development of artificial intelligence detection tools which are designed **to** identify Deepfake contents. At present, the existing laws are sparse and cannot tackle Deepfake problems effectively. There is a pressing need to have laws and regulatory methods which can effectively deal with the people and software who design or publicly share the Deepfake contents for defaming or blackmailing another person (Alanazi et al., 2024). Deepfakes also have adverse impact on legal evidence. Fake pictures or videos created by Deepfakes if submitted as evidence can impact legal decision-making (Nnamdi et al., 2023). Regulators use constitutional law, administrative law and criminal law to counter this menace. Also, other laws used are defamation, torts and copyrights. To reduce the cost of a particular activity, Deepfake is used to translate videos into multiple languages. This creates issues of copyrights and privacy, which are dealt by private laws such as contract and tort laws.

- **Ethical Implications**

Delivery of norms governing legal and ethical implications of Deepfakes is strongly influenced by communication. For countering Deepfake, there is a need for clear and concise communication from various segments. This step will encourage use of best practices to counter Deepfake. Technology experts, legal consultants and communicators can come together to fight against the rise of Deepfake menace. Misusing of Deepfake videos has negative impact on society as they affect individual's right to privacy and identity. To counter the growing concern about Deepfake misuse, there has to be collaboration between various departments so as to develop a complete framework that has all the aspects of legal, technical, ethical and innovation (Islam, 2025). Deepfakes also have other implications which include increased carbon emission, misinformation dissemination, infringement of copyrights and job displacement. All these form a part of ethical implications impacting society as a whole (Khatiwada et al., 2025).

- **Societal and Psychological impact**

Deepfake has a negative impact on society. The societal impact happens when Deepfake, without consent, posts pornographic material. A second societal impact of Deepfake is using this technology to bring a dead person to life. This is done by using Deepfake technology to create videos using the face of a dead person and using them to speak through. Another problem of this technology is manipulating

Deepfakes for financial gains. The content created through Deepfakes can be used to adversely affect share price, leading to the fall of a company. Also, a smaller use can be creating videos of family members or friends asking for money. These videos seem real like and lure innocent people to pay (Sloot & Yvette, 2022).

Deepfake is influencing fashion industry by artificial intelligence-powered hyper-personalisation. Deepfake videos are also useful tools for marketing by using Deepfake technology in storytelling. This can help in minimising video campaign costs and cater to hyper-personalised experience of the customer. Deepfake is also creating negative impact on society by ruining the trust in media. With Deepfake-tailored fake news events, people face a problem in distinguishing between true and false media news. The psychological impact on individuals about this Deepfake penetration is enlisted below:

(i) Anxiety and shame: Having oneself on Deepfake picture or video causes intense shame as this is not acceptable in societies where personal dignity is associated with social standing.
(ii) Social withdrawal and suicidal tendency: Deepfake contents make the victim an object of online and offline moral policing and mockery. Under such pressure, the victim goes into social exclusion which ultimately affects its mental health, giving rise to increased no. of suicides (Yadav et al., 2025).

- **Sectorial Impact**

Deepfake technology has a big impact on various sectors. In the field of education, Deepfake acts as a double-edged sword for education as on one hand it helps in increasing efficiency of students and provides multiple venues to explore, whereas on the other hand, it has multiple issues like plagiarism. In the education sector, teachers have to act as mentors of students, guiding them on ethical usage of Deepfake technology (Tiwari et al., 2025). Deepfake can foster higher student engagement and innovative teaching methodology. With the help of this technology, study content can be created using images of historical scholars which may seem to teach their own subject. This method of teaching will enhance student interest in the subject. These Deepfake videos of content can also be used as a tool for online presentations, making them livelier (Kokate et al., 2021). Artificial intelligence has touched various banking processes from processing of loans to personalised customer service offers in the form of Chatbots or virtual assistants. The ability of AI to handle large data volumes with speed and accuracy has paved the way for its large-scale use in the financial sector. Chatbots and virtual assistants have taken customer service to the next level by providing 24/7 customer query resolution with the slightest human intervention (Leema, 2024). Deepfake technology has found its use in advertising. The companies are using Deepfakes to engage with their clients. Deepfakes in advertising pose a challenge as this technology can be misused to deceive the consumers. This problem can only be resolved through collaboration across the industry. Transparency is another way by which the use of Deepfake technology can be stopped for malicious purposes. Brands can help in maintaining consumer trust through transparency and accountability, while maintaining creative aspects of advertising through

Deepfakes (Zafar et al., 2025). Deepfake has impacted the healthcare sector. AI in healthcare is involved in personalised patient education to help patients understand and stick to a treatment plan. It also helps in the development of skills of staff. AI can also influence areas such as the management of chronic disease and rehabilitation. Along with advantages, Deepfake also has disadvantages such as medical scams and spread of disinformation with an effect on the medical community at large (Qureshi & Khan, 2024). Deepfake is highly used in financial frauds and the country which is most impacted is the Philippines, which saw an increase of 4500% in a year. This is followed by countries like Vietnam, the United States and Belgium. Deepfake pornography has also skyrocketed recently with the use of the face of celebrity females on explicit material (Florian, 2024).

8.1 Counter Measures for Deepfake Prevention

- **Detection Methods**

Deepfakes have proved to be a menace for society by impacting the mental health and well-being of people. Various techniques employed for detection are enlisted below:

I. Frame-by-Frame analysis: In this method, the video is broken down in smaller frames to capture deviations between them.
II. Blending Analysis: Under this technique, changes in colour or texture of deepfaked face are scrutinised.
III. Blink Analysis: Under this method of scrutiny, the blinking of eyes is used to detect Deepfake videos. As a matter of fact, there is always difficulty in getting an adequate data set with closed eyes, hence it is a good method. The blinking of eyes is natural blinking, which is taken into consideration.
IV. Edge Analysis: This method utilises the analysis of borders and contours of Deepfake and real videos. Inconsistent pixelled edges and irregular lighting are the identifiers for this technique.
V. Digital Watermarking: Under this mechanism, an imperceptible piece of data known as a watermark is embedded into the media file. This watermark can be extracted at a later date to prove authenticity of data. Based on its strength, digital watermark is classified as robust, fragile and semi-fragile watermarking. Watermarking using proactive forensic technology ensures that the information in the original image, if modified, will be mismatched.

- **Forensic Techniques:** Under this technique, media files are examined scientifically, using human expertise and specialised software to exhibit tampering in audio or video.
- **Policy and Governance:** As per United Nations convention on cybercrime, member states are required to adopt legislative measures to criminalise unauthorised access to information and communication technology system.

- **Institutional Capacity Building:** This can be implemented by providing technical training to judges, police officers for accessing advanced digital forensic tools to aid in effective investigation of AI-enabled crimes.
- **Technical Counter measures:** With regular use of state of art technology and AI-driven detection systems, law enforcement agencies can easily identify Deepfake engineered synthetic and unusual behaviour patterns. This will act as gateway for prevention and proliferation of cybercrime.
- **Public–Private Information Exchange:** Technology companies having on-hand knowledge can help the common man to understand the dangers of Deepfake and ways to avoid the same.
- **Digital Literacy and education** Educating people about harmful effects of Deepfakes can be done by helping in providing digital literacy. People can be given training on identification of Deepfake videos, images and footages.
- **Cross-border Cooperation** Countries in specific region can come together in combating Deepfakes through joint operations, information sharing and extradition agreements (Amerini et al., 2025; Jordan, 2024; United Nations Brief, 2025; Zhimao et al., 2025).
- **Social Media Platform Scanning Tools and moderators:** These social media scanning tools and moderators which can help in reducing the effect of Deepfake contents are posted on social media.
- **Awareness Campaigns:** To alert public with repercussions of negative use of Deepfake, awareness campaigns can be run by media and government agencies. These campaigns can guide the people of consequences of creating and sharing such negative Deepfake content, so that people can disdain from using Deepfake technology for crimes like blackmailing, bullying and embarrassing others (Alanazi et al., 2024).
- **Synthetic Media regulations:** To control the menace of Deepfake, more stringent regulations are needed from Statutory authorities.

9 Discussion

Socio-psychological factors influence Deepfake spread. Policy-makers should consider these factors such as FOMO (Fear of missing out), DSR (Deficient Self-regulation) and cognitive ability when formulating laws for Deepfake. Intervention programmes created without considering psychological factors and learning ability of individuals will not create widespread impact on counter measuring Deepfake misinformation spread (Ahmed et al., 2023). Deepfake technology tries to destabilise a nation with falsification of news which is spread like wildfire and deeply impacts the trust and sentiments of general public on the government. To counter measure this negative use of Deepfakes, Regulatory authorities can raise public awareness against manipulative use of Deepfakes, by facilitating digital literacy programmes. Also, AI-based technology can be developed which can prevent further dissemination of Deepfakes, thereby controlling the damage caused by it. Regulators can set up a mechanism to identify malicious Deepfake developer and penetrators, so that both can be brought under law and serve as lessons to others to stop using the technology for negative purpose (Purwadi et al., 2022). Deepfake issues are increasing at a faster pace than its countermeasures. This has particularly amplified in the Southeast Asian region with rise of audio Deepfakes. In Southeast Asian countries, Deepfake detection models are not successful as there is mismatch in synthesis quality, language-specific characteristics and data scarcity. To counter this issue, SEA-Spoof is large-scale audio Deepfake detection has been developed. This technology spans 300 + hours of paired real and spoof speech in multiple Southeast Asian languages such as Tamil, Hindi, Thai, Indonesian, Malay and Vietnamese (Wu et al., 2025). Print Media has also become victim of Deepfakes. As Deepfakes are using the media information, it gives encouragement to the miscreants as the news that spread seems to be legitimate. This is affecting the good names of these print media which may lead to loss of public trust in the long run (Sunvy et al., 2023). The study is restricted to only Southeast Asia and could have included other parts of the world such as the African subcontinent. The sectorial impact analysed in the study was restricted to major fields such as education, market, healthcare, whereas the study could have been extended to impact of Deepfakes in rural and semi-urban areas where digital literacy is low and people trust the content which is posted online very easily. Future studies can be done on the impact of word of mouth and peer group on spread of Deepfakes among people. AI is spreading wrong information which is impacting the society as a whole. To counter these, strategies as mentioned below are needed:

I. Amplification of Digital Literacy: Digital literacy is crucial for combating AI. Younger population which is more susceptible to AI-generated propaganda must be educated about the AI-created content so as to avoid any transmission of misinformation.
II. Evolving a strong regulatory framework: International Regulatory authorities and technical experts have to come together to develop a strong framework which is enforceable against misuse of AI. This framework should have an understanding about innovation through Deepfakes and misuse of Deepfakes.
III. Encouraging cross-sector collaboration: Various sectors such as private, government, international bodies and society can come together to fight against this menace of Deepfakes. This collaboration can work on information sharing, setting up standards for introducing common laws and regulations for combating Deepfakes.
IV. Reinforcing human insights in identifying Deepfakes: The best countermeasure for Deepfakes is training society as well as regulators to identify malicious material posted by Deepfake. This will help in removing algorithm biases and control the spread of misinformation.
V. Election-related counter measures to be introduced: Use of Deepfakes by rival parties during election campaigns is very rampant. By introducing morphed Deepfake material, the direction elections are completely changed. To counter this false propaganda, efforts are required for using techniques such as media

watermarking, and creating standards for authentic AI-generated content is important to regain trust of people. (Tetiana, 2025).

10 Conclusion

Malicious impact of Deepfakes has tremendously increased in Asian subcontinent. This region with its linguistic and ethnic diversity has far more bearing complications due to Deepfakes. Deepfake crimes have impacted all the sectors particularly finance, healthcare, education, entertainment to name a few. In order to come out of this Deepfake jeopardy, all the nations in Asian subcontinent need to come together on common platform to have collective regulatory framework which can be applicable across the boundaries. Improvising research and development for countering Deepfake is also very crucial. Nations have to bring their technical teams on common platform to devise ways to solve this issue. In today's changing world scenario, Genz is that segment of population who are tech-savvy. This segment of population is greatly influenced by social media and word of mouth. Hence, they are most vulnerable to spread of misinformation by Deepfakes. Educating them through digital literacy programmes can help in winning a battle against Deepfakes. With the rise of Deepfake-based financial scams, organisations can protect themselves from this through increasing their investments in their employees. Better educated and tech-savvy employees can act as human firewalls against such crimes. Companies can start educating their employees for robust cyber security culture. Also, investments should be done by companies on organisation risk intelligence. Upgrading technology for countering Deepfakes is also very crucial for businesses (Matthew, 2023). Nations in Southeast Asia should leave their political differences behind and fight this Deepfake problem, which is impacting all, as one common group. With collective efforts of all this Deepfake can be channelised for beneficial purpose like health care and education instead of utilising it for pornography or financial frauds. To summarise staff education, enhance regulatory framework and greater Research and development are very crucial for stopping spread of Deepfake crime in the society. These collective efforts will be helping in developing a crime-free society which is safe for all. The creative aspect of Deepfakes can aid in improving teaching methodology and better understanding of educational matter by students. The chapter in the end will conclude with the way forward which the government and financial institutions should take to encourage and work towards educating people and enhance financial literacy. This will safely guard common man from falling in trap of such Deepfake-type cyber frauds. Also, way forward for government to develop laws and regulations to prevent rampant occurrence of cybercrime, thus protecting common man from all such financial losses.

References

Achary, R. (2021). AI transforming Indian banking sector. *International Journal of Economics and Management Systems, 6*. ISSN 2367-8925.

Adam, A., & Chen, F. F. (2025). East Asia's legal blind spot in the deepfake era, 3/10/2025. https://eastasiaforum.org/2025/10/03/east-asia-legal-blind-spot-in-the-deepfake-era/

Alanazi, S., Asif, S., & Moulitsas, I. (2024). Examining the societal impact and legislative requirements of deepfake technology: A comprehensive study. *International Journal of Social Science and Humanity, 14*(2), 58–64. https://doi.org/10.18178/ijssh.2024.14.2.1194

Amerini, et al. (2025). Deepfake media forensics: Status and future challenges. *Journal of Imaging, 11*(3), 73. https://doi.org/10.3390/jimaging11030073

Bregler, C., Covell, M., & Slaney, M. (1997). Video rewrite: Driving visual speech with audio. In *Seminal graphics papers: Pushing the boundaries* (Vol. 2). https://api.semanticsscholar.org/corpus id:2341707

Caporusso, N. (2021). Deepfakes for the good: A beneficial application of contentious artificial intelligence technology. In *Advances in intelligent systems and computing*. https://doi.org/10.1007/978-3-030-51328-3-33

Chataut, R., & Upadhyay, A. (2024). Introduction to deepfake technology and its early foundations. In *Deepfakes and their impact on business* (pp. 1–18). ISBN13: 9798369368909. https://doi.org/10.4018/979-8-3693-6890-9-ch001

Deepfake Forensics. (2023, June 26). 5 types of deepfakes-discover the different deepfakes types. https://powerhouseforensics.com

Deepfake Threats to Companies, KPMG Website. https://www.kpmg.com

Dixit, P. (25/11/2023). What are the different types of Deepfakes and how you can spot them? today.in/technology/news/story/what-are-the-different-types-of deepfakes-and -how-you-can-spot-them-407118–2023–11–25

Dymples, L.S. (2022). Deepfakes and Disinformation in Asia, Book: Deepfakes Algorithms and Society, Michael, F., Ist Edition, Chapter2, ISBN 9781003173397. https://doi.org/10.4324/978 1003173397.

Emerging Threats: The Intersection of Criminal and Technological Innovation in the Use of Automation and Artificial Intelligence in the Cybercrime Landscape of Southeast Asia, Sept 2025.United Nations Office on Drugs and Crime, Cyber Crime Technical Brief Series. https://www.undoc.org

Florian, Z. (2024). How dangerous are deepfakes and other AI powered frauds, March 13th, 2024. www.statista.com/chart/31901/countries-per-region-with-biggest-increases-in-deepfake-specific-fraud-cases/

Islam, S. (2025). Unethical use of artificial intelligence: The phenomena of deepfakes. *World Telecommunication and Information Society*, 65–77. https://www.researchgate.net/profile/Saiful-Islam-24/publication/392172229_Unethical_Use_of_Artificial_Intelligence_The_Phenomena_of_Deepfakes/links/6837

Jordan, C. (2024). Digital forensics techniques to detect deepfakes. Cyber. https://westoahu.hawaii.edu/Cyber/forensics-weeklyexecutive-summaries/digital-forensic-techniques-to-detect-deepfakes/

Khatiwada, P., Walsh, T., Hamed, A. S., & Bhatta, L. (2025). The ethical implications of AI in creative industries: A focus on AI-generated art. https://doi.org/10.48550/arXiv.2507.05549

Kokate, S., Hedge, H., Lokhande, N., & Pathak, N. (2021). An approach to education: Improvements using image animation for deepfakes. *International Research Journal of Engineering and Technology, 8*(4), 5009–5015. e ISSN 2395-0056.

Kosarkar, U., & Sakarkar, G. (2024). Unmasking deepfake advancements, challenges and ethical considerations. In A. Ghosh & S. L. Shimi (Eds.), *Innovations in electrical and electronic engineering, ICEEE2023. Lecture notes in electrical engineering* (Vol. 1115). Springer, Singapore. https://doi.org/10.1007/978-981-99-866-3_19

Leema, R. (2024). Role of artificial intelligence in banking sector. *International Journal of Cultural Studies, 20*(2), 2347–4777.

Marino, D., & Monaca, M. A. (Eds.). (2023). *Innovations and economic and social changes due to artificial intelligence: The state of the art*. ISBN 978-303133461

Matthew. (2023). Deepfakes: Real threat, reality defender. https://kpmg.com/kpmg-us/content/dam/kpmg/pdf/2023/deepfakes-real-threat.pdf

Meskys, E., Kalpokiene, J., Jurcys, P. and Aidas, L. (2020). Regulating deep fakes: Legal and ethical considerations. *Journal of Intellectual Property Law and Practices, 15*(1), 24–31. https://ssrn.com/abstract=3497144

Miriam, B., Corsaro, S., & Nguyen, P. V. (2025). Mobilising digital feminism in the age of AI: South Korean and Chinese women against pornographic deepfakes, peace human rights governance, pp. 2–40. https://doi.org/10.25430/pupj-PHRG-2025-AOF-3

Monetary Authority of Singapore. (2025). Cyber risks associated with deepfakes. https://www.mas.gov.sg/-/media/mas-media-library/regulation/circulars/trpd/cyber-risks-associated-with-deepfakes.pdf

Nnamdi, N., Oniyinde, O. A., & Abegunde, B. (2023). An approach of the implications of deepfakes: The need for urgent international legislations. *American Journal on Leadership and Governance, 8*(1), 43–70. ISSN online 2957-7284.

Purwadi, A., Serfiyani, C. Y., & Serifiyani, C. R. (2022). Legal landscape on national cybersecurity capacity in combating cyber terrorism using deepfake technology in Indonesia. *International Journal of Cyber Criminology, 16*(1). https://doi.org/10.5281/zenodo.4766560

Qureshi, J., & Khan, S. (2024). Artificial intelligence (AI) deepfakes in healthcare system: A double-edged sword, balancing opportunities and navigating risks. https://www.preprints.org/. https://doi.org/10.20944/preprints2024020176

Ramluckan, T. (2024). Deepfakes: The legal implications. In *Proceedings of the 19th international conference on cyber warfare and security, ICCWS, 2024*, pp.282–288

Sloot, B. V., & Yvette, W. (2022). Deepfakes: Regulatory challenges for the synthetic society. *Computer Law and Security Review, 46*. https://doi.org/10.1016/j.clsr.2022.105716

Sumsub Research. (2022). Global deepfake incidents surge tenfold from 2022 to 2023. https://sumsub.com/newsroom/sumsub-research-global-deepfake-incidents-surge-tenfold-from-2022-to-2023/

Sunvy, A. S., Reza, R. B., & Al Imran, A. (2023). Media coverage of deepfake disinformation: An analysis of the south Asian countries. *Informasi, 53*(2), 295–308. https://doi.org/10.21831/informasi.v53i2.66479

Surasit, N. (2024, February 29). Rogue replicants-criminal exploitation of deep fakes in Southeast Asia, Global Initiative against Transnational Organised Crime. www.globalinitiative.com. https://globalinitiative.net/analysis/deepfakes-ai-cyber-scam-south-east-asia-organized-crime/

Thies, J., Zolhofer, M., Stamminger, M., Theobalt, C., & NieBnes, M. (2018). Face to Face: Real-time face capture and re-enactment of RGB videos. *Proceedings of the IEEE Conference on Computer Vision and Pattern Recognition*, 2387–2395. https://doi.org/10.1145/2929464.2929475

Thukrul, A., Patil, S., & Anusurkar, G. (2024). Deepfake technology from positive applications to malicious exploits. *Journal of Emerging Technologies and Innovative Research, 11*(10), a806–a812. ISSN-2349-5162.

Tiwari, R., Anjum, B. Chand, K., & Bajaj, S. (2025). Dynamics of deepfake in higher education. In S. Taneja, S. Gupta, M. Kukreti, & Ozen, E. (Eds.), *Navigating the deepfake conundrum: A manager's roadmap, information systems engineering and management* (Vol. 44). Springer. https://doi.org/10.1007/978-3-031-90742-5_13

Wan Rosli, W. R. (2022). Violent extremism and artificial intelligence: A double-edged sword in the context of ASEAN. *Commonwealth Cyber Journal, 2*, 46–58.

Wu, J., Hou, N., Pan, Z., Zhang, Q., Sailor, B. H., & Mondal, S. (2025). SEA-Spoof: Bridging the gap in multilingual audio deepfake detection for southeast Asian, computer science. https://arXiv.2509.19865

Yadav, G., Sadique, M. Z., Kumar, S., Sharma, R., Sharma, M., Sharma, R., & Rattan, T. (2025). Psychological trauma and legal challenges of deepfake technology. *Sciences of Conservation and Archaeology, 37*(1), 143–150.

Yessimova, M. S., & Shevyakova, T. V. (2024). Deepfakes in the digital media age opportunities and threats. *Herald of Journalism, 73*(3), 44. https://doi.org/10.26577/HJ.2024.v73.13.4

Zafar, G., Kasheer, M., Hameed, R. M., Ullah, I., Khan, W. A., Nisar, H., & Niazi, S. (2025). Impact of deepfake advertising disclosure on purchase intention with mediating roles of perceived reality, trust, perceived ethicality and Irritation. *International Journal of Social Sciences Bulletin, 3*(4), 578–603. https://doi.org/10.5281/zenodo.15266445

Zhimao, L., Saad, A., Cong, F., Guangjun, L., & Chuntao, W. (2025). Enhancing deepfake detection: Proactive forensics techniques using digital watermarking. *Computer, Material and Continua, 82*(13), 73–102.

Part V
Circular Economy Transitions, Carbon Markets, and Global Sustainability Policy

Chapter 21
Strategizing Urban Sustainability for Emission Reduction: A Best–Worst Method (BWM) Approach

Kiran Yadav, Noman Arshed, Bhavna Yadav, and Shubham Shukla

Abstract Rapid urbanization is becoming a global and modern trend of the twenty-first century. Urbanization is causing significant economic growth; at the same time, increasing pressure on the environment is also one of the dimensions of the same, particularly carbon emissions. As cities are the major energy consumers and emission hubs, which plays a key role in events of climate mitigation. For ensuring the alignment of urban development with the sustainability goals, it becomes crucial to identify and prioritize ideal strategies so that a reduction in emissions may take place with the infrastructural growth for accommodating the population. This chapter employs the Best–Worst method (BWM), which is generally accepted as one of the best and robust multi-criterion decision-making (MCDM) techniques. This chapter aims to systematically evaluate and rank key urban sustainability strategies on the basis of expert inputs. There are mainly eight core strategies commonly identified for urban sustainability. Such core strategies include public transportations, green buildings, renewable energy integration, waste management, and urban green infrastructure. The approaches are necessary to test against metrics of carbon emission abatement cost effectiveness, scalability, and fairness. Improvements to PT networks and active travel, expanding green building codes, and a gradual shift to renewables provide the best benefits for the least effort. As they target the basic urban systems, it is only to be expected that they emerge as the best in a holistic approach to emissions reduction. This chapter summarizes the research for a range of urban planners and

K. Yadav (✉)
PGDAV College, University of Delhi, New Delhi, India
e-mail: kiran.yadav@pgdav.du.ac.in

N. Arshed
Department of Business Analytics, Sunway Business School, Sunway University, Petaling Jaya, Malaysia
e-mail: nomana@sunway.edu.my

B. Yadav
Shyam Lal College(E), University of Delhi, New Delhi, India

S. Shukla
Sri Venkateswara College, University of Delhi, New Delhi, India

K. Kumar et al. (eds.), *AI-Driven Circular Economy for Carbon Neutrality in Emerging Economies*, Environmental Footprints and Eco-design of Products and Processes,
https://doi.org/10.1007/978-3-032-30298-4_21

climate change regulators and reform advocates and translates that information into a prescription for the climate-resilient city. The framework of urban planning and policy is being endorsed priority. This provides us with the way for perspective development. The information acquired, through the study, can be used to guide urban policy. This study will enable us to move towards climate-resilient transformation in urban economies either developed or developing.

Keywords Urban sustainability · Sustainable urban planning · Carbon emissions · Best–worst method (BWM) · Multi-criteria decision-making (MCDM)

1 Introduction

Rapid urbanization has turned out as one of the most transformative global occurrence of the twenty-first century. As per the United Nations projections, about 68% of the global population is expected to have their residence in urban areas by 2050, causing exponential increment in pressures on energy systems, transportation and logistics networks, housing infrastructure, and natural resources. While cities are being considered as the engine of economic growth, hub of innovations, and means of social development and at the same time, they are simultaneously responsible for a major share of global carbon emissions. Estimates suggest that cities account for more than 70% of energy-related CO_2 emissions worldwide. This dual role is making cities to assume central positions in both of the extreme situations which is climate crisis and its potential solutions of urbanization (Liang & Gong, 2020).

The primary and significant carbon emissions in urban environment is caused by fossil fuels based energy consumption, vehicular emissions by motorized transportations, inefficient buildings, industrial activities, and under capable waste management systems. Rapid urbanization without any planning, increasing vehicles on the roads (Zhang et al., 2020), and energy-driven lifestyles are further intensifying emission statistics (Black, 2018; Kumar et al., 2021), predominantly in dynamically growing cities of emerging economies. In order to achieve various global targets, such as sustainable development goals and Paris agreement, the urban sustainability must be employed effectively. This implies that urban development and sustainability is highly correlated with the prospective climatic goals (Moroke et al., 2019).

Urban sustainability is far wider phenomena as compared to environmental protection in order to achieve economic viability, social inclusion, and institutional resilience. There are various strategies available to urban planners to ensure urban sustainability. Strategies like public and non-motorized transport coverage, green building adoption, integration of renewable energy alternatives, circular waste management systems, and urban green infrastructure development have gained global acceptance among the urban planners for sustainable development of urban areas. However, implementations of such strategies are not a cakewalk due to budgetary constraints, complexities involved in governance, technological limitations, and short-term developmental priorities (Romero-Lankao et al., 2016).

While performing the process of urban planning, policymakers face various dilemmas causing the decision-making process highly challenging. Policymakers and urban planners face the issues of prioritization of sustainability strategies with the constraints like limited resources and uncertainty. Also not all interventions fetch the consistent reductions in emission at the same time socio-economic costs of such strategies are differentiated. So, policymakers and urban planners are expected to employ systematic, transparent, and evidence-based approaches which shall be capable of evaluating multiple criteria simultaneously (Zeng et al., 2022).

There has been increased reliance on multi-criteria decision-making (MCDM) methods to address complex problems like sustainability and planning. Among MCDM, the Best–Worst method has gained prominent acceptance of scholars due to the strengths to generate consistent and reliable priority weights with a handful of pairwise comparisons as compared to traditional techniques by incorporating expert judgements and minimizing inconsistencies (Sahoo & Goswami, 2023). A vigorous framework for ranking urban sustainability strategies is being provided by BWM which is based on multiple criteria such as potential of emission reduction, cost effectiveness, scalability, and social impact (Pamučar et al., 2020).

This chapter attempts to apply the BMW framework in order to evaluate and prioritize key urban sustainability strategies aiming emission reduction systematically against the background, by connecting and bridging the methodological rigour with practical relevance.

The contribution of this chapter will be towards a growing body of literature in the preview of climate-responsive urban planning. The chapter will be ensuring the provisioning of actionable insights in the developed as well as emerging urban context.

2 Literature Review

2.1 Urbanization and Carbon Emissions

Majority of literature in the given field of study is indicating that there is a strong and positive relationship between urbanization and carbon emissions. Early studies are emphasizing that population density in cities leads to increased energy demand. This excess energy demand is being noticed predominantly in transportation, residential buildings, and industrial sectors. In the recent times, various empirical studies have produced strong set evidences that urban environments, population density, and infrastructural quality are the key factors that significantly influence the emission levels (Shukla et al., 2022). Cities equipped with efficient transportation systems will be tagged as sustainable and compact cities. Such cities will be critical to achieve lower per-capita emissions as compared to traditional and sprawling urban setup which is identified by a feature called automobile dependency (Ahmed et al., 2019; N. Zhang et al., 2017).

Recently academic studies are highlighting the heterogeneity involved in urban emissions across the different income groups. This heterogeneity can be identified and classified with unique features. The key sources of emissions, in the developed economies, are essentially affected by consumption patterns and structures consuming energy (Koca et al., 2021). In the developing economies, key sources of emissions are different. Such key sources are excessive reliance on fossil fuels, and excessive infrastructural growth. These conclusions are directing us towards context-specific and unique sustainability strategies over homogeneous solutions (Dorling, 2021; Hailemariam et al., 2020).

Why MCDM/MADM

The existing studies are reviewed and summarized in Table 1. Table X clearly explains that multi-criteria decision-making (MCDM) techniques have emerged as essential analytical tools for addressing the issues like urban sustainability and challenges of emission reduction (Zeng et al., 2023). Single criterion or conventional econometric approaches are not adequately equipped to evaluate urban emission mitigation. Urban emission mitigation involves multiple and conflicting criteria such as environmental effectiveness, economic feasibility, social acceptance, and scalability (Siksnelyte et al., 2018). Methods like MCDM enable researchers to make systematic integration of qualitative expert judgements with the quantitative assessments as shown in Table 1. It allows decision-makers to compare alternatives under uncertainty and limited resources. Earlier, researches have successfully used methods like AHP, TOPSIS, VIKOR, DEMATEL, and hybrid models as well to analyse low-carbon city performance (Brugha, 2004; Daga et al., 2025). It helped to prioritize transport and energy interventions, and assessment of smart city sustainability (Sharifi & Murayama, 2013). Traditional approaches, on the other hand, require extensive and pairwise comparisons. They may also get affected by consistency issues as criteria increase. The Best–Worst method provides a methodological advantage by reduction of cognitive burden on experts while ensuring upper level of consistency and robustness of results. Although application of BWM to prioritize integrated urban emission reduction strategies will be limited in the current body of literature. Employment of BWM will directly address such methodological gap and serves as a structured, evidence-based prioritization framework which endorses informed urban policy formulation and sustainable city planning (Oztas & Erdem, 2021; Yadav et al., 2025).

2.2 *Urban Sustainability Strategies for Emission Reduction*

The body of literature identifies multiple urban sustainability interventions which possess the capability of mitigating carbon emissions shown in Table 2:

Table 1 Summary of MCDM/MADM technique used in this field

References	Study context	Objective	MADM/MCDM method	Key criteria	Key findings
Rezaei et al. (2019)	General decision-making problems	Propose a consistent MCDM method	Best–Worst Method (BWM)	Preference intensity, consistency	BWM requires fewer comparisons and ensures higher consistency
Sharifi and Murayama (2013)	Urban sustainability assessment tools	Review neighbourhood sustainability assessment frameworks	AHP, TOPSIS	Environmental, social, economic indicators	Highlighted methodological gaps in existing urban sustainability tools
Zavadskas et al. (2016)	European urban development projects	Rank sustainable urban development alternatives	AHP–VIKOR	Energy efficiency, emissions, land use	Hybrid MCDM improves decision robustness
Wang et al. (2019)	Low-carbon cities in China	Evaluate low-carbon urban performance	TOPSIS	Carbon intensity, energy mix, innovation	Identified significant regional disparities in low-carbon transition
Siksnelyte et al. (2018)	Urban energy systems	Select sustainable urban energy technologies	AHP–COPRAS	Cost, emissions, reliability	Renewable energy options ranked as most sustainable
Kumar et al. (2021)	Smart cities in India	Assess smart city sustainability dimensions	DEMATEL–ANP	Governance, environment, technology	Governance factors strongly influence sustainability outcomes
Zhang et al. (2020)	Urban transport planning	Prioritize transport-based emission reduction strategies	AHP	Accessibility, emissions, cost	Public and non-motorized transport ranked highest

Source Authors'own creation

Table 2 Urban sustainability strategies for emission reduction

Code	Strategy	Description	Key references
S1	Promoting compact and mixed land-use urban design	Encourages high-density, mixed-use development to reduce urban sprawl, travel distance, and car dependency	Jabareen (2006), Neuman (2005)
S2	Enhancing public and non-motorized transport systems	Improves accessibility and reduces emissions through investments in public transit, cycling lanes, and walkable infrastructure	Banister (2008), Newman and Kenworthy (2015)
S3	Increasing use of renewable energy in urban infrastructure	Focuses on integrating solar, wind, and other renewables in municipal buildings, street lighting, and transport systems	Jensen et al. (2019)
S4	Implementing green building standards and retrofitting	Adoption of energy-efficient building codes, LEED/BREEAM certifications, and retrofitting of old infrastructure for energy savings	Rezaei et al. (2019)
S5	Deploying smart technologies for urban management	Use of IoT, AI, and data platforms for optimizing traffic, waste, energy, and environmental monitoring in real time	Ahvenniemi et al. (2017), Bibri and Krogstie (2017)
S6	Strengthening urban waste management and circular economy	Emphasizes waste reduction, segregation, recycling, composting, and circular material use to reduce urban emissions	Wilson et al. (2012)
S7	Expanding urban green spaces and nature-based solutions	Increases urban resilience and carbon sequestration via parks, green roofs, bioswales, and reforestation in urban settings	Kabisch et al. (2016), Haase et al. (2014)
S8	Promoting sustainable water and resource use	Enhances water conservation through rainwater harvesting, greywater reuse, and low-carbon water infrastructure	Canton (2021b)

Source Authors' own creation

Code	Strategy
S1	Promoting compact and mixed land-use urban design
S2	Enhancing public and non-motorized transport systems
S3	Increasing use of renewable energy in urban infrastructure
S4	Implementing green building standards and retrofitting
S5	Deploying smart technologies for urban management
S6	Strengthening urban waste management and circular economy
S7	Expanding urban green spaces and nature-based solutions
S8	Promoting sustainable water and resource use

S1 Compact and Mixed Land-Use Urban Design

It involves integration of residential, commercial, institutional, and recreational functions in the dense urban areas for reduction and spatial separation of daily activities.

Such approach contributes significantly towards lowering the travel demands and private vehicle dependency. Ultimately, it will lead us to lower energy consumption and carbon emissions. Some sustainable activities like walking, cycling, and public transportation systems will get promoted by such urban planning strategies (Bouckaert et al., 2021). Compact city development is critical to promote cost-effectiveness and efficiency in urban infrastructures and services (Neuman, 2005). Sharing of resources becomes an essential need when population density starts touching new heights. Sharing of resources can be adopted through the various strategies such as mixed use of land, shared transportation, and other infrastructures. These things will promote social inclusion, urban liveability, and employment (Jabareen, 2006).

S2 Public and Non-motorized Transport

Transport and vehicular emissions are the largest contributors of urban emissions (Newman & Kenworthy, 2015). Studies are consistently showing that infrastructural investments in mass transit systems, pedestrian-friendly designs, and cycling infrastructures can significantly contribute in reduction of vehicular emissions as well as enhancing social equity and livelihood in urban areas (Banister, 2008).

S3 Renewable Energy Integration

Decentralization of renewable energy systems can play a significant role in reduction of dependency on fossil fuels. Decentralized energy systems include rooftop solar photovoltaics and urban micro-grids (Jensen et al., 2019). Literature suggests that cities which are adopting localized renewable energy solutions will tend to have higher resilience alongside emission reduction (Canton, 2021a).

S4 Green Buildings and Energy-Efficient Infrastructure

Buildings are consuming nearly 40% of the global energy consumption. Standards of green building such as energy-efficient materials, passive designs, and smart energy management have been proven effective for lowering operational emissions. In order to move towards urban decarbonization retrofitting of existing building stock is increasingly gaining recognition (Jensen et al., 2019).

S5 Deploying Smart Technologies for Urban Management

It refers to the employment of digital tools in the urban management such as sensors, analytics, artificial intelligence, and internet of things (IoT). These technologies will assist in monitoring, managing, and optimizing urban systems in a real-time manner. Smart technologies will improve the dynamics of energy efficiency, reduction in emissions, and enhancement in service delivery systems (Bibri & Krogstie, 2017). Technology will enable the authorities to make intelligent traffic management, smart grids, efficient water utilization, and optimized waste management. In aggregation, technology will help in minimization of resource wastage, lower operational costs, increase in mobility, and making of evidence-based decisions (Gupta & Barua, 2017). Further, it will help to advance in the dimensions of transparency, responsiveness, and standard of living for citizens (Ahvenniemi et al., 2017).

S6 Waste Management and Circular Economy Practices

If waste management systems are inefficient, they contribute to methane emissions and environmental degradation. Circular economy-oriented strategies are getting acceptance increasingly as effective reduction tools. Some well-known circular-oriented strategies are composting, recycling, and waste-to-energy systems (MacArthur, 2017).

Although the overall effectiveness of such aforesaid strategies is well documented and widely accepted, scholars and researchers have observed that their relative importance of such strategies varies depending on the economic, institutional, and spatial conditions (Bashmakov et al., 2022). The existence and establishment of such fact compels us to reinforce, rethink, and redesign the need for structured and standardized prioritization frameworks (Wilson, 2012).

S7 Urban Green Infrastructure

Emission mitigation will be contributed by green spaces, green roofs, and urban forests by enhancing carbon sequestration, reduction in heat island effects, and cooling down the energy demand (Kabisch et al., 2016). These strategies also contribute to generate joint benefits like improved public health and biodiversity conservation (Haase et al., 2014).

S8 Promoting Sustainable Water and Resource Use

Focus of this objective is on managing urban water, energy, and material resources in a way that minimization of waste, enhancement of efficiency, and long-term availability of resources to generations (Canton, 2021b). This strategy focuses on measures such as water-efficient infrastructures, rainwater harvesting, wastewater recycling, demand side management, and promotion of circular practices to reduce the pressure on nature (Vörösmarty et al., 2010). It will not only reduce the pressure on resources, but it will be also critical to improve the resilience to climate risks. This is beneficial to the socio-economic context as it will restore the service reliability, equitable distribution, and reduced operational costs (Farrelly & Brown, 2011).

2.3 Application of MCDM Techniques in Urban Sustainability

Multi-criteria decision-making (MCDM) approaches and techniques are globally employed and accepted in the research areas like urban sustainability. This enables the researchers to address the various trade-offs involved in the context of environmental, economic, and social objectives. Methodologies and techniques like AHP, VIKOR, TOPSIS, DEMATEL, and ANP are being employed to rank the sustainability indicators, assessment of smart cities of performance, and evaluation of climate mitigation policies (Zavadskas et al., 2016). However, the traditional

MCDM methodologies often suffer from the various limitations like excessive pairwise comparisons, inconsistencies in expert judgements, and respondents' fatigue. Such challenges are motivating for the adoption of more streamlined approaches, particularly in expert-driven decision contexts (Triantaphyllou, 2000).

2.4 *Best–Worst Method (BWM) in Sustainability Research*

The efficiency and robustness of Best–Worst Method (BWM) has ensured that BWM emerges as one of the promising MCDM techniques. Experts are expected to identify the most and least important criteria in deployment of BWM (Liang et al., 2020). Further BWM requires comparison of extremes against other criteria, causing in fewer comparisons and higher consistency ratios (Oztas & Erdem, 2021).

BWM is being applied on recent studies of sustainability to the areas such as selection of renewable energy, sustainable supply chain management, climate policy evaluation, and adoption of green technology (Rezaei, 2015). The ability of BWM to concatenate the qualitative insights given by the experts with the quantitative aspects makes it perfectly applicable and suitable for studying urban sustainability planning. Applicability of this method increases in the case of increased contextual complexity (Agyekum et al., 2021).

Although the application of BWM is growing but very limited research has focussed on using BWM specifically to point out prioritized urban emission reduction strategies, especially in the context of rapidly urbanizing economies. This chapter attempts to address the gap by employing BWM to systematically rank urban sustainability strategies based on multi-dimensional criteria (MCDM) (Luo et al., 2024). It would be a step towards policy-oriented and methodological urban climatic research (Mohammadi & Rezaei, 2020).

3 Research Methodology

The process initiated with the identification of relevant performance measurement indicators (PMIs) for urban sustainability and reduction in emission. The Best–Worst Method (BWM) is applied to determine the relative importance of PMIs. A panel of experts was appointed and asked to figure out most important and least important criteria and also sub-criteria on the basis of their professional knowledge and practice experience. This stage will be minimizing the interpretational uncertainty and enhancing the internal coherence of expert evaluations. Further Best–Worst Method (BWM), as provided by (Salimi & Rezaei, 2016), is used to analyse and derive the optimal weighting schemes for established PMIs. As compared to MCDM techniques, BWM ensures a significant drop in cognitive burden involved in extensive pairwise comparisons and also simultaneously improving the robustness and consistency of derived outcomes. Due to the discussed strengths, the methodology is being extensively adopted across the various domains like sustainability appraisal, energy system planning, and environmental governance. In the given context, a systematic and transparent prioritization of mitigation of urban emission strategies is derived through application of BWM. Unlikely to the other methodologies like AHP (Saaty, 2013), which requires pairwise comparisons among all criteria, BWM compares only the best criterion with the worst criterion. In between, all other criteria are placed on the basis of a preference scale and order ranging from 1 (equal importance) to 9 (extreme importance). Such comparison-based methodology simplifies the process and also addresses the inconsistency issues appearing in conventional pairwise comparison frameworks. Lastly BWM includes a consistency measure through an optimization framework (Rezaei, 2016).

The procedural steps of the BWM applied in this study are described as follows:

Step 1: Identification of Decision Criteria

It is based on extensive literature review and consultations of experts. The set of evaluation criteria is established and identified as *n*.

All of these criteria represent the key performance measurement indicators (PMIs) which are to be critically used to evaluate all the possible decision alternatives.

Step 2: Determination of the Best and Worst Criteria

The designated experts are expected to identify and establish the best (most important) and worst (least important) criteria among the identified and selected set on the basis of their relevance to the objectives of study.

Step 3: Best-to-Others (BO) Preference Vector

The preference of the best criterion (*B*) over all the other criteria is assessed and analysed by employing a numerical scale from 1 to 9 where *aBja*_{*Bj*) *aBj* indicates that the preference of the best criterion *B* over criterion *j*. Ultimately, the final preference values are obtained by relying on expert consensus.

Step 4: Others-to-Worst (OW) Preference Vector

In the same manner, the preference of each criterion for the worst criterion (*W*) will be determined, forming the others-to-worst (OW) vector.

Step 5: Determination of Optimal Weights

In the given step, the central objective is to determine the optimal weights, i.e. *wjw_jwj*, of the criteria so the minimization between maximum absolute deviation, derived weight ratios, and the stated preferences can take place. The linear programming model provides a unique optimal solution suitable for the criteria weights. The optimal value acts as an indicator of the consistency of the judgements provided by the experts. Indicators are such that values closer to zero indicate a higher level of consistency in the case of pairwise comparisons shown in Table 3.

Table 3 BWM method ranking with weights to strategies

Strategy code	Strategy description	Final weight	Rank
S2	Enhancing public and non-motorized transport systems	0.28	1
S3	Increasing use of renewable energy in urban infrastructure	0.180	2
S4	Implementing green building standards and retrofitting	0.09	3
S7	Expanding urban green spaces and nature-based solutions	0.12	4
S6	Strengthening urban waste management and circular economy	0.12	4
S1	Promoting compact and mixed land-use urban design	0.12	4
S5	Deploying smart technologies for urban management	0.07	5
S8	Promoting sustainable water and resource use	0.020	6

Source Author's own creation

4 Analysis and Discussion

To achieve the objective of this chapter, we apply the BMW framework in order to evaluate and prioritize key urban sustainability strategies aiming at emission reduction systematically. The analysis reveals that S2—Enhancing Public and Non-Motorized Transport Systems is the best strategy with a final weight of 0.28, while S8—Promoting Sustainable Water and Resource Use, with a weight 0.02, is the worst strategy out of all these strategies. This prioritization underscores the dominant role of urban mobility in greenhouse gas emissions, as transportation accounts for approximately 25–30% of urban CO_2 emissions globally (Kikstra et al., 2022). The second-ranked strategy is S3—Increasing Use of Renewable Energy in Urban Infrastructure with final weight 0.180—highlights the imperative for energy decarbonization, aligning with findings that renewable integration can reduce urban emissions by up to 40% in high-density areas (Dechamps, 2023). The BWM's consistency ratio (CR $\approx$ 0.36) validates the results, as values below 0.1–0.3 are deemed reliable (Rezaei, 2015).

The top ranking of S2 reflects empirical evidence that shifting from private vehicles to public transit and active mobility yields the highest immediate emission reductions. Studies show that comprehensive public transport enhancements can cut urban transport emissions by 20–50% within a decade. This is particularly relevant in rapidly urbanizing contexts, where modal shifts offer co-benefits like reduced congestion and improved public health. S3's high weight (0.180) emphasizes that electrifying urban grids with renewables is foundational as shown in Fig. 1. Without this, efficiency gains in buildings or transport are undermined by fossil-based electricity. Recent analyses confirm that solar and wind integration in urban settings can achieve 30% emission cuts by 2030 (IRENA, 2024). Mid-tier strategies are S1, S6,

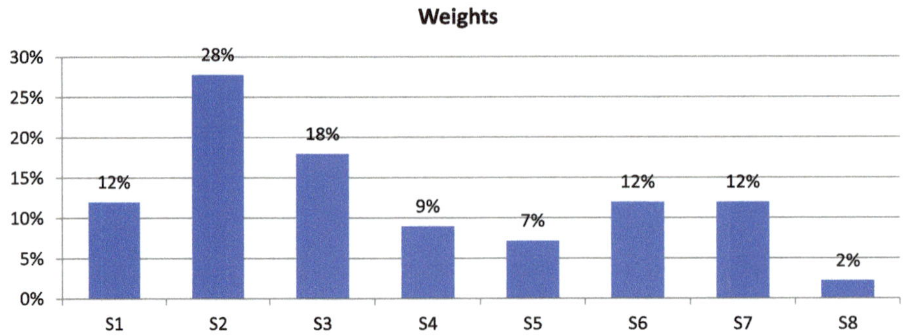

Fig. 1 Weights of all strategies after BWM analysis. *Source* Author's own creation

and S7, each with a weight of 0.12, form a balanced cluster. Compact land use (S1) reduces travel distances, synergizing with S2. Circular economy practices in waste management (S6) divert methane emissions, while urban greening (S7) sequesters carbon and cools cities. Their tied ranks suggest expert consensus on their complementary roles, whereas green retrofitting (S4, 0.09) and smart tech (S5, 0.07) rank lower due to upfront costs and scalability challenges, despite long-term potential. S8's minimal weight (0.020) indicates its indirect link to emissions; water efficiency primarily conserves energy embedded in treatment rather than directly curbing CO_2 (Vörösmarty et al., 2010).

5 Implications of the Study

5.1 Theoretical Implications

This study will be contributing to the growing set of literature on urban sustainability and climate mitigation by employing and reinforcing the applicability of multi-attribute decision-making (MADM) techniques and approaches in the environmental context having complex scenarios. Through the employment of Best–Worst Method (BWM), the study intends to make advancement in methodological rigour in the research of sustainability. The study will be demonstrating how expert-driven weighting techniques will effectively handle multidimensional and conflicting criteria associated with urban emission reduction. Unlikely to the traditional evaluation approaches which depend on single indicators or aggregate indices, the BWM framework enables us to have a framework which has structured mechanism for capturing the relative priorities among all sustainability dimensions. This feature of BWM enhances the theoretical discussions on urban climate governance by providing a sound replicable and consistency-oriented decision-support model that can be extended to the other sustainability domains.

5.2 Policy Implications

From the perspective of policy, the findings of this chapter offer some valuable insights that are helpful for urban policymakers and local governments who are willing to design effective emission reduction strategies under the constraints of resources. Policymakers should prioritize S2 through investments in bus rapid transit and bike networks, integrating S3 via renewable-powered charging infrastructure. Mid-tier strategies offer resilience dividends, such as flood mitigation from S7. The prioritized measurement indicators (PMIs) highlight the areas where policy interventions can move us towards the maximized policy impact and reduction in emissions. Policymakers and regulators can use the outcomes based on rankings to align with the urban development plans with the national climate commitments and international sustainability targets, such as the Sustainable Development Goals (SDGs), and the Paris Agreement. Further, the BWM-oriented prioritization framework enables us to make transparent and evidence-based formulation of policy, reduction of dependency on politically driven or ad hoc decision-making. Strategic sequencing of urban sustainability investments is being supported through this policy. This also ensures that high-impact initiatives get priority funding and institutional backing.

5.3 Practical and Managerial Implications

This study is going to provide practical and critical insights to urban planners, municipal authorities, and sustainability practitioners. The systematic and structural prioritization of strategies of reduction in urban emissions will assist the decision-makers in identifying critical intervention areas, such as sustainable transportation systems, energy-efficient buildings, and renewable energy integration. The BWM framework can be readily adopted and implemented by city administrations to evaluate the alternatives in urban projects, assessment of policy trade-offs, and monitoring the performance of sustainability over time. Further, BWM offers simplicity and reduces the burden of comparison. Such features of BWM make it suitable for real-world applications involving diversified stakeholders, also facilitating participatory and inclusive decision-making processes.

5.4 Methodological Implications

In methodological manner, our study demonstrates the effectiveness of BWM in addressing and pointing out the inconsistency issues which are commonly associated with the traditional MCDM techniques. Inculcation and incorporation of the consistency indicator enhances the credibility of expert judgements. It also strengthens the reliability of the derived weights. Further researchers and scholars can adopt and extend this framework by integrating BWM with other relevant analytical tools, which includes sensitivity analysis or scenario analysis and evaluation, further it will be enriching and enhancing urban sustainability assessments. The aforesaid approach can also be replicated in the whole set of different contexts of urban development. This technique also enables comparative analysis between cities in the emerging and developed economies.

6 Conclusion

This study examines the sustainability strategies in the urban context, predominantly a reduction in urban emissions by using Best–Worst method (BWM) which is a robust multiattribute decision-making approach. It is well capable of addressing complex, multidimensional policy challenges. The purpose is achieved by systematically prioritizing performance measure indicators (PMIs), which is based on the expert judgements. This study provides a systematic and structured framework for the evaluation of strategies in reduction in urban emission that too under economic, environmental, technical, and social considerations. The implications of the findings necessitate efforts to target the critical urban systems of sustainable transport, energy-efficient building, and renewable energy for the achievement of the ambitious urban carbon emissions reduction goal. The BWM framework's focus on transparency, consistency, and analytical rigour will thus enable the continued use of evidence-based decision-making for urban planning and climate governance. Our research contributes to the conversation around urban sustainability by providing a concrete and replicable decision-making framework for policymakers and urban planners. There will be contributions from the studies, yet also limitations. The analysis depends on expert judgement, and this is systematically achieved using the BWM framework. However, it is possible that this study may reflect subjective perceptions that arise from this contextual expertise in these urban and emissions systems. The study does not consider any dynamic or temporal aspects of urban emissions, which may limit the long-term effectiveness of the prioritized mitigation strategies. A focus on the framework means that there is a relative rather than absolute focus on the anticipated emission reduction results, which may limit the applicability of findings to predicting environmental outcomes. The findings of the study may be context-dependent to various degrees, and may not extend to any city.

There are several avenues along which this study can be extended. Extensions of this study can build on its findings by employing fuzzy extensions of the BWM framework. The BWM framework can be combined with other decision-making approaches, such as DEMATEL, ISM or system dynamics modelling, to explore the relationships between indicators of urban sustainability. Future developments of the BWM framework can be tested with real-world emissions datasets to further strengthen the robustness of the findings that emerge from its use in this and other studies. Other studies may compare cities or even countries with an aim to revealing insights into the contextual factors that impact sustainable priorities. The BWM framework may be extended by future studies to incorporate stakeholder participation and scenario analysis in the urban climate governance process.

References

Agyekum, E. B., Kumar, N. M., Mehmood, U., Panjwani, M. K., Alhelou, H. H., Adebayo, T. S., & Al-Hinai, A. (2021). Decarbonize Russia—A Best-Worst method approach for assessing the renewable energy potentials, opportunities and challenges. *Energy Reports, 7*, 4498–4515.

Ahmed, Z., Wang, Z., & Ali, S. (2019). Investigating the non-linear relationship between urbanization and CO_2 emissions: An empirical analysis. *Air Quality, Atmosphere & Health, 12*(8), 945–953. https://doi.org/10.1007/s11869-019-00711-x

Ahvenniemi, H., Huovila, A., Pinto-Seppä, I., & Airaksinen, M. (2017). What are the differences between sustainable and smart cities? *Cities, 60*, 234–245.

Banister, D. (2008). The sustainable mobility paradigm. *Transport Policy, 15*(2), 73–80.

Bashmakov, I. A., Nilsson, L. J., Acquaye, A., Bataille, C., Cullen, J. M., Fischedick, M., Geng, Y., & Tanaka, K. (2022). *Climate change 2022: Mitigation of climate change. Contribution of working group III to the sixth assessment report of the intergovernmental panel on climate change* (Chap. 11). https://escholarship.org/uc/item/23n103jv

Bibri, S. E., & Krogstie, J. (2017). Smart sustainable cities of the future: An extensive interdisciplinary literature review. *Sustainable Cities and Society, 31*, 183–212.

Black, J. (2018). *Urban transport planning: Theory and practice*. Routledge. https://www.taylorfrancis.com/books/mono/10.4324/9781351068604/urban-transport-planning-john-black

Bouckaert, S., Pales, A. F., McGlade, C., Remme, U., Wanner, B., Varro, L., D'Ambrosio, D., & Spencer, T. (2021). *Net zero by 2050: A roadmap for the global energy sector*. https://trid.trb.org/View/1856381

Brugha, C. M. (2004). Structure of multi-criteria decision-making. *Journal of the Operational Research Society, 55*(11), 1156–1168. https://doi.org/10.1057/palgrave.jors.2601777

Canton, H. (2021a). International energy agency—IEA. In *The Europa directory of international organizations 2021* (pp. 684–686). Routledge. https://www.taylorfrancis.com/chapters/edit/10.4324/9781003179900-103/international-energy-agency%E2%80%94iea-helen-canton

Canton, H. (2021b). United Nations human settlements programme—UN-habitat. In *The Europa directory of international organizations 2021* (pp. 234–240). Routledge. https://www.taylorfrancis.com/chapters/edit/10.4324/9781003179900-32/united-nations-human-settlements-programme%E2%80%94un-habitat-helen-canton

Daga, S., Yadav, K., Singh, D., Pamucar, D., & Simic, V. (2025). Unveiling Greenwashing: Analyzing the interaction of factors discouraging ESG Greenwashing through TISM and MICMAC. *Journal of Environmental Management, 380*, 124850. https://www.sciencedirect.com/science/article/pii/S0301479725008266

Dechamps, P. (2023). The IEA World Energy Outlook 2022–a brief analysis and implications. *European Energy & Climate Journal, 11*(3), 100–103.

Dorling, D. (2021). World population prospects at the UN: Our numbers are not our problem? In *The struggle for social sustainability* (pp. 129–154). Policy Press. https://bristoluniversitypressdigital.com/edcollchap/book/9781447356127/ch007.xml

Farrelly, M., & Brown, R. (2011). Rethinking urban water management: Experimentation as a way forward? *Global Environmental Change, 21*(2), 721–732.

Gupta, H., & Barua, M. K. (2017). Supplier selection among SMEs on the basis of their green innovation ability using BWM and fuzzy TOPSIS. *Journal of Cleaner Production, 152*, 242–258.

Haase, D., Larondelle, N., Andersson, E., Artmann, M., Borgström, S., Breuste, J., Gomez-Baggethun, E., Gren, Å., Hamstead, Z., Hansen, R., Kabisch, N., Kremer, P., Langemeyer, J., Rall, E. L., McPhearson, T., Pauleit, S., Qureshi, S., Schwarz, N., Voigt, A., … Elmqvist, T.

(2014). A quantitative review of urban ecosystem service assessments: Concepts, models, and implementation. *AMBIO, 43*(4), 413–433. https://doi.org/10.1007/s13280-014-0504-0

Hailemariam, A., Dzhumashev, R., & Shahbaz, M. (2020). Carbon emissions, income inequality and economic development. *Empirical Economics, 59*(3), 1139–1159. https://doi.org/10.1007/s00181-019-01664-x

International Renewable Energy Agency. (2024). *Renewable energy statistics 2024*. https://www.irena.org/Publications/2024/Jul/Renewable-energy-statistics-2024

Jabareen, Y. R. (2006). Sustainable urban forms: Their typologies, models, and concepts. *Journal of Planning Education and Research, 26*(1), 38–52. https://doi.org/10.1177/0739456X05285119

Jensen, U. T., Andersen, L. B., Bro, L. L., Bøllingtoft, A., Eriksen, T. L. M., Holten, A.-L., Jacobsen, C. B., Ladenburg, J., Nielsen, P. A., Salomonsen, H. H., Westergård-Nielsen, N., & Würtz, A. (2019). Conceptualizing and measuring transformational and transactional leadership. *Administration & Society, 51*(1), 3–33. https://doi.org/10.1177/0095399716667157

Kabisch, N., Frantzeskaki, N., Pauleit, S., Naumann, S., Davis, M., Artmann, M., Haase, D., Knapp, S., Korn, H., & Stadler, J. (2016). Nature-based solutions to climate change mitigation and adaptation in urban areas: Perspectives on indicators, knowledge gaps, barriers, and opportunities for action. *Ecology and Society, 21*(2). https://www.jstor.org/stable/26270403

Kikstra, J. S., Nicholls, Z. R., Smith, C. J., Lewis, J., Lamboll, R. D., Byers, E., Sandstad, M., Meinshausen, M., Gidden, M. J., & Rogelj, J. (2022). The IPCC Sixth Assessment Report WGIII climate assessment of mitigation pathways: From emissions to global temperatures. *Geoscientific Model Development, 15*(24), 9075–9109.

Koca, G., Egilmez, O., & Akcakaya, O. (2021). Evaluation of the smart city: Applying the dematel technique. *Telematics and Informatics, 62*, Article 101625.

Kumar, P. G., Lekhana, P., Tejaswi, M., & Chandrakala, S. (2021). Effects of vehicular emissions on the urban environment-a state of the art. *Materials Today: Proceedings, 45*, 6314–6320.

Liang, F., Brunelli, M., & Rezaei, J. (2020). Consistency issues in the best worst method: Measurements and thresholds. *Omega, 96*, Article 102175.

Liang, L., & Gong, P. (2020). Urban and air pollution: A multi-city study of long-term effects of urban landscape patterns on air quality trends. *Scientific Reports, 10*(1), 18618.

Luo, Y., Yang, Z., & Qin, Y. (2024). A hybrid BWM–CRITIC–VIKOR approach for assessing oil and gas risk scenarios in probabilistic linguistic term set. *Heliyon, 10*(19). https://www.cell.com/heliyon/fulltext/S2405-8440(24)14545-8

MacArthur, E. (2017). Beyond plastic waste. *Science, 358*(6365), 843–843. https://doi.org/10.1126/science.aao6749

Mohammadi, M., & Rezaei, J. (2020). Bayesian best-worst method: A probabilistic group decision making model. *Omega, 96*, Article 102075.

Moroke, T., Schoeman, C., & Schoeman, I. (2019). Developing a neighbourhood sustainability assessment model: An approach to sustainable urban development. *Sustainable Cities and Society, 48*, Article 101433.

Neuman, M. (2005). The compact city fallacy. *Journal of Planning Education and Research, 25*(1), 11–26. https://doi.org/10.1177/0739456X04270466

Newman, P., & Kenworthy, J. (2015). The end of automobile dependence: How cities are moving beyond car-based planning. *Island Press/center for Resource Economics*. https://doi.org/10.5822/978-1-61091-613-4

Oztas, G. Z., & Erdem, S. (2021). Framework selection for developing optimization algorithms: Assessing preferences by conjoint analysis and best–worst method. *Soft Computing, 25*(5), 3831–3848. https://doi.org/10.1007/s00500-020-05411-8

Pamučar, D., Ecer, F., Cirovic, G., & Arlasheedi, M. A. (2020). Application of improved best worst method (BWM) in real-world problems. *Mathematics, 8*(8), 1342.

Rezaei, J. (2015). Best-worst multi-criteria decision-making method. *Omega, 53*, 49–57.

Rezaei, J. (2016). Best-worst multi-criteria decision-making method: Some properties and a linear model. *Omega, 64*, 126–130.

Rezaei, F., Bulle, C., & Lesage, P. (2019). Integrating building information modeling and life cycle assessment in the early and detailed building design stages. *Building and Environment, 153*, 158–167.

Romero-Lankao, P., Gnatz, D. M., Wilhelmi, O., & Hayden, M. (2016). Urban sustainability and resilience: From theory to practice. *Sustainability, 8*(12), 1224.

Saaty, T. L. (2013). Analytic hierarchy process. In *Encyclopedia of operations research and management science* (pp. 52–64). Springer.

Sahoo, S. K., & Goswami, S. S. (2023). A comprehensive review of multiple criteria decision-making (MCDM) methods: Advancements, applications, and future directions. *Decision Making Advances, 1*(1), 25–48.

Salimi, N., & Rezaei, J. (2016). Measuring efficiency of university-industry Ph.D. projects using best worst method. *Scientometrics, 109*(3), 1911–1938. https://doi.org/10.1007/s11192-016-2121-0

Sharifi, A., & Murayama, A. (2013). A critical review of seven selected neighborhood sustainability assessment tools. *Environmental Impact Assessment Review, 38*, 73–87.

Shukla, P. R., Skea, J., Slade, R., Al Khourdajie, A., van Diemen, R., McCollum, D., Pathak, M., Some, S., Vyas, P., & Fradera, R. (2022). Climate change 2022: Mitigation of climate change. *Contribution of Working Group III to the Sixth Assessment Report of the Intergovernmental Panel on Climate Change, 10*, 9781009157926.

Siksnelyte, I., Zavadskas, E. K., Streimikiene, D., & Sharma, D. (2018). An overview of multi-criteria decision-making methods in dealing with sustainable energy development issues. *Energies, 11*(10), 2754.

Triantaphyllou, E. (2000). Multi-criteria decision making methods. In *Multi-criteria decision making methods: A comparative study* (pp. 5–21). Springer. https://link.springer.com/chapter/%DB%B1%DB%B0.%DB%B1%DB%B0%DB%B0%DB%B7/%DB%B9%DB%B7%DB%B8-%DB%B1-%DB%B4%DB%B7%DB%B5%DB%B7-%DB%B3%DB%B1%DB%B5%DB%B7-%DB%B6_%DB%B2

Vörösmarty, C. J., McIntyre, P. B., Gessner, M. O., Dudgeon, D., Prusevich, A., Green, P., Glidden, S., Bunn, S. E., Sullivan, C. A., & Liermann, C. R. (2010). Global threats to human water security and river biodiversity. *Nature, 467*(7315), 555–561.

Wang, Z., Hao, H., Gao, F., Zhang, Q., Zhang, J., & Zhou, Y. (2019). Multi-attribute decision making on reverse logistics based on DEA-TOPSIS: A study of the Shanghai End-of-life vehicles industry. *Journal of Cleaner Production, 214*, 730–737.

Wilson, G. (2012). *Community resilience and environmental transitions*. Routledge. https://www.taylorfrancis.com/books/mono/10.4324/9780203144916/community-resilience-environmental-transitions-geoff-wilson

Wilson, D. C., Rodic, L., Scheinberg, A., Velis, C. A., & Alabaster, G. (2012). Comparative analysis of solid waste management in 20 cities. *Waste Management & Research, 30*(3), 237–254.

Yadav, K., Daga, S., & Saini, A. (2025). Structural analysis of sustainable fashion adoption: An integrated TISM and MICMAC approach. *Journal of Sustainable Marketing*, 1–24. https://doi.org/10.51300/jsm-2025-149

Zavadskas, E. K., Govindan, K., Antucheviciene, J., & Turskis, Z. (2016). Hybrid multiple criteria decision-making methods: A review of applications for sustainability issues. *Economic Research-Ekonomska Istraživanja, 29*(1), 857–887.

Zeng, S., Gu, J., & Peng, X. (2023). Low-carbon cities comprehensive evaluation method based on Fermatean fuzzy hybrid distance measure and TOPSIS. *Artificial Intelligence Review, 56*(8), 8591–8607. https://doi.org/10.1007/s10462-022-10387-y

Zeng, X., Yu, Y., Yang, S., Lv, Y., & Sarker, M. N. I. (2022). Urban resilience for urban sustainability: Concepts, dimensions, and perspectives. *Sustainability, 14*(5), 2481.

Zhang, L., Li, Z., Jia, X., Tan, R. R., & Wang, F. (2020). Targeting carbon emissions mitigation in the transport sector–A case study in Urumqi, China. *Journal of Cleaner Production, 259*, Article 120811.

Zhang, N., Yu, K., & Chen, Z. (2017). How does urbanization affect carbon dioxide emissions? A cross-country panel data analysis. *Energy Policy, 107*, 678–687.

Chapter 22
Carbon Markets in the Global South: Neoliberal Foundations Amid Neocolonial Critiques

Sana Fatima and Nabeel Ahmad

Abstract This study elucidates the architectural principles and conceptual foundations essential for developing carbon markets in the Global South, focusing on emerging mechanisms under Article 6 of the Paris Agreement. These market-based instruments internalize greenhouse gas emission externalities via mitigation outcome trading, optimizing abatement costs across heterogeneous economies and harmonizing Nationally Determined Contributions with development goals. Specifically, Internationally Transferred Mitigation Outcomes (Article 6.2) and the centralized Article 6.4 framework amplify mitigation ambition, sustainable development, and environmental integrity within nascent systems. Robust accounting, averting double counting through corresponding adjustments and adaptation proceeds anchor global emission reductions, while neoliberal governance facilitates bilateral cooperation, channeling finance into renewables and nature-based solutions. Nevertheless, political economy critiques illuminate asymmetries: Northern offsets may entrench green capitalism, compromising sovereignty and elevating exchange-value above ecological priorities. Compounding these issues are institutional hurdles, such as regulatory deficits, weak monitoring/reporting/verification, and data gaps obstructing baselines and additionality. Neocolonial lenses depict Article 6 as extracting, appropriating Southern carbon sinks despite differentiated responsibilities; yet hybrid models—fusing domestic emissions trading with international ties and South-South collaboration—promise equitable pathways. Finally, scalability via compliance-voluntary integration, bolstered by stringent pricing and equity measures, renders carbon markets either efficient global tools or arenas of unequal exchange, demanding vigilant safeguards.

S. Fatima (✉)
Symbiosis School for Online and Digital Learning, Symbiosis International (Deemed University), Pune, India
e-mail: fsana3683@gmail.com

N. Ahmad
Al-Barkaat College of Graduate Studies, Aligarh, India

K. Kumar et al. (eds.), *AI-Driven Circular Economy for Carbon Neutrality in Emerging Economies*, Environmental Footprints and Eco-design of Products and Processes,
https://doi.org/10.1007/978-3-032-30298-4_22

Keywords Carbon market · Carbon footprints · Climate change · Paris agreement · Sustainable development · Global south

1 Introduction

The interest in cooperative mechanisms that could assist in enhancing mitigation ambition and be flexible to meet country-specific circumstances has increased with the adoption of the Paris Agreement, thus bringing back carbon markets as one of the main instruments of international climate policy. In comparison with the previous top-down regimes, this agreement is based on nationally determined contributions (NDCs), where market-based cooperation is established as a voluntary instrument in ensuring a successful achievement of the emissions cuts in the heterogeneous economies (Asian Development Bank, 2024). Carbon markets are being set up in the Global South as a mechanism of striking a balance between climate mitigation and development agendas by marshaling funds, facilitating transfer of technology, and supporting low-carbon growth pathways.

Under the Paris Agreement, Article 6 introduces decentralized and centralized market mechanisms as a legal and institutional basis of such cooperation. Article 6.2 permits bilateral or plurilateral transfer of mitigation performance between Parties, while Article 6.4 is a centrally operated mechanism, focused on reduction of the emissions, as well as the provision of the sustainable development benefits (Kumar et al., 2025; Prakash & Tiwari, 2025; Schneider & La Hoz Theuer, 2018; Singh et al., 2025a, 2025b). These requirements are a great improvement compared to the offset regimes in the past as they realize the market cooperation of national accounting systems that are focused on environmental integrity and greater transparency (Agrawal, 2025; Groenewoudt & Romijn, 2022; Joselin et al., 2025; Singh et al., 2024). As the practical implementation of Article 6 continues to be refined over the years, the topic of scholarly analysis regarding the implications of such mechanisms in the developing economies has drawn enormous discussion.

Global South holds a distinguishing position in the emerging carbon market landscape. The mitigation potential of the developing countries is high due to the growing energy systems, development of infrastructure, land-use patterns, often at lower marginal abatement cost (Singh et al., 2025b; Stek et al., 2025). At the same time, such nations also have structural constraints such as insufficiency of fiscal resources, unbalanced institutionalization, and pressing socio-economic demands, such as access to energy and reduction of poverty. Carbon markets are often advocated as tools injected to international funds into mitigation efforts that support national development goals, especially the deployment of renewable energy and nature-driven solutions (UNFCCC, 2024; Zhao & Shaw, 2025).

While some scholars advocate the benefits of efficiency and the potential of higher mitigation ambitions of carbon markets, others highlight unequal bargaining power, methodological complexities, and the fact that the responsibility of mitigation could be transferred to the developing economies instead of high-emitting ones (Kaith &

Sachdeva, 2024; Redvers et al., 2025). Such discussions are supported by historical experience with previous market mechanisms that demonstrated the presence of endemic issues similar to additionality, baselines, and distribution of benefits unevenly across host nations (Minas, 2023). Consequently, the modern debate is more concerned with the institutional and governance provisions, according to which carbon markets can help in emissions reductions without weakening decarbonizing strategies.

Thus, this chapter concentrates on designing and functioning of carbon market mechanisms covered by Article 6 of the Paris Agreement, the accounting and institutional basis necessary to protect the integrity of the environment and the developmental consequences of market involvement to developing economies. Simultaneously, it interacts with political economy and neocolonial critique that has emerged alongside international carbon trading, keeping these views within the changing structure of climate cooperation. The next section initially follows the role of carbon markets in global climate collaboration and as a tool of economic-efficient mitigation.

2 Carbon Markets and International Climate Cooperation

Market-based mechanisms have made numerous contributions to international climate policy. Their initial institutionalization under the Kyoto Protocol and in particular by their emissions trading and project-based approaches attempted to strike a balance between their objectives of reduction of emissions and economic efficiency by allowing mitigation to occur in the countries where these costs would be the lowest. Despite these mechanisms contributing to the investment funding and experimentation with carbon pricing, the lack of equal participation, low environmental additionality, and lack of integration into domestic climate policy frameworks recurrently became a problem (Minas, 2023).

One significant redefinition of this strategy was marked by the Paris Agreement through cooperative mechanisms part of NDCs. It offers the Parties discretion in policy instruments instead of mandatory levels of emissions, while necessitating transparency, consistency of accounting, and periodic ambition. In accordance with this, carbon markets have turned into not external compliance but linked to the national policy and system of mitigation (Asian Development Bank, 2024). This transition is particularly in the scenario of the developing economies that focus on the needs of national development. Recent literature describes this change as a continuation of a larger trend toward flexibility and pluralism in climate governance. The Paris framework enables various national solutions with the perspective of realizing consistency through reporting and transparency principles (Mukhia et al., 2024). In this changing governance environment, carbon markets are important depending on how they are operationalized and combined with other policy tools.

Carbon markets help to lower the aggregate cost of emissions reduction through redistribution of abatement effort in jurisdictions with different marginal abatement costs. The differences in the technological maturity, energy endowments, and developmental level create strong disparities in mitigation costs, which opens possibilities of the cooperative mechanisms enhancing the efficiency outcomes (Wang et al., 2025). Carbon markets are geared toward exploiting such disparities by allowing the emissions benefits realized in the less-costly settings to be relocated to the more-costly environments.

This logic can be linked in the Global South with opportunities of mitigation in the deployment of renewable energy, improvement in energy efficiency, and land-use change, in which emissions reductions are often cost-effective. Consequently, market-based cooperation is often introduced as the tool that facilitates mitigation and provides improvement in the areas related to development such as the increase in infrastructure, employment, and access to better energy sources (Alsamawi, 2024). These expected synergies make carbon markets attractive to the development-based climate policies.

Further, researchers also point out that cost-effectiveness is not a decisive factor of the desirability and efficiency of carbon markets. The design features in these markets are setting baselines, crediting rules, and eligibility criteria that determine incentives and distributional outcomes. In cases of not well-defined elements, carbon markets can prefer low-cost mitigation on a short-term basis, instead of structural transformation over the long-term, especially in areas that need sustained regulatory and investment support (Gold Standard, 2024). These reflect the significance of efficiency claims in the context of more comprehensive evaluations of decarbonization pathways.

2.1 Position of the Global South in Cooperative Climate Frameworks

The Global South has a significant part in the global carbon market dealings, both as a potential provider of mitigation accomplishment and a more active site of domestic climate policy structure. Several developing countries have initiated or tested carbon pricing instruments, including emissions trading systems (ETS) and carbon taxes, as a broader policy to curb the rise in emissions and support climate policy objectives (Asian Development Bank, 2024). This dynamic policy landscape poses a challenge of determining the host and acquiring countries, which presupposes a more diversified role of the developing economies in further collaboration on the market.

Despite these changes, there are high structural asymmetries. Successful participation in carbon markets requires institutional capacities in regard to inventory management of the emissions, regulation and monitoring, reporting and verification (MRV). According to the empirical research, capacity limitations are one of the major impediments to most developing states that restrict their operations in the field of cooperation and access to financial and technological benefits (Michaelowa &

Keßler, 2024). According to these constraints, there can be unequal participation of Global South in the carbon market.

Another consideration is that of dealing with the relationship between the carbon market and long-term mitigation strategies. Internationalization of mitigation outcomes deprives nations in which such initiatives are undertaken the privilege to balance such efforts against national commitments. NDCs would pose a trade-off between the short-term financial incentives and long-term mitigation requirements since they are likely to become more ambitious (Stek et al., 2025). The choices made on the permissibility and magnitude of market entry hence have significant connotations on policy space maintenance in third world economies.

These dynamics highlight the necessity to study the organization and the regulation of international carbon market mechanisms under the Paris Agreement. The institutional framework of cooperative strategies is the key to the evaluation of the operations of carbon markets and the relationship with domestic mitigation policies. Expanding on this argument, the following section considers Article 6 of the Paris Agreement, its design, and its operationalization in more detail.

3 Article 6 of the Paris Agreement

The international carbon markets in the post-2020 climate regime get the form of institutional foundation in Article 6 of the Paris Agreement. It incorporates collaboration mechanism into the structure of the national determination of the climate commitments, thereby connecting the international transfer of mitigation outcomes (ITMOs) directly to domestic emission accounting. Such integration makes Article 6 more advanced than the previous market mechanisms as it does not assume carbon market activity as an extrinsic compliance measure but rather an integral part of the national policy (Global Green Growth Institute, 2022).

Article 6 consists of three parts, which are interrelated as follows: cooperative approaches to Article 6.2, a centralized mechanism under Article 6.4, and a framework of approaches to non-market under Article 6.8. Although non-market approaches are complementary, the main usage of carbon markets is based on Articles 6.2 and 6.4. These provisions fix the legal framework within which the results of mitigation can be cross-border transferred, and, at the same time, introduce accounting provisions aimed at protecting the integrity and uniformity of the environment with NDCs (Groenewoudt & Romijn, 2022).

3.1 Article 6.2: Cooperative Approaches and Internationally Transferred Mitigation Outcomes

Article 6.2 allows Parties the opportunity to participate in voluntary cooperative methods that include ITMOs. By design, these arrangements are decentralized to enable the countries to adopt bilateral or plurilateral agreements that are based on their priorities regarding policies and mitigation measures. Instead of standardizing methodologies or types of the project, Article 6.2 promotes national discretion under the transparency and reporting considerations in the Paris Agreement (Global Green Growth Institute, 2022).

The main characteristic of the Article 6.2 is the necessity to ensure that the participating Parties implement relevant changes to their emission accounts to avoid the risks of double counting. In case of transfer of a mitigation outcome, it is placed on the host country to increase its emissions balance and the acquiring country makes an equivalent requirement of a downward adjustment. This accounting system guarantees that transferred mitigation results can be claimed only once in relation to NDCs achievement (Groenewoudt & Romijn, 2022). The success of Article 6.2, thus, lies heavily on the soundness of the country inventories and accounting systems.

Article 6.2 provides Global South countries with a leeway in organizing cooperation according to the domestic development priorities. Bilateral agreements can be industry or technology-specific, such as renewable energy, transport electrification or industrial efficiency. Concurrently, researchers observe that decentralized cooperation exposes national institutions with a lot of responsibility to guarantee integrity that has raised concerns on the distributed implementation competence among countries (Michaelowa & Keßler, 2024). Technical knowledge and regulatory controls may also be used to negotiate outcomes differentially and can lead to the allocation of benefits and costs of ITMO transfers.

3.2 Article 6.4: Centralized Market Mechanism and Sustainable Development Requirement

Article 6.4 provides a centralized system that works with international oversight and that aims at creating some certified emission cuts that are beneficial in mitigation practices besides ensuring sustainable growth in host nations. Article 6.4 mechanism, unlike the Article 6.2, is based on standardized methodologies, centralized approval systems, and international governance to bring consistency and plausibility to projects and initiatives (Global Green Growth Institute, 2022).

Its importance lies in the fact that it directly contributes to the overall mitigation of global emissions. This is operationalized with provisions like the compulsory cancellation of a portion of generated mitigation outcomes that are to be used to ensure that market activity leads to a net reduction of global emissions as opposed to mere offsetting. This quality reflects experience of the prior market mechanisms that have been denounced to cause a displacement of emissions with no aggregate climate advantages (Minas, 2023).

The centrality of Article 6.4 has its possibilities and restrictions to the Global South. Standardized processes and controlled centralization can eradicate uncertainty among the investors, enhancing accessibility to international markets. However, in a location with limited data access, the same characteristics may pose severe administrative and technical demands on host nations. Researchers also note the importance of ensuring that the host countries have significant influence over project selection, and benefits allocation to enable the recommendation to align market participation with the NDCs (UNFCCC, 2024).

3.3 Environmental Integrity and Overall Mitigation in Global Emissions

Environmental integrity is one of the fundamental principles the design of Article 6 mechanisms is based on. The credibility of international carbon markets has been questioned on matters of non-additional projects, overstated bases, and poor monitoring facilities. Article 6 aims to mitigate these issues by proposing improved accounting principles, disclosure principles, and institutional protection.

The overall mitigation in global emissions concept comes in as another effort to go beyond offsetting into net atmospheric benefit. Practically, this goal is conditional upon conservative setting of a baseline, restrictions of credit issuing, and a permanent cancellation of a part of mitigation successes. According to the scholars, the rigor of these design characteristics would prove to be a defining factor in deciding whether carbon markets under Article 6 can help make a significant contribution to global emissions reductions (Schneider & La Hoz Theuer, 2018).

In developing countries, the process of achieving environmental integrity interrelates with the issue of development and policies. Stricter regulations could increase credibility level in the market but also decrease the amount of tradable mitigating outcomes, which has a bearing on the possible revenue streams. Conversely, there might be short-term financial gains on lax regulation of climate objectives in the long term. The importance of these trade-offs is to identify the accounting systems, data quality, and institutional capacity to guide the outcomes of Article 6 implementation in the Global South (Jain & Deshmukh, 2025).

These design areas are concerned with the technical and institutional foundations on which plausible carbon markets are anchored. The operation of both decentralized and centralized mechanisms is dependent on quantifying emissions reductions and the way in which accounting rules are practiced. The next section consequently looks at various preconditions needed to operationalize carbon markets under Article 6.

4 Accounting, Measurement, and Institutional Preconditions

The accounting frameworks, measurement practices, and institutional capacity are the keys to the operational credibility of carbon markets under the Paris Agreement. Although Articles 6.2 and 6.4 create the legislative framework of the cooperative market mechanisms, their success in providing actual and supplementary emissions reductions depends on the level of the quantification of the mitigation outcomes, their reporting, and their inclusion in national emissions accounts. To the nations of the Global South, these technical and institutional preconditions do not only offer opportunities but also impose limitations that define the possibility and the attractiveness of participation in the market.

4.1 Corresponding Adjustments and the Avoidance of Double Counting

The key accounting process entailed by the Paris Agreement is the adjustments which are made corresponding to retain the integrity of the environment of international carbon markets. In case the mitigation results are carried across the borders, the resultant adjustments ensure that both the host and the acquiring nations have the same level of emissions reflected in the emissions balances of the respective countries. Such adjustments are necessary to ensure that a single mitigation outcome is counted against many NDCs and compromises the integrity of the collective mitigation efforts (Groenewoudt & Romijn, 2022).

The corresponding adjustments to be implemented must involve detailed emissions inventories, express definition of the transferable outcomes, and their uniformity across the sectors and over the time. According to scholars, these requirements are very complex, especially to developing countries whose inventory systems might not be fully developed (Global Green Growth Institute, 2022).

In the Global South, some of these adjustments can increase strategic considerations related to long-term policy space. When emissions reductions are authorized for international transfer and adjusted accordingly, host countries lose the right to claim the reductions to form part of the domestic quota in the future. Since mitigation ambition is expected to increase over successive NDC cycles, the transfers in an early or high-volume can limit the possibilities of ambitious commitments in the future (Stek et al., 2025). These factors demonstrate that there is a necessity to carefully sequence and selectively authorize market participation.

4.2 Baselines, Additionality, and Permanence

In addition to accounting changes, baselines and additionality determine the credibility of carbon markets. Baselines determine the counterfactual situation whereby the reduction in emissions is measured and additionality whereby the reduction outcome would not have come up without the carbon market incentives. Weak baseline construction or ineffective additionality testing may lead to issuance of credits which are not related to actual emissions reductions, an issue which has held significant focus in reviews of previous market mechanisms (Minas, 2023).

Recent research highlights that it is especially difficult to set the baseline in fast-evolving economic and policy contexts, which is typical of most Global South situations. The changes in the structure of energy systems, developing regulatory frameworks, and shifting patterns of investments make it hard to create plausible counterfactual scenarios (Michaelowa & Keßler, 2024). Such environments have risks of over-incentivizing mitigation results that represent general developmental changes as opposed to market-driven action.

Permanence is another issue that is particularly associated with land-use and nature-based mitigation efforts. The emissions that are captured by afforestation, reforestation, or soil carbon sequestration can be reversed by natural disturbances, change in land use or even by failure of governance. The relevance of buffer mechanisms, long-term monitoring, and liability provisions in eliminating the risks of permanence is pointed out by the scholars, indicating the lack of those measures may erode the credibility of the market (Michaelowa et al., 2025).

These technical problems are not just methodological problems but have distributional and strategic consequences. Stringent baselines and extreme additionality thresholds may lead to reduction in the scale of ITMOs, limiting sources of revenue to host nations. Contrarily, the supply of short-term credit can be increased by the lenient measures at the cost of market integrity. These issues help in designing carbon markets which are both credible and development-compatible in the Global South.

4.3 Monitoring, Reporting, and Verification Capacity in the Global South

Monitoring, reporting, and verification (MRV) systems are the institutional frameworks of plausible carbon markets. Effective MRV helps to quantify the decrease in emissions, making it easy to introduce corresponding adjustments and providing transparency and confidence among the Parties. The MRV requirements under the Paris Agreement are incorporated into the improved transparency framework, which aims to harmonize reporting and accommodate differing national capacities (Asian Development Bank, 2024).

Despite this, there are still large gaps in MRV capacity among countries. Data accessibility, technical know-how, and institutional alignment are some of the challenges encountered in many developing economies, especially where emissions are diffuse and challenging to quantify, as is the case in the agricultural, land use, and small-scale energy system sectors (Jain & Deshmukh, 2025). Such restrictions may raise transaction costs, slow up projects, and discourage the use of international carbon markets. Governance and accountability are also affected by the capacity constraints. Weak MRV systems can reduce the capacity of host countries to monitor the activities of project developers, create compliance and make mitigation measures in line with national priorities. Researchers argue long-term investment in the institutional capacity-building is thus a requirement in achieving equitable participation in the carbon markets, especially the least developed countries and the small island developing states (UNFCCC, 2024).

Notably, the MRV capacity is not fixed. Over the last few years, a number of countries in the Global South have made significant contributions in creating national emissions inventories, online monitoring platforms, and regulatory frameworks. The combination of carbon markets involvement and the wider scale of capacity-building processes under the Paris Agreement can thus create positive feedbacks, enhancing domestic climate governance in the long run. The manner in which this integration can be attained is based on the policy decisions about how the market is designed, how the institutions can coordinate with each other, and the international support.

The technical basis of carbon markets comprises these accounting, measurement, and institutional considerations. Their solution is at the core of deciding whether cooperation mechanisms under Article 6 can be viable instruments of mitigation in the Global South. On this technical basis, the next section shifts to the relations between carbon markets, development aims, and climate finance, exploring how market engagement cuts across more economic and social concerns in developing economies.

5 Carbon Markets, Development, and Climate Finance

Carbon markets in the Global South are often justified on grounds of mitigation as well as their contribution toward the wider development goals. Monetizing the emission of reduction is projected to marshal global climate finance, decrease the risk of investments, and assist in the dissemination of low-carbon technologies in settings where capital limitation persists to be insignificant. The degree to which the expectations are fulfilled depends on the relationships of carbon markets with domestic policies on development, sectoral investment patterns, as well as finance structures.

5.1 Mobilization of Climate Finance and Investment Flows

One of the key arguments in support of carbon markets is their ability to draw the private investment in mitigation actions to help in emissions reductions. Carbon revenues are regarded as a way to enhance the project bankability and mobilize the private capital toward low-carbon investment, particularly in developing economies where there is lack of public climate financing and fiscal space (Asian Development Bank, 2024). This logic has been especially prevalent in the fields of renewable energy, food waste management, and industrial efficiency, where the initial capital expenses may present a substantial impediment to implementation (Srivastava et al., 2018, 2021)).

Empirical studies have, however, indicated that the intensity and dependability of carbon finance fail to meet initial expectations. The prices of the carbon market have been highly volatile, and in most instances, the revenues were not high enough to make investment choices without any other incentives in the form of feed-in tariff, subsidies, or concessional finance (Michaelowa et al., 2025). Thus, carbon markets are complementary and not core funding tools, thereby supplementing broader policy frameworks instead of replacing them.

This complementarity is of significance to the Global South. In places where carbon markets are incorporated into consistent policy packages that are supplemented by regulation and government investment, they can help to reduce the cost of capital and improve deployment. In the absence of such integration, carbon finance can potentially provide incentive to develop more fragmented or short-term projects instead of a full-scale structural change (Gold Standard, 2024).

5.2 Renewable Energy, Nature-Based Solutions, and Sectoral Distribution

The sectoral distribution of carbon market activity in the Global South exhibits cost factor and methodological viability. Traditionally, renewable energy projects, specifically wind, solar, and small hydro projects, have garnered much interest in the carbon market because of comparatively low abatement goals and quantifiable emissions decrease. Alongside mitigation potential and co-benefits in the form of biodiversity and ecosystem services, nature-based solutions, such as afforestation, reforestation, and avoided deforestation, have also become increasingly popular categories (Carton et al., 2021; Griscom et al., 2020).

As scholars observe, sectoral concentration presents critical issues on the effects of development and distribution of risks. Renewable energy projects are often capital-intensive and may be dominated by large developers, limiting local value capture. Although nature-based solutions offer mitigation potential, they are also associated with land tenure and permanence challenges and community involvement, especially in areas with weak governance or disputed property rights (Redvers et al., 2025).

In terms of development, these trends emphasize the need to align the carbon market involvement with the national priorities and social protection. Without deliberate policy guidance, carbon markets can strengthen sectoral and spatial inequalities by directing investment toward regions with a good cost and risk situation instead of the sectors that are significantly associated with inclusive developmental outcomes (Michaelowa et al., 2025).

5.3 Policy Space, Sovereignty, and Long-Term Mitigation Pathways

The participation in global carbon markets is also a concern regarding policy autonomy and long-term mitigation policy in the Global South. Host nations lose the right to offset ITMOs on future domestic target when mitigation outcomes are authorized to be transferred to other countries. Since NDCs are predicted to become progressive, this dynamic creates possible trade-offs between short-term financial profits and long-term decarbonization requirements (Stek et al., 2025). Recent studies stressed that these trade-offs are not uniform across countries and sectors. Economies, while in the early phases of industrialization, can experience an increasing pattern of emissions and growing mitigation costs over the years, retaining low-cost mitigation options as strategically important. Under these conditions, high dependency on international transfer would limit future policy alternatives or high prices of domestic mitigation (Asian Development Bank, 2024).

Simultaneously, institutional learning and policy formulation can be facilitated through carbon market participation, when strategically incorporated. Carbon market revenues can be used to fund domestic climate initiatives, and involvement in global mechanisms can enhance administrative capacities, data infrastructure, and regulations. The potential benefits are accomplished depending on the governance decisions in terms of authorization, revenue application, and higher-level planning integration with the development planning (UNFCCC, 2024).

These deliberations indicate that carbon markets are neither necessarily good nor bad in developing the consequences in the Global South. Their effects are determined by the ways they are embedded in the national policy frameworks and trade-offs between financial, sovereignty, and long-term mitigation aims. Such interactions provide an insight for such critical political economy interpretations of carbon markets, questioning the broad rationalities and power structures behind the market-based climate cooperation, as elaborated in the next section.

6 Political Economy Perspectives on Carbon Markets

Research related to political economy provides an interpretive lens of analyzing carbon markets that does not simply focus on their design and professed efficiency goals. The literature does not see markets as neutral instruments, but rather they are embedded in more general institutional rationalities that influence the organization, valuation, and exchange of mitigation. Such views shed light in the context of international climate cooperation in how the structure participation of carbon markets, assign risks, and distribute benefits across the actors and regions of the Global South.

6.1 Market-Based Coordination and Institutional Rationalities

The focus of the political economy study is that carbon markets are based on the dependence on price-based coordination, tradability, and involvement of the private sector. These characteristics are reminiscent of what the literature defines as neoliberal foundations, as far as shared climate ambitions are being pursued using commodification and exchange instead of a direct regulatory distribution (Kus & Jackson, 2025). The emissions cuts are converted to homogeneous units of value, allowing cross-jurisdictional comparisons, transfer, and accumulation.

Notably, detecting neoliberal foundations does not presuppose that carbon markets are post-ideological or homogenous in their impacts. Instead, it emphasizes a formal reasoning of how mitigation is structured. Carbon markets give preference to market indicators and voluntary cooperation, diverting mitigation actions to activities that can be measured, priced, and traded (Gold Standard, 2024). This logic can be applied to the nature of projects that should be given priority and the actors that are in the best position to contribute.

6.2 Value Distribution, Risk Allocation, and Asymmetries

The issue of value distribution and risk distribution in carbon markets is also a subject of interest in the political economy perspectives. Participation is frequently associated with a need to access capital, technical know-how, and regulatory capability, providing benefits to actors who have access to pre-existing financial and institutional resources. Consequently, researchers observe that market-based mechanisms can concentrate benefits among multinational companies, financial intermediaries, and countries with developed administrative systems, while pushing some risks such as price fluctuation or project failure, onto less-resourced entities (Mukhia et al., 2024; Shen et al., 2023; Tripathi et al., 2022).

In the Global South, these asymmetries are especially relevant. Although mitigation activities might be hosted in the developing nations, the downstream value created through credit trading, aggregation, and financialization might be realized in other locations. Political economy analysis is therefore concerned with distributional effects of market participation and how economic rents are captured and exposed to market fluctuations. These questions overlap with the issues of equity, development, and long-term policy autonomy without assuming that similar results would be reached in different contexts.

6.3 State Authority, Hybrid Governance, and Policy Mediation

Contrary to interpretations that suggest that market mechanisms reduce the state power, recent research gives importance to the hybridity of the carbon market governance. States still maintain strong powers of making authorization decisions, eligibility criteria, and integration of market mechanisms into national policy frameworks. In this respect, neoliberal foundations refer to coordination rather than the withdrawal of a public authority (Kus & Jackson, 2025).

For the countries in Global South, this hybridity gives space to policy mediation. Governments can influence the interactions of carbon markets with national development policies, policy tools, and public investment priorities. The approaches on political economy focus on contingency: the impacts of market-based mechanisms rely on governance decisions, institutional capacity, and overall policy landscape in which markets are incorporated (Shen et al., 2023; Srivastava et al., 2024). These insights elucidate the manner in which carbon markets can be adjusted by certain institutional rationality while remaining susceptible to the politics of bargaining and regulation. They also provide a conceptual array to more critical perspectives, challenging the historical and structural implication of market-based climate cooperation. This analysis leads to the further discussion engaging the neocolonial critiques of carbon markets by putting forward arguments that market processes can reproduce unequal patterns of pre-existing exchange in the Global South.

7 Neocolonial Critiques of Carbon Markets in the Global South

Neocolonial critiques of carbon markets examine the global climate partnership through historical power asymmetries and exchange inequality. Rather than focusing on the technical design of market mechanisms, this analysis questions the re-establishment of structural relationships between industrialized and developing economies with the help of market-based mitigation. Under this strategy, carbon markets are appraised under the distributive effects, sovereignty, and potential to re-develop the existing North–South relations into a new institutional form (Durmaz & Schroeder, 2025).

7.1 *Mitigation Export and Unequal Exchange*

A central issue in neocolonial criticisms is the export of mitigation responsibility from high-emitting countries to that of the Global South. When emissions reductions are transferred from developing countries to other countries, critiques argue that mitigation transfers without changing carbon-intensive production and consumption processes in the acquiring countries (Fuhr, 2021). This dynamic is viewed as the privilege cost responsibility over historical responsibility, enabling wealthier economies to delay structural decarbonization.

Based on this, carbon markets are prone to entrenching a novel division of labor whereby developing nations provide mitigation results and industrialized nations retain the activities that emit higher emissions. These are considered to be the unequal exchange where environmental value is taken away from the Global South on market conditions that exhibit ambiguity in economic power, bargaining, and finance accessibility (Durmaz & Schroeder, 2025). Neocolonial critiques hence point out that allocation of mitigation effort cannot be judged just by aggregate efficiency.

7.2 *Land, Carbon Sinks, and Socio-Ecological Appropriation*

Carbon markets are criticized by neocolonialists based on land-based mitigation activities that are related to nature-based solutions. The activities like afforestation, reforestation, or avoided deforestation usually overlap with the intricate land tenure, customary rights, and livelihood systems. With weak or exclusionary governance structures, these activities may limit access to land and resources, causing displacement of local population or reorganize land use that favors carbon value over socio-ecological requirements (Carton et al., 2021).

Critiques, further, believe that turning forests, soils, and ecosystems into standardized units of carbon value eases socio-ecological appropriation. By this, local landscapes are registered in global markets, provided external valuation and domination. In neocolonial terms, this abstraction runs the risk of replicating extractive relationships by segregating ecological value from local social contexts and channeling benefits towards external actors (Asiyanbi & Lund, 2020).

7.3 Limits, Contestations, and Contextual Variation

Although neocolonial critiques raise significant issues, recent studies warn against seeing such consequences as uniform or general across countries. Empirical research has noted variability in the manner in which carbon market projects are governed, regulated, and contested in the Global South. The decisive elements in defining the market participation or mitigation unequal results are host-country authorization procedures, benefit-sharing arrangements, and legal safeguards (Minas, 2023).

Further, the Global South is a heterogeneous actor. Governments, policymakers, and local communities look at carbon markets differently, with diverse development and political priorities. The market interactions developed as the subgroup of climate and development policies in developing countries is actively pursued, challenging narratives of carbon markets being externally enforced practices (Stek et al., 2025).

The neocolonial critiques function less as categorical denials of carbon markets than as diagnostic tools to determine which market-based mitigation can erode equity, sovereignty, or social inclusion. Their analytical utility is highlighted by their capability to governance lapses, power imbalance, and distributive risks that can exist despite technical integrity. These issues can be addressed through institutional structures which go beyond accounting accuracy and market efficiency. Proposals, like domestic policy, host-country dominating over authorization, and alternative patterns of collaboration like South–South exchanges, stand as pathways of reconciling market-based mitigation to equity and development goals. The next section discusses hybrid and reform-oriented methods for redesigning carbon markets to address the critiques.

8 Toward Hybrid and Equitable Carbon Market Pathways

The discussions analyzed in the previous sections state that the role of the carbon markets in the Global South is not about their existence itself, but rather about the way they are designed and managed. By addressing the issue of equity, sovereignty, and inequality of values, research is increasingly focusing on hybrid patterns to complement market mechanisms with robust domestic policies and alternative international level cooperation. These strategies aim to maintain the efficiency and coordination advantages of markets while overcoming the structural hazard of the political economy and neocolonial critiques.

8.1 Domestic Carbon Pricing with International Linkages

Combining international carbon market participation with domestic carbon pricing instruments is one of the key pathways. Instead of being autonomous mechanisms,

ITMOs may be connected selectively with national ETS or carbon taxes. Such integration enables governments to maintain control over authorization decisions, credit use, and revenue distribution, maintaining policy space as well as participating in cooperative mitigation (Asian Development Bank, 2024).

In the Global South, this approach assists in prioritizing the sequence of market involvement based on the decarbonization policies of the country. A negative risk of early market involvements can be minimized by governments focusing on local mitigation needs and international transfers being limited to surplus reductions. Researchers insist that this paradigm should be accompanied by long-term plans and multilateral coordination by institutions to transform international relations to supplement the domestic policy objectives (Groenewoudt & Romijn, 2022).

8.2 South–South Cooperation and Regional Carbon Markets

Another reform-oriented pathway that the literature focuses on is the development of South–South collaboration in carbon markets. Regional or plurilateral arrangements between developing countries in comparable economies and mitigation concerns could have a negative effect on power imbalances associated with North–South in the market transactions (Berger, 2020). This may help in knowledge sharing, harmonization of methods, and shared capacity-building while enabling participating countries to negotiate terms on a more equal footing.

Regional initiatives of carbon market can also facilitate the implementation of the common standards and infrastructure, reducing transaction costs and raising market liquidity. Although the empirical data regarding large-scale South–South carbon markets is scarce, analysts suggest that such arrangements have potential to transform the geography of carbon trading to be more responsive to development priorities in the Global South (UNFCCC, 2024).

8.3 Safeguards, Equity Measures, and Institutional Reform

Across both domestic–international hybrid approaches and South–South cooperation, there is a need to incorporate carbon markets in more extensive models of social and institutional safeguards. Such actions as open authorization processes, inclusive decision-making, and income distribution to domestic climate and development initiatives can enhance legitimacy and distributive justice.

Rather than considering safeguards as externalities to the market efficiency, scholars are viewing them as sustainable and politically viable carbon market conditions. Reforms such as these can build trust among the participants since they help in solving equity, sovereignty, and accountability problems, consequently reducing opposition to cooperation based on markets. These considerations result in redesigning of carbon markets as part of a multi-layered climate governance.

These hybrid and reform-based ways do not resolve all the tensions on the carbon markets of the Global South. However, they portray that market processes are accessible to political decisions. Under the condition of evaluating carbon markets by the prism of these alternatives, these markets can provide a collection of criteria to bring them to efficiency considerations and develop into more balanced tools to align the mitigation objectives to meet the developmental and equity goals.

References

Agrawal, D. (2025). A comprehensive review of investing and financing for sustainable tourism projects. *Environment, Sustainability, and Governance Insights, 1*(1), 53–70.

Alsamawi, A. (2024). *The role of carbon markets in transitions*. International Energy Forum. https://ercst.org/wp-content/uploads/2025/01/20240301-Carbon-Markets-Report_IEF.pdf

Asian Development Bank. (2024). *Carbon pricing as a key policy for climate change mitigation.* https://www.adb.org/sites/default/files/publication/1008086/ap-climate-report-2024-chapter5.pdf

Asiyanbi, A., & Lund, J. F. (2020). Policy persistence: REDD+ between stabilization and contestation. *Journal of Political Ecology, 27*(1), preprint. https://doi.org/10.2458/v27i1.23493

Berger, T. (2020). The "Global South" as a relational category—Global hierarchies in the production of law and legal pluralism. *Third World Quarterly, 42*(9), 1–17. https://doi.org/10.1080/01436597.2020.1827948

Carton, W., Lund, J. F., & Dooley, K. (2021). Undoing equivalence: Rethinking carbon accounting for just carbon removal. *Frontiers in Climate, 3*. https://doi.org/10.3389/fclim.2021.664130

Durmaz, Z., & Schroeder, H. (2025). Indigenous contestations of carbon markets, carbon colonialism, and power dynamics in international climate negotiations. *Climate, 13*(8), 158–158. https://doi.org/10.3390/cli13080158

Fuhr, H. (2021). The rise of the Global South and the rise in carbon emissions. *Third World Quarterly, 42*(11), 1–23. https://doi.org/10.1080/01436597.2021.1954901

Global Green Growth Institute. (2022). *GGGI technical report no. 25 developing carbon markets based on article 6 of the Paris Agreement: Challenges and opportunities.*

Gold Standard. (2024). *A practitioner's guide: Aligning the voluntary carbon market with the Paris Agreement test.* Gold Standard. https://www.goldstandard.org/publications/a-practitioners-guide-aligning-the-voluntary-carbon

Griscom, B. W., Busch, J., Cook-Patton, S. C., Ellis, P. W., Funk, J., Leavitt, S. M., Lomax, G., Turner, W. R., Chapman, M., Engelmann, J., Gurwick, N. P., Landis, E., Lawrence, D., Malhi, Y., Schindler Murray, L., Navarrete, D., Roe, S., Scull, S., Smith, P., & Streck, C. (2020). National mitigation potential from natural climate solutions in the tropics. *Philosophical Transactions of the Royal Society b: Biological Sciences, 375*(1794), 20190126. https://doi.org/10.1098/rstb.2019.0126

Groenewoudt, A. C., & Romijn, H. A. (2022). Limits of the corporate-led market approach to off-grid energy access: A review. *Environmental Innovation and Societal Transitions, 42*, 27–43. https://doi.org/10.1016/j.eist.2021.10.027

Jain, G., & Deshmukh, S. (2025, September 11). *How to fully operationalize article 6 of the Paris Agreement*. Center on Global Energy Policy at Columbia University SIPA. https://www.energypolicy.columbia.edu/publications/how-to-fully-operationalize-article-6-of-the-paris-agreement/

Joselin, D., Singh, A., & Tiwary, K. S. (2025). The nexus between CSR practices and financial performance: Empirical evidence from the manufacturing sector in India. *Environment, Sustainability, and Governance Insights, 1*(1), 1–17.

Kaith, A., & Sachdeva, G. (2024). Bibliometric insights into green human resource management: A connection with business ethics and values. *PURUSHARTHA-A Journal of Management, Ethics and Spirituality, 17*(1), 51–63. https://doi.org/10.21844/16202117104

Kumar, K., Singh, R., Choudhary, M., & Sharma, A. (2025). Climate change mitigation strategies in environmentally sensitive industries. In *Climate change and social responsibility* (Vol. 25, pp. 185–204). Emerald Publishing Limited.

Kus, B., & Jackson, G. (2025). Green transitions: Rethinking political economy in the context of climate change. *Regulation & Governance, 19*(2), 287–302. https://doi.org/10.1111/rego.70013

Michaelowa, A., & Keßler, J. (2024). *Methodological challenges of policy crediting under Article 6 of the Paris Agreement* [Discussion Paper].

Michaelowa, A., Keßler, J., Dalfiume, S., & Ahonen, H.-M. (2025). *Reversal risk and buffer pool contribution analysis Discussion paper*. Perspectives Climate Group. https://www.carbon-mechanisms.de/fileadmin/media/dokumente/PCG_Reversal_Risk_Paper_20250616.pdf

Minas, S. (2023). Market making for the planet: The Paris Agreement Article 6 decisions and transnational carbon markets*. *Transnational Legal Theory*, 1–34. https://doi.org/10.1080/20414005.2023.2174690

Mukhia, A., Shen, Q., & Xiaolong, Z. (2024). Climate governance pathway for BRICS in the post-paris era. *Global Journal of Emerging Market Economies*. https://doi.org/10.1177/09749101241244456

Prakash, A., & Tiwari, A. K. (2025). The role of institutional factors in achieving SDG 16: A thematic review of drivers of corporate anti-corruption action. *Environment, Sustainability, and Governance Insights, 1*(1), 34–52.

Redvers, N., Chan, J., Odochao, S., Pratt, V., Sim, J., Gougsa, S., Kobei, D. M., & Willetts, L. (2025). Carbon markets: A new form of colonialism for indigenous peoples? *The Lancet Planetary Health, 9*(5), e421–e430. https://doi.org/10.1016/S2542-5196(25)00086-5

Schneider, L., & La Hoz Theuer, S. (2018). Environmental integrity of international carbon market mechanisms under the Paris Agreement. *Climate Policy, 19*(3), 386–400. https://doi.org/10.1080/14693062.2018.1521332

Shen, X., Gatto, P., & Pagliacci, F. (2023). Unravelling the role of institutions in market-based instruments: A systematic review on forest carbon mechanisms. *Forests, 14*(1), 136. https://doi.org/10.3390/f14010136

Singh, R., Kumar, K., & Khan, S. (2024). A comprehensive view of artificial intelligence (AI)–based technologies for sustainable development goals (SDGs). *Artificial Intelligence Enabled Management: An Emerging Economy Perspective, 183*, 9783111172408–012.

Singh, R., Sharma, A., & Kumar, K. (2025a). Navigating carbon neutrality: Strategies for sustainable future and environmental challenges. In *Climate neutrality through smart eco-innovation and environmental sustainability* (pp. 69–80). Springer Nature Switzerland.

Singh, R., Tiwary, K. S., Kumar, K., & Kumar, A. (2025b). Empowering a sustainable future: Blockchain's role in economy, organizations, and finance towards smart circular economy. In *Blockchain technologies for smart circular economy and organisational sustainability* (pp. 271–283). Springer Nature Switzerland.

Srivastava, S. K., Anshul, A., Pathak, P., & Bansal, J. (2018). Responsible consumption for curbing food wastage: An exploratory enquiry. *PURUSHARTHA-A Journal of Management, Ethics and Spirituality, 11*(2), 104–116.

Srivastava, S. K., Singh, S., & Srivastava, M. (2021). Food waste at household and social gatherings: Drivers and possible remedies. *PURUSHARTHA-A Journal of Management, Ethics and Spirituality, 14*(2), 117–133.

Srivastava, S. K., Sinha, A. K., & Kumar, S. (2024). Workplace spirituality empowered by spiritual leadership to promote organizational effectiveness: A bibliometric analysis. *PURUSHARTHA-A journal of Management, Ethics and Spirituality, 17*(1), 1–22. https://doi.org/10.21844/16202117101

Stek, P. E., Lima-de-Oliveira, R., & Vasudhevan, T. (2025). The development of carbon markets in upper-middle-income countries. *Regulation & Governance, 19*(2), 482–495. https://doi.org/10.1111/rego.70010

Tripathi, V., Singh, K., Parihar, S., & Srivastava, S. K. (2022). A bibliometric analysis on supply chain collaboration. *PURUSHARTHA-A journal of Management, Ethics and Spirituality, 15*(2), 66–103. https://doi.org/10.21844/16202115205

UNFCCC. (2024). *National capacity-building workshop on carbon markets and article 6 implementation for guatemala FINAL REPORT*. https://unfccc.int/sites/default/files/resource/National_Capacity-Building_Workshop_on_Carbon_Markets_and_Article%206_Implementation_for_Guatemala.pdf

Wang, Z., Sun, Y., Kong, H., & Xia-Bauer, C. (2025). An in-depth review of key technologies and pathways to carbon neutrality: Classification and assessment of decarbonization technologies. *Carbon Neutrality, 4*(1). https://doi.org/10.1007/s43979-025-00129-8

Zhao, Z., & Shaw, R. (2025). Nature-based ESG solutions: An analytical overview. *Peace and Sustainability, 1*(2), 100015–100015. https://doi.org/10.1016/j.nerpsj.2025.100015

Index

A

Accountability, 3, 4, 10, 11, 14–19, 26, 27, 29–33, 40, 46, 51, 61, 62, 65–68, 76, 79, 84, 85, 100, 135, 161, 201, 205, 270–272, 275, 280, 287, 289, 307, 313, 326, 364, 371

Adaptive capacity, 167, 169, 171, 173–175, 233

AI governance, 3, 10, 13–16, 19, 21, 49, 67

AI systems, 10, 12, 23, 26–33, 78, 113, 128, 135, 201, 234, 236, 251, 271, 273, 279, 307, 310, 311

Algorithmic bias, 3, 10, 39, 67, 77, 78, 128, 130, 135, 273, 275, 279, 280, 304, 311

Algorithmic ethics, 12

Artificial Intelligence (AI), 1, 2, 9, 10, 12, 14, 25, 30, 60, 63, 67, 68, 75, 77–79, 83, 94, 96, 114, 128, 132, 146, 183, 184, 189, 195, 200, 217, 231–234, 238, 240, 242, 248–250, 252, 255, 269, 271–276, 279, 280, 304, 305, 307, 310, 318, 319, 325, 326, 343

Automated trading, 272, 273, 279

B

Best-Worst Method (BWM), 339–341, 345–350

Bibliometric analysis, 161

Blockchain ethics, 65

BRICS, 4, 112–118, 120–122

C

Capabilities, 29, 65, 80, 82, 94–97, 100, 101, 103, 128, 132, 136–138, 236–241, 248, 272, 307

Carbon accounting, 2, 4, 65, 131, 144–146, 150, 151, 155, 158, 161

Carbon emissions, 5, 38, 114, 115, 131, 134, 136, 137, 145, 155, 159, 184, 191, 202, 232, 235, 239, 338–340, 343, 350

Carbon footprints, 66, 137, 239, 306

Carbon market, 5, 144, 145, 156, 157, 356–359, 361–372

Carbon neutrality, 2, 3, 51, 94, 115, 130–133, 136, 137, 184, 185, 190, 194

CIMO framework, 168–170, 172, 174–176

Circular, 1, 3–6, 60, 61, 64, 93–98, 100–103, 112–114, 127–133, 135–137, 183–190, 192–195, 200, 201, 203, 204, 206–209, 211, 217–219, 221, 223, 224, 226, 239, 247, 248, 251–256, 259, 261, 262, 297, 317, 323, 338, 342, 344, 347, 348

Circular Economy (CE), 2–6, 60, 61, 64, 94, 96–103, 112–114, 128–133, 135–137, 184, 185, 190, 192, 194, 195, 200, 209, 218, 219, 221–224, 226, 227, 248, 250–252, 254, 255, 262, 318, 323, 342, 344, 348

Circular economy practices, 251, 344

Circular value chains, 65, 184, 185, 187, 190, 194, 195

K. Kumar et al. (eds.), *AI-Driven Circular Economy for Carbon Neutrality in Emerging Economies*, Environmental Footprints and Eco-design of Products and Processes,
https://doi.org/10.1007/978-3-032-30298-4

Climate adaptation, 166, 168–171, 173, 174, 176, 177, 235, 313
Climate change, 58, 124, 134, 145, 154, 155, 159, 161, 166, 168, 172, 184, 203, 204, 235, 248, 306
Climate change Vulnerability, 167–169, 174–176
Climate governance, 4, 144, 145, 151, 175, 313, 348, 350, 351, 357, 364, 371
Climate technology, 143
Content analysis, 288, 289
Context-sensitive policy, 175
Corporate Responsibility (CR), 3, 38, 39, 51, 58, 59, 64–68
Corporate sustainability, 38, 59, 60, 308, 310

D

Decentralised Finance (DeFi), 269–272, 274, 275, 277–280, 286–296, 299
Deepfakes, 318–330
Digital economy, 38, 39, 58, 59, 61, 62, 65–68
Digital transformation, 3, 15, 18, 61, 62, 66–68, 70, 94–98, 103, 195, 255

E

Economy, 1, 2, 4, 5, 37–39, 55, 57–62, 65–69, 71, 93, 96, 98–103, 111–113, 115–117, 127, 129–132, 135, 137, 144, 183–185, 188, 190, 192–194, 200, 201, 207–209, 217, 219, 221, 222, 224, 226, 247, 248, 250, 251, 254, 255, 290, 317, 324, 342, 344, 347, 355, 357, 366–368, 370
Emerging, 2–4, 23, 49, 52, 67, 85, 86, 94, 97, 103, 124, 150, 154, 162, 194, 207, 211, 219, 220, 287, 293, 296, 298, 339
Emerging economies, 2–5, 12, 94–98, 101, 103, 112–114, 116, 123, 124, 128, 130–138, 184, 186, 188, 189, 193–195, 218, 226, 227, 286, 288, 290, 292, 294, 296, 338
Emerging markets, 38, 185, 190, 192, 194, 291, 292, 295
Environmental sustainability, 41, 43, 200, 222, 224, 226
Ethical and governance, 130, 270
Ethical dilemmas, 26, 28, 83, 310
Ethical leadership, 10–14, 16, 20–23, 276
Ethics, 12–16, 23, 26, 28–30, 32, 33, 38, 39, 44–46, 48, 50, 52, 53, 61, 65–68, 83, 99, 100, 270, 274, 287, 295, 304, 312

F

Financial inclusion, 51, 95, 114, 286, 287, 291, 293
Financial sector, 45, 52, 63, 271, 272, 290, 326

G

Generative AI (GenAI), 4, 85, 86, 232–234, 236–241, 318, 319
Global south, 5, 6, 210, 286, 287, 356, 358–372
Governance, 2–5, 12–17, 21, 26, 28, 30, 31, 40, 46, 49, 50, 59, 61–63, 65–67, 78, 79, 85, 95, 98–101, 103, 130, 131, 135–138, 144–146, 156, 158–161, 166, 167, 170–173, 175, 176, 194, 200, 202–205, 207, 209, 270, 272, 274, 275, 277–280, 288, 289, 291, 295, 296, 306, 307, 312, 314, 338, 341, 346, 360, 363, 364, 366, 368–370
Green growth, 4, 112, 113, 123, 124, 360
Greta Effect (GE), 248–250, 255, 256, 258–261
Guest engagement, 258, 259, 261

H

Horizon scanning, 200, 203, 206–208
Human-Centric Artificial Intelligence (HCAI), 76–85
Human Resource Management (HRM), 76, 78, 79, 85
Human Resource Sustainability, 75

I

ICT, 4, 39, 112–118, 120–124, 290
Intelligence, 76, 77, 85, 195, 209, 210, 220, 224, 236, 241, 250, 271, 330

K

Knowledge structure, 144, 162

M

Market fairness, 4, 270, 273

Multi-Criteria Decision-Making (MCDM), 339–341, 344, 345, 349

N

Nifty50, 41, 43, 44

P

Paris agreement, 130, 144, 157, 175, 224, 338, 349, 356, 357, 359, 360, 362–364
Policy and strategy, 172

R

Renewable energy adoption, 112, 113, 117, 118, 121, 122
Research evolution, 143
Resilience, 40, 94, 98, 133, 167, 169, 173–176, 203, 204, 232, 233, 236–239, 241, 242, 251, 277, 342–344, 349
Resilient, 11, 12, 60, 76, 77, 83, 137, 172, 194, 200, 204, 226, 232, 233, 235, 238, 242
Resource efficiency, 103, 132, 133, 251, 254
Responsible innovation, 11, 13, 23, 38

S

Scenario planning, 200, 203, 205–208, 210
Social sustainability, 4, 63, 94–98, 103, 311
Southeast Asia, 186, 319, 320, 329, 330
Strategic Foresight (SF), 3, 200, 202–204, 207
Supply Chain Resilience (SCR), 232–236, 238, 240–242
Sustainability, 2–5, 9–16, 18–20, 22, 23, 30, 37–41, 43, 49–53, 55, 57, 59, 61, 63–70, 75–77, 79, 82, 83, 93–98, 101, 103, 111–115, 123, 124, 131, 137, 146, 149–151, 155, 159, 161, 162, 166, 169, 175, 176, 183–186, 189, 191, 194, 195, 200–204, 207, 209, 211, 217, 219, 220, 222, 224–226, 231–242, 247–256, 258, 259, 261, 262, 280, 286, 288, 289, 303–314, 317, 324, 337, 338, 340–342, 344–346, 348–350
Sustainability-Oriented Innovation (SOI), 248, 251–256, 258–261
Sustainability-Oriented Leadership, 13
Sustainable Development Goals (SDGs), 4, 22, 38, 41, 44, 45, 47–49, 51, 52, 54, 55, 58, 59, 64, 65, 79, 112, 127–129, 131, 132, 135–137, 218, 232, 239
Sustainable finance, 4, 304–307, 309–312, 314
Sustainable procurement, 185, 191, 235
Sustainable Urban planning, 134

T

Technology Acceptance Model (TAM), 249, 250
Technology adoption, 99, 101
Tokenised economy, 286
Transformation, 2, 11, 23, 44, 66, 77, 83, 85, 86, 96, 97, 99, 103, 167, 194, 211, 251, 252, 304, 358
Trustworthy AI, 14

U

Urban sustainability, 338–341, 344, 345, 347–350

W

Waste management, 2, 38, 96, 132, 135, 185, 187, 192, 195, 206, 208, 209, 218–227, 250, 338, 342, 343, 347, 348, 365

Kolibri 360 GmbH
Ferdinand-Jühlke-Straße 7
99095 Erfurt, Deutschland
produktsicherheit@kolibri360.de